谨以此书献给中山大学世纪华诞暨材料科学与工程学院十周年庆。

液相激光烧蚀
及其纳米材料制备应用

杨国伟 著

科学出版社

北京

内 容 简 介

本书介绍一种独特的纳米制备技术即液相激光烧蚀（laser ablation in liquids，LAL）技术，以及其纳米材料制备的应用。LAL 技术的优点表现在：①化学上"简单、干净"，不需要任何催化剂，属于绿色合成，可以在温和环境中进行诸如高温高压相等亚稳相纳米材料制备；②特殊的反应空间使得合成的纳米材料的组分来自固体和液体，为科学家基于基础或应用的需求所合成的纳米材料进行组分设计提供了可能；③可以通过调节激光参数、液体种类和靶材，对制备的纳米材料的尺寸、形貌、结构等进行调控；④可以通过电场、磁场、温度场、电化学的引入，原位对功能纳米结构组装进行操纵。

本书可供高等院校的学生及从事纳米材料研究的科研人员参考。

图书在版编目（CIP）数据

液相激光烧蚀及其纳米材料制备应用 / 杨国伟著. -- 北京：科学出版社，2024.9. -- ISBN 978-7-03-079402-4

Ⅰ.TB383

中国国家版本馆 CIP 数据核字第 20248PD778 号

责任编辑：郭勇斌　邓新平　常诗尧 / 责任校对：高辰雷
责任印制：徐晓晨 / 封面设计：义和文创

科学出版社 出版
北京东黄城根北街 16 号
邮政编码：100717
http://www.sciencep.com

北京天宇星印刷厂印刷
科学出版社发行　各地新华书店经销
*
2024 年 9 月第 一 版　开本：720×1000　1/16
2024 年 9 月第一次印刷　印张：20 插页：14
字数：396 000
定价：168.00 元
（如有印装质量问题，我社负责调换）

序

  在世界上第一台激光器问世之后，人们将它应用于材料的制备与加工，激光烧蚀（laser ablation）是激光材料制备与加工的最基本的过程，也就是通过激光烧蚀固体靶进行材料制备与加工。早期的激光材料制备与加工主要是在真空和气体（包括自然大气）环境中进行的，例如，激光切割、焊接、打孔、清洗，薄膜的脉冲激光沉积（pulsed laser deposition，PLD）和材料表面改性等。后来，由于水下材料加工的需求和激光技术在医学领域中的应用，水下激光切割、焊接、打孔、烧蚀液体（液态生物组织）等激光技术应运而生，极大拓展了激光技术在材料加工方面的应用。20世纪80年代后期，国际上纳米材料的研究风起云涌，人们开始应用激光技术制备纳米粉体。主要的方法是在气体环境中采用脉冲激光烧蚀固体靶，然后在气体环境中冷却并获得纳米粉体，这种技术的优点是可以连续且大量地制备各种纳米粉体。在激光材料制备中，无论是薄膜的沉积还是粉体的合成，激光烧蚀固体靶都是在真空或气体环境中进行的。然而，这种激光材料制备过程发生在液体环境中会怎么样呢？一个偶然的发现不仅回答了这个问题，而且发展了液相激光烧蚀纳米制备技术。20世纪90年代后期，我们的实验室需要搭建一个PLD系统用于碳及碳氮薄膜制备研究，当固体脉冲激光器安装调试完毕，PLD系统却迟迟没有做好。看着昂贵的激光器闲置在那里我们心里很不是滋味，就一直想着暂时能用它做些什么。当时突发奇想，就顺手把一块小石墨靶放进装有水的烧杯里用激光去打，结果水的颜色迅速发生变化，并且我们在激光作用后的水中发现了金刚石纳米晶！这在当时是还没有人做到的事情。于是，我们将脉冲激光与液-固界面的相互作用引入到亚稳相纳米材料的制备中，开发并发展了系列基于LAL的常温常压下制备亚稳相纳米材料的新方法。后来，经过国际同仁们近三十年的努力，将LAL技术发展成为国际公认的一种重要的纳米制备技术，而我们的研究组也被国际同行公认为LAL领域的"开拓者"（pioneer）。在本书中，我们将阐述LAL技术的基本物理和化学过程，介绍它在纳米材料合成和纳米结构组装方面的最新成果，以及LAL基纳米材料的应用领域，包括光学、磁学、环境、能源和生物医学等，并总结LAL纳米制备技术的主要优势，存在的问题及解决方案，展望未来LAL纳米制备技术的广阔前景。

目 录

第1章 引言 ··· 1
    参考文献 ·· 8
第2章 气体环境中固体靶的激光烧蚀 ··· 16
    参考文献 ··· 19
第3章 液体环境中固体靶的激光烧蚀 ··· 22
  3.1 基本物理过程 ·· 22
      3.1.1 液体环境中毫秒激光烧蚀固体靶及熔化过程 ························ 22
      3.1.2 液体环境中纳秒激光烧蚀固体靶及等离子体羽形成 ··············· 24
      3.1.3 液体环境中皮秒激光烧蚀固体靶及空泡产生 ························ 28
      3.1.4 液体环境中飞秒激光烧蚀固体靶及多形态过程 ····················· 29
  3.2 基本化学过程 ·· 33
      3.2.1 等离子体中的化学反应 ·· 33
      3.2.2 固体与液体界面的化学反应 ·· 36
  3.3 热力学特征 ··· 37
  3.4 动力学特征 ··· 40
    参考文献 ··· 43
第4章 LAL 中纳米晶形成的物理化学过程 ··· 49
  4.1 成核热力学 ··· 49
  4.2 相变热力学 ··· 51
  4.3 生长动力学 ··· 53
    参考文献 ··· 55
第5章 LAL 合成纳米金刚石及新碳相纳米材料 ··································· 58
  5.1 纳米金刚石 ··· 58
      5.1.1 Bottom-up 合成纳米金刚石 ·· 59
      5.1.2 Top-down 合成纳米金刚石 ··· 62
      5.1.3 来自煤的纳米金刚石 ··· 67
      5.1.4 纳米金刚石的荧光起源 ·· 73
  5.2 LAL 中金刚石的相变 ··· 81
      5.2.1 金刚石-石墨相变的中间相 ··· 82

5.2.2　金刚石-碳葱可逆相变 ·· 86
　　　5.2.3　金刚石-新金刚石相变 ·· 95
　5.3　新碳相纳米材料 ·· 102
　　　5.3.1　C_8 纳米晶 ·· 103
　　　5.3.2　C_8-like 碳纳米方块 ·· 110
　　　5.3.3　白碳纳米晶 ·· 117
　参考文献 ·· 128

第 6 章　新颖亚稳相纳米材料的 LAL 探索 ···································· 139
　6.1　闪锌矿硅纳米晶 ·· 140
　6.2　四方相锗纳米晶 ·· 148
　6.3　双层六角密堆积铁纳米晶 ·· 152
　6.4　立方氮化硼纳米晶 ·· 156
　6.5　立方碳氮纳米晶 ·· 159
　6.6　立方氮化镓纳米晶 ·· 161
　参考文献 ·· 166

第 7 章　LAL 纳米制备技术 ·· 171
　7.1　电场辅助 LAL 用于金属氧化物纳米晶形貌控制 ······················· 171
　　　7.1.1　多种形貌金属氧化物半导体纳米晶 ·························· 171
　　　7.1.2　电场对纳米晶形貌形成的影响 ································ 173
　7.2　电场辅助 LAL 用于纳米结构组装 ···································· 174
　　　7.2.1　金属氧化物功能纳米结构 ···································· 174
　　　7.2.2　纳米结构组装中的定向附着机制 ······························ 175
　7.3　温度场辅助 LAL 用于纳米结构组装 ·································· 176
　　　7.3.1　一维和二维纳米结构 ·· 176
　　　7.3.2　纳米结构组装中的 Ostwald 熟化机制 ························ 177
　7.4　电化学辅助 LAL 用于复杂纳米结构组装 ···························· 179
　　　7.4.1　多金属氧酸盐纳米结构 ·· 179
　　　7.4.2　纳米结构组装中的化学反应 ···································· 181
　7.5　磁场辅助 LAL 用于磁性纳米链组装 ·································· 182
　　　7.5.1　磁性纳米颗粒一维链束 ·· 182
　　　7.5.2　磁场感应定向附着机制 ·· 185
　参考文献 ·· 188

第 8 章　液体环境中纳米图案 LAL 组装 ······································ 193
　8.1　液体环境中脉冲激光沉积用于纳米图案组装 ·························· 193
　　　8.1.1　液体环境中的脉冲激光沉积 ···································· 193

8.1.2　在透明基片上组装纳米颗粒图案 193
　8.2　液体环境中功能纳米结构图案的激光直写 195
　　8.2.1　异质结构纳米图案的组装 195
　　8.2.2　液体环境中激光诱导的相变 196
　参考文献 199

第9章　LAL 制备的纳米材料 200
　9.1　纳米颗粒-聚合物复合材料 200
　9.2　掺杂半导体纳米晶体 202
　9.3　亚微米球形颗粒 204
　9.4　单分散胶体量子点 208
　参考文献 212

第10章　LAL 基纳米材料的应用 216
　10.1　光学功能纳米结构 216
　　10.1.1　荧光发射 216
　　10.1.2　可见光散射 223
　　10.1.3　非线性光学 226
　　10.1.4　光热转换 229
　10.2　磁性功能纳米结构 234
　10.3　应用于环境科学的功能纳米结构 237
　　10.3.1　吸附 237
　　10.3.2　光催化降解 239
　　10.3.3　传感 245
　10.4　应用于绿色能源中的功能纳米结构 251
　　10.4.1　超级电容器 251
　　10.4.2　锂离子电池 253
　　10.4.3　太阳电池 255
　　10.4.4　混合发光二极管 257
　　10.4.5　光催化分解水产氢 258
　　10.4.6　电化学催化剂 260
　10.5　生物医学功能纳米结构 262
　　10.5.1　生物分子载体 262
　　10.5.2　阳性造影剂 264
　　10.5.3　生物识别 266
　　10.5.4　纳米酶 267
　参考文献 272

第 11 章 结论和展望 287
11.1 LAL 纳米制备技术的主要优势 288
11.1.1 清洁表面 288
11.1.2 亚稳纳米相 290
11.2 LAL 纳米制备技术的不足和解决方案 291
11.2.1 产额 291
11.2.2 尺寸和分散度控制 293
11.3 LAL 纳米制备技术的未来发展 294
11.3.1 物理化学机制探索 294
11.3.2 应用领域拓展 295
参考文献 298
附录 杨国伟研究组发表 LAL 纳米制备技术论文目录 302
后记 309
彩图

第 1 章 引 言

20 世纪 60 年代初，当世界上第一台激光器即红宝石激光器问世时，人们就报道了通过激光烧蚀（ablation）固体靶制备材料的激光技术应用研究。不久之后，脉冲激光烧蚀（pulsed laser ablation，PLA）作为一种固体材料加工技术得到了迅速的发展。PLA 因其在材料加工（包括固体薄膜制备、晶体生长、表面清洁和微电子器件制造等）方面的潜在应用而备受关注。由于固体靶的激光烧蚀很容易在真空或气体环境的传统沉积室（chamber）中进行，所以，针对如上所述的各种应用，大多数研究人员都专注于在真空或稀释气体环境中进行固体靶的激光烧蚀研究[1-4]。例如，用于制备固体薄膜的脉冲激光沉积、在气体环境中固体靶的 PLA 合成超细粉末、材料表面的激光清洁等[5-7]。然而，与气-固界面上的 PLA 研究相比，沉浸在液体环境中的固体靶的 PLA 研究仅限于脉冲激光与固体靶相互作用的基础和应用基础研究[8]。一个合理的解释是，脉冲激光和液体环境中的固体靶的相互作用比其和真空或气体环境中的固体靶的相互作用要复杂得多[9]。尽管近几十年来研究人员采用多种光谱技术对脉冲激光和液体环境中固体靶的相互作用进行了研究[10, 11]，但是，针对 PLA 在液体环境中材料加工应用的系统研究并没有得到很好地开展，主要的原因就是液体环境中的 PLA 表征更困难。

在现有的文献中，关于液体环境中的激光烧蚀存在两种不同的定义。一种定义是在气体或液体环境中对液体进行激光烧蚀。详细地说，就是激光在气-液界面或液-固界面上烧蚀液体。由于其在高温化学合成和材料加工等应用中的巨大潜在应用价值，尤其是在人体内引导烧蚀"软"组织的医学应用价值，这种液体的激光烧蚀一直备受关注。此外，也为研究激光与结构更复杂的软物质的相互作用提供了一条途径。另一种定义是在液体环境中对固体靶的激光烧蚀，即激光在液-固界面烧蚀固体靶。大多数情况下，所使用的液体在辐照波长下是透明的。然而，与液体的激光烧蚀相比，关于液体环境中固体靶的激光烧蚀研究并不多，人们对于其基本的物理和化学过程还是知之甚少。

近年来，由于先进光谱技术的发展，研究人员可以采用多种技术手段来研究液体环境中液体和固体靶的激光烧蚀所涉及的基本物理和化学过程，所以，这方面的工作取得了很大的进展。研究人员已经发表了许多综述论文，总结了液相激光烧蚀的基础和应用基础的研究成果[12]，包括生物组织激光烧蚀的机制[13]、固-

液界面激光辐照的水下钻孔.和焊接[14]及液-固界面的激光辐照用于材料表面清洁和水下刻蚀和切割[15]等。另外，还有出版聚焦于材料表面的激光清洁的学术专著[16]。因此，从液相激光烧蚀的上述应用不难看出，液相激光烧蚀在生物医学领域得到了很好的发展，而液体环境中激光烧蚀固体靶技术主要应用于特殊环境中的材料加工如水下刻蚀、钻孔、焊接等领域。然而，液体环境中激光烧蚀固体靶在材料制备中的应用起步较晚。1987 年，Patil 等[17]报道了将 PLA 作用于液体环境中的固体靶，作为新材料制备的开创性工作，他们通过水中铁靶的 PLA 在块体铁表面合成了具有亚稳相的氧化铁。随后，Ogale[18]研究了液体环境中固体靶的 PLA 在金属表面改性如氧化、氮化和碳化等方面应用的潜力，并且他还利用不同液体合成不同的亚稳结构，例如，他通过脉冲红宝石激光辐照沉浸在液态苯中的石墨靶，合成了许多碳的小颗粒并观察到了金刚石相[19]。这些开创性工作为 PLA 在液体环境中材料制备与加工开辟了新的途径，涌现了一系列新颖的液相激光材料制备与加工技术。研究人员通过湿刻蚀技术实现了激光诱导的表面图案化，该技术的物理机制源于激光辐照液体-衬底界面处的可吸收溶液。人们利用激光辐照液-固界面在衬底上制备表面涂层。基于激光在液体中的反冲效应，人们开发了激光蒸气方法去除材料表面的小颗粒。在不同液体中，人们通过不同固体靶的 PLA 合成了多种亚稳结构。

在液体环境中激光烧蚀固体靶的基础研究方面，Devaux 等应用发射光谱技术和冲击波（shock wave，SW）法等实验手段，讨论了 PLA 产生的等离子体羽的热力学机制[20-27]。Yavas 等使用光学技术详细描述了空泡成核、生长和坍塌的热力学及动力学行为[28-33]。Sakka 等论述了发生在等离子体羽和液体的界面上的重要化学反应，并测量了等离子体羽的热力学参数[34-39]。研究人员发现，在液体环境中，PLA 诱导的等离子体羽的温度可能达到几千开（K），压强可能达到 GPa 数量级。这些研究初步揭示了液体环境中 PLA 在材料制备方面的独特性和优越性，尤其是亚稳结构制备方面，意味着它可以为亚稳相的形成提供一个高温高压（high-temperature and high-pressure，HTHP）环境。Yang 等[40]将这一过程称为脉冲激光诱导液-固界面反应（pulsed laser induced liquid-solid interface reaction，PLIR）。

20 世纪 90 年代末，Yang[41]将液体环境中的激光烧蚀即液相激光烧蚀引入纳米尺度亚稳相材料的制备中，即在温和环境中如常温常压下产生局域的极端条件进行亚稳相纳米材料的制备，发展了一系列基于 LAL 的纳米材料制备和纳米结构组装新技术，这些研究工作极大地推动了 LAL 纳米制备技术的发展。如今，LAL 已经被国际公认为是一种功能强大的纳米制备新技术而得到广泛应用[42]，并且作为一种在极端条件下的纳米制备技术，在许多有着重大应用需求前景的新型（特种）纳米材料制备和功能纳米结构组装方面发挥着越来越大的作用。近年来，应用 LAL 纳米制备技术，研究人员合成了众多且具有重要潜在技术应用的纳米材料和纳米结构，如金属、半导体、复杂金属氧化物、氮化物材料和有机-无机杂化结构等。由于纳米尺寸效应，LAL 合成的这些纳米材料常常表现出优于传统材料的光、电、磁、热、力、机械、生物等性能，从而展现出它们在微电子学、光电子学、催化化学、生物医学等领域的巨大应用前景。

尽管近几十年来纳米材料制备研究取得了长足的进展，但是仍然面临着一些根本性的挑战。例如，如何控制纳米材料合成中结构单元的相（结构）、尺寸和形状，如何轻松地利用这些结构单元组成功能纳米结构，以及如何实现从纳米材料制备或合成到功能纳米结构组装的转变等[43-46]。为了应对这些纳米材料制备中的关键科学技术挑战，研究人员开发了一系列实用技术。例如，气-液-固机制的热化学气相传输制备一维纳米结构[47-50]、溶液化学反应合成纳米晶[51-53]，以及 DNA 模板组装复杂纳米结构[54-56]等[57]。而在这些技术中，LAL 纳米制备技术由于其独特的合成环境引起了人们极大的关注[41, 58-62]，其中包括合成具有极高稳定性的新型亚稳纳米相和极高纯度的纳米颗粒（nanoparticle，NP）胶体。

传统的纳米材料制备技术存在一定的不足。例如，气相合成法形成的微米颗粒或纳米颗粒容易发生团聚，难以形成均匀的单分散，而湿化学合成法通常会引入来自添加剂和前驱体的杂质。相比而言，LAL 纳米制备技术具有以下优点：①在化学上，LAL 简单且干净，该过程几乎没有副产物形成，而且起始原料简单，不需要催化剂。这就使得 LAL 可以确保合成出的纳米材料具有高表面活性、高纯度、高清洁表面[63]。②LAL 是在常温常压下进行的，不需要施加额外极端温度或压强。LAL 独特的激光与材料的相互作用使其能够产生局域的极端环境，从而为亚稳相的形成提供了好的热力学环境。③LAL 简便且通用，几乎适用于所有的材料和溶剂。由于 LAL 生成的新相涉及液体环境和固体靶，因此研究人员可以根据不同的需求来设计液体种类和固体靶组分，合成所需要纳米材料[64]。④LAL 可以通过调节激光参数、液体种类和靶材，较容易地控制所组装纳米结构的相、尺寸和形状，从而完成从纳米晶的合成到纳米结构功能性的调控。例如，Wang 等[65]利用液相激光选择性加热合成了多尺度的亚微米球，Liang 等[66]使用磁场辅助 LAL（magnetic field-assisted laser ablation in liquids，MFLAL）组装 FeC 复合微纳纤维、亚微米 Co_3C 颗粒链[67]和 Fe 基双金属合金纳米颗粒链[68]等。因此，LAL 纳米制备技术作为一种功能强大且应用广泛的技术受到了充分的关注，并取得了飞跃性的发展。下面，我们回顾 LAL 纳米制备技术的主要进展，简单总结它的发展历程。

图 1-1 显示了 2000~2016 年 LAL 领域文章发表数量和引用数量两项指标的情况。这些数据是通过在 Web of Science 数据库中搜索 "laser ablation in liquids"（液相激光烧蚀）条目获得的，使用不同的搜索字符串和数据集细化会得到不同的

结果[69]。这两项指标多年来稳步上升,表明了 LAL 技术的快速发展。图 1-2 展示了 LAL 技术使用的各种先进光学原位表征手段,例如,用于表征等离子体的发射光谱、用于探测空泡和等离子体动力学的快速阴影法(fast shadowgraph method)、用于确定所产生的材料在液体中的传递机制的激光散射法、用于观察空泡中纳米颗粒形成的小角度 X 射线散射等[70, 71]。此外,研究人员还开发了各种新型设备以在极端条件下实现 LAL,图 1-2(c)描绘了专为在 400 K 的温度和 30 MPa 的压强下进行原位实验而设计的装置[72]。这些表征手段和实验装置极大地推动了 LAL 的机理探索、LAL 纳米材料合成的研究等。

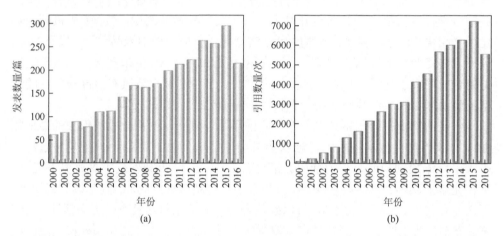

图 1-1　2000~2016 年以"液相激光烧蚀"搜索的文章(a)发表数量和(b)引用数量
数据收集自 Web of Science 数据库

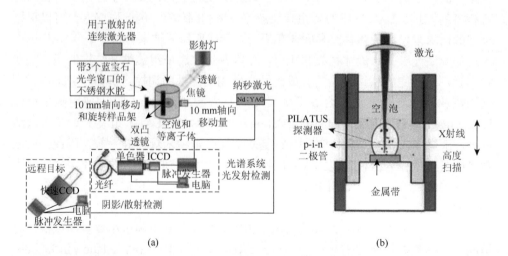

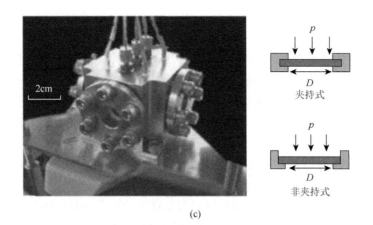

图 1-2　LAL 技术使用的各种先进光学原位表征手段

（a）用于光发射检测和阴影/散射检测的检测系统；（b）小角度 X 射线散射检测空泡中纳米颗粒的形成；
（c）高压光学池的照片及用于固定窗口的夹持式和非夹持式法兰盘的示意图

新的激光技术、表征手段和先进实验装置的引入使得 LAL 成为更加强大且有效的制备新纳米材料的方法。由于 LAL 适合几乎所有的材料和液体溶剂，因此它已经被用来合成各种先进的纳米材料[65, 73]，例如，高功率短脉冲的皮秒（ps）激光器可以显著提高 LAL 合成纳米材料的产率[74, 75]。此外，每两年一次的 LAL 专题国际会议（EOS Conference on Laser Ablation and Nanoparticles Generation in Liquid）极大地促进了 LAL 技术的国际发展，该会议将世界各地的同行们聚集在一起讨论该领域的关键科学技术问题。第一届会议于 2010 年在日本举办，2024 年会议在美国举办。该会议已经成为 LAL 领域最具影响力的国际会议，涵盖了 LAL 理论和实验方面的几乎所有重要进展，如光谱、影像和辐射成像[76-79]、空泡动力学[80, 81]、纳米颗粒宏观量合成方法[74, 75]、能源和生物医学领域的应用等[82-84]。

LAL 纳米制备技术最重要的因素是激光器的选择。以往，LAL 研究主要使用纳秒激光器，纳米材料的合成也集中在此类激光器上。相比纳秒激光器，那时人们对其他脉宽的激光器了解有限。如今，各种脉宽的激光器都已经用于 LAL 纳米制备技术中，包括毫秒（ms）[85]、微秒（μs）[86]、纳秒（ns）[87]、皮秒（ps）[88]、飞秒（fs）[89]等激光器。研究人员已经证明，在 LAL 纳米制备技术中，不同脉宽的激光在不同的应用场景都有其独特的优势，这极大地拓宽了所制备纳米材料的前驱体的可选范围。注意，液体中的材料与不同脉宽的激光的相互作用有很大差别，例如，纳米液滴由微秒激光产生，而等离子体羽由纳秒激光产生。

在LAL纳米制备技术中,液体种类是仅次于激光器的重要因素。水和常见的有机溶剂,包括酒精、丙酮和十二烷基硫酸钠(SDS),已经取得广泛应用[90-95]。目前,寡核苷酸水溶液[96]、聚合物[97]、液氮[98]、超临界二氧化碳和极低温度下的液氦[99-101]等可作为独特的溶剂进行纳米晶合成和纳米结构组装。这些新溶剂的使用极大地拓展了LAL技术在催化化学和生物医学中的应用。特别是在局域极端环境下的LAL常常会产生一些非常有趣的物理现象,而它们在其他过程中是很难发生的。

额外附加的辅助环境是能极大地拓展LAL技术功能的因素。例如,最近发展的各种LAL技术,温度场辅助LAL(temperature field-assisted LAL,TFLAL)[102, 103]、电场辅助LAL(electric field-assisted LAL,EFLAL)[104, 105]、MFLAL[67, 68],以及电化学辅助LAL(electrochemistry-assisted LAL,ECLAL)[106, 107]等。重要的是,实验证明,LAL制备的纳米材料的形态、组分和结构可以通过辅助环境很容易地操控。此外,许多基于LAL的微米/纳米加工技术,如液体中的脉冲激光沉积和液体中纳米图案的激光诱导刻蚀等[108, 109],都已经被开发出来,这些技术在纳米材料合成和纳米结构组装方面显示出了巨大的潜在应用价值。

综合考虑以上三个因素,我们可以看出,LAL可以充分实现纳米材料的形貌控制和组分调控[110]。例如,多种形态纳米结构包括纳米棒[111]、纳米片[112-114]、纳米层状[115]、纳米纺锤体[104, 105]、纳米管[115]、空心纳米颗粒[85, 116]和纳米立方体[117, 118]等都可以采用LAL技术进行制备。还有多组分纳米材料,如金属[119, 120]、氧化物[121]、氮化物[122]、碳化物[123, 124]、硫化物[125]、硒化物[126]、合金[106, 107]和多金属氧酸盐(polyoxometalate,POM)[107, 111]等。我们在图1-3中总结了形貌较为独特的若干类微米/纳米结构。由于近年的研究积累,LAL纳米制备技术已逐渐成为一种众所周知的纳米材料合成和纳米结构组装的有效方法,所制备的纳米材料广泛应用于光学、磁学、能源、环境和生物医学等领域[57, 126-139],尤其是其独特的洁净、无杂质吸附的高活性表面,赋予了LAL合成的纳米材料在环境和生物医学等领域应用的巨大优势[96, 133]。如图1-4所示。

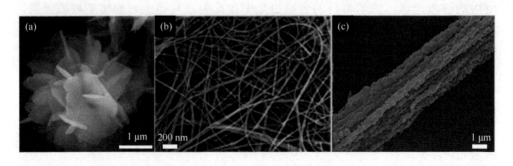

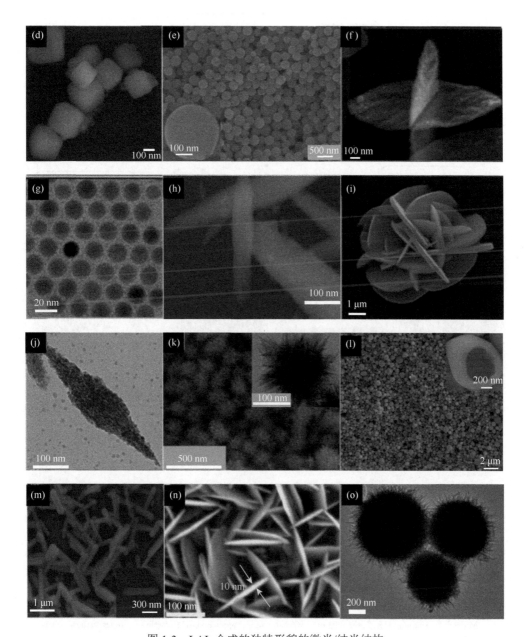

图 1-3 LAL 合成的独特形貌的微米/纳米结构

(a) 氢氧化锌/十二烷基硫酸盐纳米结构；(b) MnOOH 纳米线；(c) Fe_3C 微纳纤维；(d) C_8 纳米立方体；(e) CuO 亚微米球；(f) $H_2WO_4·H_2O$ 纳米片；(g) Cd 单分散量子点；(h) GeO_2 纳米纺锤体；(i) $ZnMoO_4$ 纳米花；(j) CuO 纳米纺锤体；(k) 海胆状 $ZnSnO_3$ 纳米颗粒；(l) Co_3O_4 空心纳米球；(m) $Cu_3Mo_2O_9$ 纳米棒；(n) Ge 掺杂 $α-Fe_2O_3$ 纳米片；(o) 栗子状 $Fe_3O_4@C@ZnSnO_3$ 核-壳纳米结构

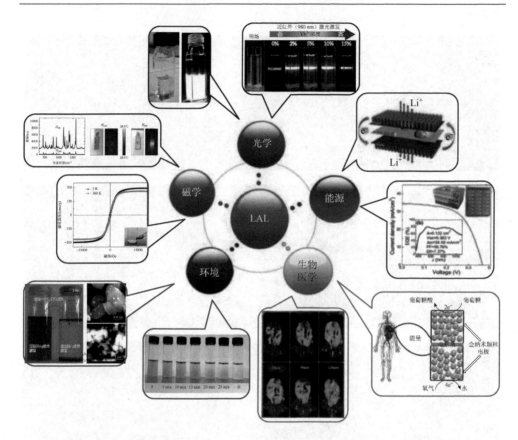

图 1-4　LAL 所制备的纳米材料在光学、磁学、能源、环境及生物医学等领域的应用

参 考 文 献

[1] Srinivasan R，Braren B. Ultraviolet laser ablation of organic polymers[J]. Chemical Reviews，1989，89（6）: 1303-1316.

[2] Lazare S，Granier V. Ultraviolet laser photoablation of polymers：A review and recent results[J]. Laser Chemistry，1989，10: 25-40.

[3] Root R G. Laser-induced plasma and applications，chapter 2[M]. New York：Marcel Dekker，1989.

[4] Bäuerle D. Laser processing and chemistry[M]. Berlin：Springer-Verlag，2000.

[5] Miller J C，Haglung Jr R. Laser ablation：Mechanisms and applications[M]. Berlin：Sprnger-Verlag，1991.

[6] Fogarassy E，Lazare S. Laser ablation of electronic materials：Basic mechanisms and applications[M]. Amsterdam：Elsevier，1992.

[7] Chrisey D B，Hubler G K. Pulsed laser deposition of thin solid films[M]. New York：Wiley-Interscience，1994.

[8] Ashfold M N R，Claeyssens F，Fuge G M，et al. Pulsed laser ablation and deposition of thin films[J]. Chemical Society Reviews，2004，33（1）: 23-31.

[9] Georgiou S，Hillenkamp F. Introduction：Laser ablation of molecular substrates[J]. Chemical Reviews，2003，

103 (2): 317-320.

[10] Fabbro R, Fournier J, Ballard P, et al. Physical study of laser-produced plasma in confined geometry[J]. Journal of Applied Physics, 1990, 68 (2): 775-784.

[11] Yavas O, Leiderer P, Park H K, et al. Optical reflectance and scattering studies of nucleation and growth of bubbles at a liquid-solid interface induced by pulsed laser heating[J]. Physical Review Letters, 1993, 70 (12): 1830.

[12] Georgiou S, Koubenakis A. Laser-induced material ejection from model molecular solids and liquids: Mechanisms, implications, and applications[J]. Chemical Reviews, 2003, 103 (2): 349-394.

[13] Vogel A, Venugopalan V. Mechanisms of pulsed laser ablation of biological tissues[J]. Chemical Reviews, 2003, 103 (2): 577-644.

[14] Fabbro R, Peyre P, Berthe L, et al. Physics and applications of laser-shock processing[J]. Journal of Laser Applications, 1998, 10 (6): 265-279.

[15] Kruusing A. Underwater and water-assisted laser processing: Part 2: Etching, cutting and rarely used methods[J]. Optics and Lasers in Engineering, 2004, 41 (2): 329-352.

[16] Luk'yanchuk B. Laser cleaning[M]. Singapore: World Scientific, 2000.

[17] Patil P P, Phase D M, Kulkarni S A, et al. Pulsed-laser-induced reactive quenching at liquid-solid interface: Aqueous oxidation of iron[J]. Physical Review Letters, 1987, 58 (3): 238.

[18] Ogale S B. Pulsed-laser-induced and ion-beam-induced surface synthesis and modification of oxides, nitrides and carbides[J]. Thin Solid Films, 1988, 163: 215-227.

[19] Ogale S B, Malshe A P, Kanetkar S M, et al. Formation of diamond particulates by pulsed ruby laser irradiation of graphite immersed in benzene[J]. Solid State Communications, 1992, 84 (4): 371-373.

[20] Devaux D, Fabbro R, Tollier L, et al. Generation of shock waves by laser-induced plasma in confined geometry[J]. Journal of Applied Physics, 1993, 74 (4): 2268-2273.

[21] Peyre P, Fabbro R. Laser shock processing: A review of the physics and applications[J]. Optical and Quantum Electronics, 1995, 27: 1213-1229.

[22] Berthe L, Fabbro R, Peyre P, et al. Shock waves from a water-confined laser-generated plasma[J]. Journal of Applied Physics, 1997, 82 (6): 2826-2832.

[23] Peyre P, Berthe L, Scherpereel X, et al. Laser-shock processing of aluminium-coated 55C1 steel in water-confinement regime, characterization and application to high-cycle fatigue behaviour[J]. Journal of Materials Science, 1998, 33 (6): 1421-1429.

[24] Berthe L, Fabbro R, Peyre P, et al. Wavelength dependent of laser shock-wave generation in the water-confinement regime[J]. Journal of Applied Physics, 1999, 85 (11): 7552-7555.

[25] Berthe L, Sollier A, Peyre P, et al. The generation of laser shock waves in a water-confinement regime with 50 ns and 150 ns XeCl excimer laser pulses[J]. Journal of Physics D: Applied Physics, 2000, 33 (17): 2142.

[26] Peyre P, Berthe L, Fabbro R, et al. Experimental determination by PVDF and EMV techniques of shock amplitudes induced by 0.6~3 ns laser pulses in a confined regime with water[J]. Journal of Physics D: Applied Physics, 2000, 33 (5): 498.

[27] Sollier A, Berthe L, Fabbro R. Numerical modeling of the transmission of breakdown plasma generated in water during laser shock processing[J]. The European Physical Journal-Applied Physics, 2001, 16 (2): 131-139.

[28] Yavas O, Leiderer P, Park H K, et al. Enhanced acoustic cavitation following laser-induced bubble formation: Long-term memory effect[J]. Physical Review Letters, 1994, 72 (13): 2021.

[29] Yavas O, Leiderer P, Park H K, et al. Optical and acoustic study of nucleation and growth of bubbles at a

liquid-solid interface induced by nanosecond-pulsed-laser heating[J]. Applied Physics A, 1994, 58: 407-415.

[30] Park H K, Grigoropoulos C P, Poon C C, et al. Optical probing of the temperature transients during pulsed-laser induced boiling of liquids[J]. Applied Physics Letters, 1996, 68 (5): 596-598.

[31] Park H K, Zhang X, Grigoropoulos C P, et al. Transient temperature during the vaporization of liquid on a pulsed laser-heated solid surface[J]. ASME Joural of Heat and Mass Transfer, 1996, 118 (3): 702-708.

[32] Park H K, Kim D, Grigoropoulos C P, et al. Pressure generation and measurement in the rapid vaporization of water on a pulsed-laser-heated surface[J]. Journal of Applied Physics, 1996, 80 (7): 4072-4081.

[33] Kim D, Park H K, Grigoropoulos C P. Interferometric probing of rapid vaporization at a solid-liquid interface induced by pulsed-laser irradiation[J]. International Journal of Heat and Mass Transfer, 2001, 44(20): 3843-3853.

[34] Sakka T, Iwanaga S, Ogata Y H, et al. Laser ablation at solid-liquid interfaces: An approach from optical emission spectra[J]. The Journal of Chemical Physics, 2000, 112 (19): 8645-8653.

[35] Sakka T, Takatani K, Ogata Y H, et al. Laser ablation at the solid-liquid interface: Transient absorption of continuous spectral emission by ablated aluminium atoms[J]. Journal of Physics D: Applied Physics, 2001, 35 (1): 65.

[36] Sakka T, Saito K, Ogata Y H. Emission spectra of the species ablated from a solid target submerged in liquid: Vibrational temperature of C_2 molecules in water-confined geometry[J]. Applied Surface Science, 2002, 197-198: 246-250.

[37] Saito K, Takatani K, Sakka T, et al. Observation of the light emitting region produced by pulsed laser irradiation to a solid-liquid interface[J]. Applied Surface Science, 2002, 197-198: 56-60.

[38] Saito K, Sakka T, Ogata Y H. Rotational spectra and temperature evaluation of C_2 molecules produced by pulsed laser irradiation to a graphite-water interface[J]. Journal of Applied Physics, 2003, 94 (9): 5530-5536.

[39] Furusawa H, Sakka T, Ogata Y H. Characterization of ablated species in laser-induced plasma plume[J]. Journal of Applied Physics, 2004, 96 (2): 975-982.

[40] Yang G W, Wang J B, Liu Q X. Preparation of nano-crystalline diamonds using pulsed laser induced reactive quenching[J]. Journal of Physics: Condensed Matter, 1998, 10 (35): 7923.

[41] Yang G W. Laser ablation in liquids: Applications in the synthesis of nanocrystals[J]. Progress in Materials Science, 2007, 52 (4): 648-698.

[42] Xiao J, Liu P, Wang C X, et al. External field-assisted laser ablation in liquid: An efficient strategy for nanocrystal synthesis and nanostructure assembly[J]. Progress in Materials Science, 2017, 87: 140-220.

[43] Peng X G, Manna L, Yang W D, et al. Shape control of CdSe nanocrystals[J]. Nature, 2000, 404: 59-61.

[44] Puntes V F, Krishnan K M, Alivisatos A P. Colloidal nanocrystal shape and size control: The case of cobalt[J]. Science, 2001, 291 (5511): 2115-2117.

[45] Tao A R, Habas S, Yang P D. Shape control of colloidal metal nanocrystals[J]. Small, 2008, 4 (3): 310-325.

[46] Xia Y N, Xiong Y J, Lim B, et al. Shape-controlled synthesis of metal nanocrystals: Simple chemistry meets complex physics? [J]. Angewandte Chemie International Edition, 2009, 48 (1): 60-103.

[47] Huang M H, Mao S, Feick H, et al. Room-temperature ultraviolet nanowire nanolasers[J]. Science, 2001, 292 (5523): 1897-1899.

[48] Dasgupta N P, Sun J W, Liu C, et al. 25th anniversary article: Semiconductor nanowires-synthesis, characterization, and applications[J]. Advanced Materials, 2014, 26 (14): 2137-2184.

[49] Farrell A C, Lee W J, Senanayake P, et al. High-quality InAsSb nanowires grown by catalyst-free selective-area metal-organic chemical vapor deposition[J]. Nano Letters, 2015, 15 (10): 6614-6619.

[50] Xiang B, Wang P W, Zhang X Z, et al. Rational synthesis of p-type zinc oxide nanowire arrays using simple chemical vapor deposition[J]. Nano Letters, 2007, 7 (2): 323-328.

[51] Zhang H, Jin M S, Xiong Y J, et al. Shape-controlled synthesis of Pd nanocrystals and their catalytic applications[J]. Accounts of Chemical Research, 2013, 46 (8): 1783-1794.

[52] Zhang H, Jin M S, Xia Y N. Noble-metal nanocrystals with concave surfaces: Synthesis and applications[J]. Angewandte Chemie International Edition, 2012, 51 (31): 7656-7673.

[53] Xia Y N, Xia X H, Wang Y, et al. Shape-controlled synthesis of metal nanocrystals[J]. MRS Bulletin, 2013, 38 (4): 335-344.

[54] Zhang G M, Surwade S P, Zhou F, et al. DNA nanostructure meets nanofabrication[J]. Chemical Society Reviews, 2013, 42 (7): 2488-2496.

[55] Qi H, Ghodousi M, Du Y N, et al. DNA-directed self-assembly of shape-controlled hydrogels[J]. Nature Communications, 2013, 4: 2275.

[56] Howorka S. DNA nanoarchitectonics: Assembled DNA at interfaces[J]. Langmuir, 2013, 29 (24): 7344-7353.

[57] Liang D W, Wu S L, Wang P P, et al. Recyclable chestnut-like $Fe_3O_4@C@ZnSnO_3$ core-shell particles for the photocatalytic degradation of 2, 5-dichlorophenol[J]. RSC Advances, 2014, 4 (50): 26201-26206.

[58] Yang G W. Laser ablation in liquids: Principles and applications in the preparation of nanomaterials[M]. New York: CRC Press, 2012.

[59] Amendola V, Meneghetti M. Laser ablation synthesis in solution and size manipulation of noble metal nanoparticles[J]. Physical Chemistry Chemical Physics, 2009, 11 (20): 3805-3821.

[60] Amendola V, Meneghetti M. What controls the composition and the structure of nanomaterials generated by laser ablation in liquid solution?[J]. Physical Chemistry Chemical Physics, 2013, 15 (9): 3027-3046.

[61] Liu P, Cui H, Wang C X, et al. From nanocrystal synthesis to functional nanostructure fabrication: Laser ablation in liquid[J]. Physical Chemistry Chemical Physics, 2010, 12 (16): 3942-3952.

[62] Zeng H B, Du X W, Singh S C, et al. Nanomaterials via laser ablation/irradiation in liquid: A review[J]. Advanced Functional Materials, 2012, 22 (7): 1333-1353.

[63] Barcikowski S, Compagnini G. Advanced nanoparticle generation and excitation by lasers in liquids[J]. Physical Chemistry Chemical Physics, 2013, 15 (9): 3022-3026.

[64] Asahi T, Mafuné F, Rehbock C, et al. Strategies to harvest the unique properties of laser-generated nanomaterials in biomedical and energy applications[J]. Applied Surface Science, 2015, 348: 1-3.

[65] Wang H Q, Pyatenko A, Kawaguchi K, et al. Selective pulsed heating for the synthesis of semiconductor and metal submicrometer spheres[J]. Angewandte Chemie International Edition, 2010, 49 (36): 6361-6364.

[66] Liang Y, Liu P, Xiao J, et al. A microfibre assembly of an iron-carbon composite with giant magnetisation[J]. Scientific Reports, 2013, 3: 3051.

[67] Liang Y, Liu P, Xiao J, et al. A general strategy for one-step fabrication of one-dimensional magnetic nanoparticle chains based on laser ablation in liquid[J]. Laser Physics Letters, 2014, 11 (5): 056001.

[68] Liang Y, Liu P, Yang G W. Fabrication of one-dimensional chain of iron-based bimetallic alloying nanoparticles with unique magnetizations[J]. Crystal Growth & Design, 2014, 14 (11): 5847-5855.

[69] Barcikowski S, Devesa F, Moldenhauer K. Impact and structure of literature on nanoparticle generation by laser ablation in liquids[J]. Journal of Nanoparticle Research, 2009, 11: 1883-1893.

[70] De Giacomo A, Dell'Aglio M, Santagata A, et al. Cavitation dynamics of laser ablation of bulk and wire-shaped metals in water during nanoparticles production[J]. Physical Chemistry Chemical Physics, 2013, 15 (9):

3083-3092.

[71] Ibrahimkutty S, Wagener P, Menzel A, et al. Nanoparticle formation in a cavitation bubble after pulsed laser ablation in liquid studied with high time resolution small angle X-ray scattering[J]. Applied Physics Letters, 2012, 101 (10): 103104.

[72] Wei S Y, Saitow K. In situ multipurpose time-resolved spectrometer for monitoring nanoparticle generation in a high-pressure fluid[J]. Review of Scientific Instruments, 2012, 83 (7): 073110.

[73] Yang J, Ling T, Wu W T, et al. A top-down strategy towards monodisperse colloidal lead sulphide quantum dots[J]. Nature Communications, 2013, 4: 1695.

[74] Streubel R, Bendt G, Gökce B. Pilot-scale synthesis of metal nanoparticles by high-speed pulsed laser ablation in liquids[J]. Nanotechnology, 2016, 27 (20): 205602.

[75] Streubel R, Barcikowski S, Gökce B. Continuous multigram nanoparticle synthesis by high-power, high-repetition-rate ultrafast laser ablation in liquids[J]. Optics Letters, 2016, 41 (7): 1486-1489.

[76] Matsumoto A, Tamura A, Kawasaki A, et al. Comparison of the overall temporal behavior of the bubbles produced by short-and long-pulse nanosecond laser ablations in water using a laser-beam-transmission probe[J]. Applied Physics A, 2016, 122: 234.

[77] Fischer M, Hormes J, Marzun G, et al. In situ investigations of laser-generated ligand-free platinum nanoparticles by X-ray absorption spectroscopy: How does the immediate environment influence the particle surface? [J]. Langmuir, 2016, 32 (35): 8793-8802.

[78] Tanabe R, Nguyen T T P, Sugiura T, et al. Bubble dynamics in metal nanoparticle formation by laser ablation in liquid studied through high-speed laser stroboscopic videography[J]. Applied Surface Science, 2015, 351: 327-331.

[79] Ibrahimkutty S, Wagener P, Rolo T S, et al. A hierarchical view on material formation during pulsed-laser synthesis of nanoparticles in liquid[J]. Scientific Reports, 2015, 5: 16313.

[80] Soliman W, Nakano T, Takada N, et al. Modification of rayleigh-plesset theory for reproducing dynamics of cavitation bubbles in liquid-phase laser ablation[J]. Japanese Journal of Applied Physics, 2010, 49 (11R): 116202.

[81] Kohsakowski S, Gökce B, Tanabe R, et al. Target geometry and rigidity determines laser-induced cavitation bubble transport and nanoparticle productivity-a high-speed videography study[J]. Physical Chemistry Chemical Physics, 2016, 18 (24): 16585-16593.

[82] Gamrad L, Rehbock C, Westendorf A M, et al. Efficient nucleic acid delivery to murine regulatory T cells by gold nanoparticle conjugates[J]. Scientific Reports, 2016, 6: 28709.

[83] Krawinkel J, Richter U, Torres-Mapa M L, et al. Optical and electron microscopy study of laser-based intracellular molecule delivery using peptide-conjugated photodispersible gold nanoparticle agglomerates[J]. Journal of Nanobiotechnology, 2016, 14: 2.

[84] Streich C, Akkari L, Decker C, et al. Characterizing the effect of multivalent conjugates composed of Aβ-specific ligands and metal nanoparticles on neurotoxic fibrillar aggregation[J]. ACS Nano, 2016, 10 (8): 7582-7597.

[85] Niu K Y, Yang J, Kulinich S A, et al. Hollow nanoparticles of metal oxides and sulfides: Fast preparation via laser ablation in liquid[J]. Langmuir, 2010, 26 (22): 16652-16657.

[86] Luo N Q, Yang C, Tian X M, et al. A general top-down approach to synthesize rare earth doped-Gd_2O_3 nanocrystals as dualmodal contrast agents[J]. Journal of Materials Chemistry B, 2014, 2 (35): 5891-5897.

[87] Tsuji T, Okazaki Y, Tsuboi Y, et al. Nanosecond time-resolved observations of laser ablation of silver in water[J]. Japanese Journal of Applied Physics, 2007, 46 (4R): 1533.

[88] Barchanski A, Funk D, Wittich O, et al. Picosecond laser fabrication of functional gold-antibody nanoconjugates

for biomedical applications[J]. The Journal of Physical Chemistry C, 2015, 119 (17): 9524-9533.

[89] Petersen S, Barcikowski S. In situ bioconjugation: Single step approach to tailored nanoparticle-bioconjugates by ultrashort pulsed laser ablation[J]. Advanced Functional Materials, 2009, 19 (8): 1167-1172.

[90] Liang C H, Shimizu Y, Masuda M, et al. Preparation of layered zinc hydroxide/surfactant nanocomposite by pulsed-laser ablation in a liquid medium[J]. Chemistry of Materials, 2004, 16 (6): 963-965.

[91] Liang C H, Shimizu Y, Sasaki T, et al. Preparation of ultrafine TiO_2 nanocrystals via pulsed-laser ablation of titanium metal in surfactant solution[J]. Applied Physics A, 2005, 80: 819-822.

[92] Kabashin A V, Meunier M. Synthesis of colloidal nanoparticles during femtosecond laser ablation of gold in water[J]. Journal of Applied Physics, 2003, 94 (12): 7941-7943.

[93] Tsuji T, Hamagami T, Kawamura T, et al. Laser ablation of cobalt and cobalt oxides in liquids: Influence of solvent on composition of prepared nanoparticles[J]. Applied Surface Science, 2005, 243 (1-4): 214-219.

[94] Dolgaev S I, Simakin A V, Voronov V V, et al. Nanoparticles produced by laser ablation of solids in liquid environment[J]. Applied Surface Science, 2002, 186 (1-4): 546-551.

[95] Usui H, Shimizu Y, Sasaki T, et al. Photoluminescence of ZnO nanoparticles prepared by laser ablation in different surfactant solutions[J]. The Journal of Physical Chemistry B, 2005, 109 (1): 120-124.

[96] Petersen S, Barcikowski S. Conjugation efficiency of laser-based bioconjugation of gold nanoparticles with nucleic acids[J]. The Journal of Physical Chemistry C, 2009, 113 (46): 19830-19835.

[97] Stelzig S H, Menneking C, Hoffmann M S, et al. Compatibilization of laser generated antibacterial Ag-and Cu-nanoparticles for perfluorinated implant materials[J]. European Polymer Journal, 2011, 47 (4): 662-667.

[98] Xiao J, Liu P, Liang Y, et al. Super-stable ultrafine beta-tungsten nanocrystals with metastable phase and related magnetism[J]. Nanoscale, 2013, 5 (3): 899-903.

[99] Kato T, Stauss S, Kato S, et al. Pulsed laser ablation plasmas generated in CO_2 under high-pressure conditions up to supercritical fluid[J]. Applied Physics Letters, 2012, 101 (22): 224103.

[100] Lebedev V, Moroshkin P, Grobety B, et al. Formation of metallic nanowires by laser ablation in liquid helium[J]. Journal of Low Temperature Physics, 2011, 165: 166.

[101] Saitow K, Yamamura T, Minami T. Gold nanospheres and nanonecklaces generated by laser ablation in supercritical fluid[J]. The Journal of Physical Chemistry C, 2008, 112 (47): 18340-18349.

[102] Xiao J, Liu P, Liang Y, et al. High aspect ratio β-MnO_2 nanowires and sensor performance for explosive gases[J]. Journal of Applied Physics, 2013, 114 (7): 073513.

[103] Xiao J, Liu P, Liang Y, et al. Porous tungsten oxide nanoflakes for highly alcohol sensitive performance[J]. Nanoscale, 2012, 4 (22): 7078-7083.

[104] Liu P, Wang C X, Chen X Y, et al. Controllable fabrication and cathodoluminescence performance of high-index facets GeO_2 micro-and nanocubes and spindles upon electrical-field-assisted laser ablation in liquid[J]. The Journal of Physical Chemistry C, 2008, 112 (35): 13450-13456.

[105] Lin X Z, Liu P, Yu J M, et al. Synthesis of CuO nanocrystals and sequential assembly of nanostructures with shape-dependent optical absorption upon laser ablation in liquid[J]. The Journal of Physical Chemistry C, 2009, 113 (40): 17543-17547.

[106] Liang Y, Liu P, Li H B, et al. $ZnMoO_4$ micro-and nanostructures synthesized by electrochemistry-assisted laser ablation in liquids and their optical properties[J]. Crystal Growth & Design, 2012, 12 (9): 4487-4493.

[107] Liang Y, Liu P, Li H B, et al. Synthesis and characterization of copper vanadate nanostructures via electrochemistry assisted laser ablation in liquid and the optical multi-absorptions performance[J].

CrystEngComm, 2012, 14 (9): 3291-3296.

[108] Cui H, Liu P, Yang G W. Noble metal nanoparticle patterning deposition using pulsed-laser deposition in liquid for surface-enhanced Raman scattering[J]. Applied Physics Letters, 2006, 89 (15): 153124.

[109] Liu P, Wang C X, Chen J, et al. Localized nanodiamond crystallization and field emission performance improvement of amorphous carbon upon laser irradiation in liquid[J]. The Journal of Physical Chemistry C, 2009, 113 (28): 12154-12161.

[110] Yan Z J, Chrisey D B. Pulsed laser ablation in liquid for micro-/nanostructure generation[J]. Journal of Photochemistry and Photobiology C: Photochemistry Reviews, 2012, 13 (3): 204-223.

[111] Liu P, Liang Y, Lin X Z, et al. A general strategy to fabricate simple polyoxometalate nanostructures: Electrochemistry-assisted laser ablation in liquid[J]. Acs Nano, 2011, 5 (6): 4748-4755.

[112] Liang C H, Sasaki T, Shimizu Y, et al. Pulsed-laser ablation of Mg in liquids: Surfactant-directing nanoparticle assembly for magnesium hydroxide nanostructures[J]. Chemical Physics Letters, 2004, 389 (1-3): 58-63.

[113] Yan Z J, Bao R Q, Chrisey D B. Generation of Ag_2O micro-/nanostructures by pulsed excimer laser ablation of Ag in aqueous solutions of polysorbate 80[J]. Langmuir, 2011, 27 (2): 851-855.

[114] He C, Sasaki T, Zhou Y, et al. Surfactant-assisted preparation of novel layered silver bromide-based inorganic/organic nanosheets by pulsed laser ablation in aqueous media[J]. Advanced Functional Materials, 2007, 17 (17): 3554-3561.

[115] Nistor L C, Epurescu G, Dinescu M, et al. Boron nitride nano-structures produced by pulsed laser ablation in acetone[C]//IOP Conference Series: Materials Science and Engineering, 2010, 15 (1): 012067.

[116] Yan Z J, Bao R Q, Busta C M, et al. Fabrication and formation mechanism of hollow MgO particles by pulsed excimer laser ablation of Mg in liquid[J]. Nanotechnology, 2011, 22 (26): 265610.

[117] Yan Z J, Compagnini G, Chrisey D B. Generation of AgCl cubes by excimer laser ablation of bulk Ag in aqueous NaCl solutions[J]. The Journal of Physical Chemistry C, 2011, 115 (12): 5058-5062.

[118] Liu P, Cao Y L, Wang C X, et al. Micro-and nanocubes of carbon with C_8-like and blue luminescence[J]. Nano Letters, 2008, 8 (8): 2570-2575.

[119] Muñetón Arboleda D, Santillán J M J, Mendoza Herrera L J, et al. Synthesis of Ni nanoparticles by femtosecond laser ablation in liquids: Structure and sizing[J]. The Journal of Physical Chemistry C, 2015, 119 (23): 13184-13193.

[120] Xu X X, Duan G T, Li Y, et al. Fabrication of gold nanoparticles by laser ablation in liquid and their application for simultaneous electrochemical detection of Cd^{2+}, Pb^{2+}, Cu^{2+}, Hg^{2+}[J]. ACS Applied Materials & Interfaces, 2014, 6 (1): 65-71.

[121] Blakemore J D, Gray H B, Winkler J R, et al. Co_3O_4 nanoparticle water-oxidation catalysts made by pulsed-laser ablation in liquids[J]. ACS Catalysis, 2013, 3 (11): 2497-2500.

[122] Liu H, Jin P, Xue Y M, et al. Photochemical synthesis of ultrafine cubic boron nitride nanoparticles under ambient conditions[J]. Angewandte Chemie, 2015, 127 (24): 7157-7160.

[123] Barmina E V, Serkov A A, Shafeev G A, et al. Nanostructuring of single-crystal silicon carbide by femtosecond laser irradiation in a liquid[J]. Physics of Wave Phenomena, 2014, 22: 15-18.

[124] Yang S K, Kiraly B, Wang W Y, et al. Fabrication and characterization of beaded SiC quantum rings with anomalous red spectral shift[J]. Advanced Materials, 2012, 24 (41): 5598-5603.

[125] Aneesh P M, Shijeesh M R, Aravind A, et al. Highly luminescent undoped and Mn-doped ZnS nanoparticles by liquid phase pulsed laser ablation[J]. Applied Physics A, 2014, 116: 1085-1089.

[126] Feng G Y, Yang C, Zhou S H. Nanocrystalline Cr^{2+}-doped ZnSe nanowires laser[J]. Nano Letters, 2013, 13 (1): 272-275.

[127] Mafuné F, Kohno J, Takeda Y, et al. Formation and size control of silver nanoparticles by laser ablation in aqueous solution[J]. The Journal of Physical Chemistry B, 2000, 104 (39): 9111-9117.

[128] Sylvestre J P, Kabashin A V, Sacher E, et al. Stabilization and size control of gold nanoparticles during laser ablation in aqueous cyclodextrins[J]. Journal of the American Chemical Society, 2004, 126 (23): 7176-7177.

[129] Sylvestre J P, Kabashin A V, Sacher E, et al. Femtosecond laser ablation of gold in water: Influence of the laser-produced plasma on the nanoparticle size distribution[J]. Applied Physics A, 2005, 80: 753-758.

[130] Liu J, Deng H W, Huang Z Y, et al. Phonon-assisted energy back transfer-induced multicolor upconversion emission of Gd_2O_3 : Yb^{3+}/Er^{3+} nanoparticles under near-infrared excitation[J]. Physical Chemistry Chemical Physics, 2015, 17 (23): 15412-15418.

[131] Amendola V, Scaramuzza S, Agnoli S, et al. Laser generation of iron-doped silver nanotruffles with magnetic and plasmonic properties[J]. Nano Research, 2015, 8: 4007-4023.

[132] Xin Y Z, Nishio K, Saitow K. White-blue electroluminescence from a Si quantum dot hybrid light-emitting diode[J]. Applied Physics Letters, 2015, 106 (20): 201102.

[133] Xiao J, Wu Q L, Liu P, et al. Highly stable sub-5 nm $Sn_6O_4(OH)_4$ nanocrystals with ultrahigh activity as advanced photocatalytic materials for photodegradation of methyl orange[J]. Nanotechnology, 2014, 25 (13): 135702.

[134] Yan Z J, Bao R Q, Chrisey D B. Self-assembly of zinc hydroxide/dodecyl sulfate nanolayers into complex three-dimensional nanostructures by laser ablation in liquid[J]. Chemical Physics Letters, 2010, 497 (4-6): 205-207.

[135] Wang H Q, Kawaguchi K, Pyatenko A, et al. General bottom-up construction of spherical particles by pulsed laser irradiation of colloidal nanoparticles: A case study on CuO[J]. Chemistry: A European Journal, 2012, 18 (1): 163-169.

[136] Luo R C, Li C, Du X W, et al. Direct conversion of bulk metals to size-tailored, monodisperse spherical non-coinage-metal nanocrystals[J]. Angewandte Chemie International Edition, 2015, 54 (16): 4787-4791.

[137] Tian Z F, Liang C H, Liu J, et al. Zinc stannate nanocubes and nanourchins with high photocatalytic activity for methyl orange and 2, 5-DCP degradation[J]. Journal of Materials Chemistry, 2012, 22 (33): 17210-17214.

[138] Wang H Q, Miyauchi M, Ishikawa Y, et al. Single-crystalline rutile TiO_2 hollow spheres: Room-temperature synthesis, tailored visible-light-extinction, and effective scattering layer for quantum dot-sensitized solar cells[J]. Journal of the American Chemical Society, 2011, 133 (47): 19102-19109.

[139] Liu J, Cai Y Y, Tian Z F, et al. Highly oriented Ge-doped hematite nanosheet arrays for photoelectrochemical water oxidation[J]. Nano Energy, 2014, 9: 282-290.

第 2 章 气体环境中固体靶的激光烧蚀

自 20 世纪 60 年代初,将红宝石激光器用于固体靶的 PLA 以来,随着先进脉冲激光器的发展,固体 PLA 已经广泛应用于激光材料的制备与加工。众所周知,目前,基于 PLA 的材料制备有两种常用的技术。一种是用于制备固体薄膜的脉冲激光沉积(PLD)[1],另一种是气体环境中的固体 PLA 用于合成纳米颗粒[2-14]。此外,固体 PLA 还用于在聚合物、宽带隙半导体和清洁材料表面上制造功能微纳结构和纳米图案[15-21]。本书讲述的是薄膜材料制备和纳米颗粒合成,所以,如果读者对固体 PLA 在微纳结构和图案组装方面的应用感兴趣的话,读者可参考 Ashfold 等[22]的相关综述。由于材料制备中大部分固体 PLA 通常在真空或稀释气体环境中进行,所以研究人员系统深入地研究了脉冲激光与气-固界面处的固体靶的相互作用[23],并给出了固体 PLA 的清晰的物理图像[24-25]。为了更好地理解液体脉冲激光烧蚀,我们需要比较真空环境与稀释气体环境中的固体 PLA。为此,我们简要介绍激光材料制备与加工中气体环境中的固体 PLA 的基本过程。

通常,在激光材料制备与加工中,固体 PLA 是在真空或稀释气体环境中进行的,激光烧蚀的产物直接由脉冲激光辐照固体靶表面所产生的等离子体羽凝结生成。脉冲激光辐照固体靶表面产生等离子体羽的基本过程分为三步:等离子体羽的产生、等离子体羽的相变、等离子体羽的凝结,如图 2-1 所示。等离子体羽在激光材料制备中起着重要作用。值得注意的是,在激光烧蚀固体的过程中,短脉冲(纳秒)和超短脉冲(例如,皮秒和飞秒)激光产生等离子体羽的物理机制是不同的。由于目前国际上激光材料制备领域应用最广泛的是短脉冲激光(纳秒激光),因此本章中大多数实例用的都是纳秒激光。

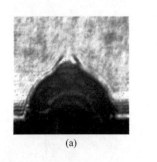 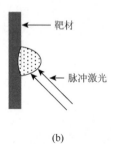 

(a)   (b)   (c)

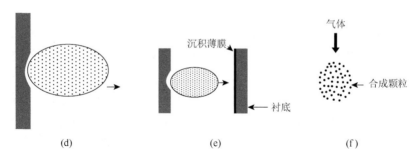

图 2-1 脉冲激光辐照固体靶表面产生等离子体羽的基本过程

(a) 1200 ps 延迟后，在空气中使用脉冲激光辐照固体靶表面产生的等离子体羽的阴影图像。(b)～(d) 在真空或稀释气体环境中产生的等离子体羽的三个典型演化阶段，其中靶材为纯 Cu，激光器为脉宽 35 ps、波长 1064 nm、能量 4.5 mJ 的 Nd：YAG 激光器；(b) 当脉冲激光的前端烧蚀靶材时，在其表面产生等离子体羽；(c) 等离子体羽吸收脉冲激光的后端能量，在真空或气体环境中自由膨胀；(d) 等离子体羽从固体靶表面喷射而出。(e) 和 (f) 分别为等离子体羽的两种不同的凝结方式：一种在衬底上沉积为薄膜，另一种在气体中形成纳米颗粒

对于纳秒激光的固体材料烧蚀，图 2-1 所示的材料粒子喷射是以热过程为主的[22]。在纳秒激光烧蚀过程中，光子可以与靶材的电子模式和振动模式耦合，并且，光子-电子耦合会导致电子温度迅速升高，并最终使得靶材瞬时加热蒸发。当蒸发开始时，靶材表面蒸发的材料粒子会受热膨胀，这样，蒸发粒子组成的羽状气团（蒸气羽）就会和环境气体发生相互作用，使得蒸气羽的体积受到限制，而环境气体则会被进一步推离固体靶表面。由于蒸气羽进一步吸收脉冲激光的能量被加热至极高的温度（超过蒸发粒子的离化温度），蒸气羽就会转变为等离子体羽，也就是脉冲激光辐照固体靶表面的前期将产生等离子体羽，如图 2-1（a）和图 2-1（b）所示。等离子体羽的激发和电离主要是由脉冲激光引起的多光子吸收、电离和轫致辐射逆过程的结果[26]。因此，等离子体羽含有大量来自固体靶的中性原子、离子和电子。

在纳秒激光烧蚀过程中，电子的影响通常主导了等离子体羽的产生，因为光子-电子耦合可以导致电子分布的快速激发（a rapid excitation of the electron distribution）。需要注意的是，在脉冲激光烧蚀过程中，由于等离子体羽的高温和脉冲激光辐照，固体靶与等离子体羽之间的界面通常处于气态。同时，等离子体羽中的离子和电子相互作用将加速它们在等离子体羽中的运动和碰撞。因此，等离子体羽的温度可以迅速达到热平衡[21, 22]。这样，随着等离子体羽的产生，入射的脉冲激光后端不仅直接辐照等离子体羽，而且还透过等离子体羽不断地辐照并烧蚀固体靶。所以，脉冲激光后端的辐照在等离子体羽的相变阶段起着两个作用。一是激光辐照增强了等离子体羽中中性粒子的激发和电离，其中光子的诱导机制非常重要。二是脉冲激光后端不断辐照等离子体羽和固体靶之间的界面，将更多的来自靶材的蒸发粒子溶入等离子体羽中，从而导致等离子体羽的进一步膨胀。

因此，脉冲激光后端辐照的这些效应不仅导致了真空（气体环境）中等离子体羽的快速持续膨胀，而且将等离子体羽推向一个更高温度、更大物质密度的状态，如图2-1（c）所示[27]。由于反冲效应[28]，等离子体羽从固体靶表面喷射出来，如图2-1（d）所示。

超短脉冲激光（例如，皮秒激光、飞秒激光）烧蚀固体靶的机制与短脉冲激光不同，超短脉冲激光与固体靶的相互作用通常是非平衡的，上述的加热、等离子体羽产生和粒子喷射过程通常发生在脉冲激光作用一段时间之后[29-32]。电子在雪崩电离和光电离作用下，在数十飞秒内[33]被激发到几或几十电子伏特[34-36]，随后从电子到离子的能量转移则是在皮秒量级内[37,38]。所以，在吸收超短脉冲激光期间靶材的晶格温度保持不变。超短脉冲激光能量在光子-电子相互作用过程中主要沉积在靶材的表面层，因此，决定加热体积的是光子吸收深度而不是热扩散深度。在脉冲激光持续时间之后，从电子到晶格的热传递取决于电子-声子耦合强度，并最终导致瞬时加热的靶材发生气化而蒸发。所以，电子主要贡献于超短脉冲激光烧蚀的能量转移。此外，由于极高的温度会使得蒸发气体激发、离化，因而导致等离子体羽的产生、膨胀。但是，与纳秒激光烧蚀中的等离子体羽膨胀是由于吸收了脉冲激光后端的能量不同，超短脉冲激光中等离子体羽在膨胀时没有任何其他加热过程。对于超短脉冲激光烧蚀固体靶，等离子体羽膨胀会很快失去温度（压强），寿命短。重要的是，由于激光（例如，飞秒激光）的作用时间"超短"，其作用时间比体积膨胀的弛豫时间更短，因此激光的加热发生在几乎恒定的体积下，这样就会导致高热弹性压强的积累。由压强松弛引起的光力效应在粒子喷射中发挥着重要作用，因为在脉冲激光持续时间之后，吸收区域的压强显著提高，由激光引起的压强梯度的弛豫驱动了喷射。由此可见，我们需要区分纳秒激光和飞秒激光烧蚀固体靶的两种不同的反冲效应。当飞秒激光烧蚀固体靶时，反冲效应（喷射）通常发生在脉冲激光持续时间之后。

等离子体羽中粒子的剧烈热扩散会导致它们的碰撞、聚集及凝聚，因此在等离子体羽中会发生相变，例如，新分子的形成和成核的发生。显然，等离子体羽的温度、压强和组分对于等离子体羽相变的激发至关重要。著名的萨哈方程（Saha's equation）可以用来估算PLA固体靶产生的等离子体羽的组分：

$$n_i = (2.4 \times 10^{15} T^{3/2} n_n \mathrm{e}^{-E_i/kT})^{1/2} \tag{2-1}$$

式中，n_i和n_n分别是单电荷离子和中性粒子的数密度；T是温度；E_i是电离势。Ashfold等[22]提供了真空（10^{-6} Torr①）环境下PLA石墨靶的情况。假设E_i = 11.26 eV，n_n = 1018 cm^{-1}，设定温度为4500 K，则在能量为20 J/cm的20 ns、193 nm激光烧蚀中，n_i/n_n约为10^{-5}。假设等离子体羽膨胀速度为20 km/s，当20 ns激光

---

① 1 Torr = 1.333 22×10^2 Pa。

烧蚀结束时，10^{15} 个烧蚀原子将被限制在 0.13 mm^3 的体积内，因此，等离子体羽中的局部压强预估为几个巴[①]的数量级。这些结果意味着中性粒子可能是等离子体羽发生相变的主要原因。然而，从热力学的角度来看，新分子的形成应该源于较高能态离子的碰撞，而成核的发生可能是中性粒子聚集的结果。等离子体羽中的压强是一个重要的热力学参数，但是，由于它在等离子体羽膨胀过程中的值较小，因此对真空和稀释气体环境中 PLA 过程的等离子体羽相变影响较小。

PLA 固体靶表面产生的等离子体羽相变的最后阶段是在真空和气体中冷却、凝结，不同的凝结方式对应着不同的材料制备。图 2-1（e）和图 2-1（f）显示了两种常见的等离子体羽凝结方式，一种是等离子体羽在衬底上沉积，形成薄膜，这种方式也称为脉冲激光沉积（PLD），如图 2-1（e）所示。另一种是等离子体羽在气体中自由凝结，形成纳米颗粒，如图 2-1（f）所示。当然，衬底的状态（例如，表面结构和温度）及气体的特性（例如，温度和电离程度）都会对 PLA 产物的晶体结构产生显著的影响。此外，对于特定应用需求，当选择在稀释气体环境中进行 PLA 时，其核心反应将发生在等离子体羽的相变和凝结过程中，这个时候等离子体羽和气体的界面为这些化学反应提供了可能。

因此，在等离子体羽与气体的界面区域可能发生两种化学反应。一种化学反应是，由于界面附近等离子体羽的高温激发，气体的分子首先变成离子，然后这些气体离子与等离子体羽中的离子发生反应，形成新的分子。另一种化学反应是，等离子体羽中的离子通过界面扩散到气体中，并且受界面处等离子体羽中的离子密度梯度影响与气体分子碰撞，发生化学反应。进而，这些扩散离子会通过化学反应从等离子体羽中的高能态降低到气体中相对低的能态，形成新的分子。因此，通常使用氧气和氮气作为环境气体，实现通过 PLD 制备氧化物薄膜或氮化物薄膜[1]。此外，额外的离子束可以有效地帮助 PLD 沉积氧化物薄膜和氮化物薄膜。例如，射频等离子体羽有助于 PLD 在室温下制备高质量的立方氮化硼（c-BN）薄膜[39]。

参 考 文 献

[1] Chrisey D B, Hubler G K. Pulsed laser deposition of thin solid films[M]. New York: Wiley-Interscience, 1994.

[2] El-Shall M S, Li S, Turkki T, et al. Synthesis and photoluminescence of weblike agglomeration of silica nanoparticles[J]. The Journal of Physical Chemistry, 1995, 99 (51): 17805-17809.

[3] Burr T A, Seraphin A A, Werwa E, et al. Carrier transport in thin films of silicon nanoparticles[J]. Physical Review B, 1997, 56 (8): 4818.

[4] Geohegan D B, Puretzky A A, Duscher G, et al. Time-resolved imaging of gas phase nanoparticle synthesis by laser ablation[J]. Applied Physics Letters, 1998, 72 (23): 2987-2989.

---

① 1 巴（bar）= 10^5 Pa。

[5] Geohegan D B, Puretzky A A, Duscher G, et al. Photoluminescence from gas-suspended SiO$_x$ nanoparticles synthesized by laser ablation[J]. Applied Physics Letters, 1998, 73 (4): 438-440.

[6] Becker M F, Brock J R, Cai H, et al. Metal nanoparticles generated by laser ablation[J]. Nanostructured Materials, 1998, 10 (5): 853-863.

[7] Geohegan D B, Puretzky A A, Rader D J. Gas-phase nanoparticle formation and transport during pulsed laser deposition of Y$_1$Ba$_2$Cu$_3$O$_{7-d}$[J]. Applied Physics Letters, 1999, 74 (25): 3788-3790.

[8] Lowndes D H, Rouleau C M, Thundat T G, et al. Silicon and zinc telluride nanoparticles synthesized by low energy density pulsed laser ablation into ambient gases[J]. Journal of Materials Research, 1999, 14 (2): 359-370.

[9] Wu K T, Yao Y D, Wang C R C, et al. Magnetic field induced optical transmission study in an iron nanoparticle ferrofluid[J]. Journal of Applied Physics, 1999, 85 (8): 5959-5961.

[10] Vyazovkin S. Kinetic concepts of thermally stimulated reactions in solids: A view from a historical perspective[J]. International Reviews in Physical Chemistry, 2000, 19 (1): 45-60.

[11] Marine W, Patrone L, Luk'yanchuk B, et al. Strategy of nanocluster and nanostructure synthesis by conventional pulsed laser ablation[J]. Applied Surface Science, 2000, 154-155: 345-352.

[12] Hata K, Fujita M, Yoshida S, et al. Selective adsorption and patterning of Si nanoparticles fabricated by laser ablation on functionalized self-assembled monolayer [J]. Applied Physics Letters, 2001, 79 (5): 692-694.

[13] Belomoin G, Therrien J, Smith A, et al. Observation of a magic discrete family of ultrabright Si nanoparticles[J]. Applied Physics Letters, 2002, 80 (5): 841-843.

[14] Nakata Y, Muramoto J, Okada T, et al. Particle dynamics during nanoparticle synthesis by laser ablation in a background gas[J]. Journal of Applied Physics, 2002, 91 (3): 1640-1643.

[15] Tornari V, Zafiropulos V, Bonarou A, et al. Modern technology in artwork conservation: A laser-based approach for process control and evaluation[J]. Optics and Lasers in Engineering, 2000, 34 (4-6): 309-326.

[16] Lu Y F, Takai M, Shiokawa T, et al. Excimer-laser removal of SiO$_2$ patterns from GaAs substrates[J]. Japanese Journal of Applied Physics, 1994, 33 (3A): L324.

[17] Lu Y F, Takai M, Komuro S, et al. Surface cleaning of metals by pulsed-laser irradiation in air[J]. Applied Physics A, 1994, 59: 281-288.

[18] Lu Y F, Aoyagi Y A Y. Acoustic emission in laser surface cleaning for real-time monitoring[J]. Japanese Journal of Applied Physics, 1995, 34 (11B): L1557.

[19] Elg A P, Andersson M, Rosén A. REMPI as a tool for studies of OH radicals in catalytic reactions[J]. Applied Physics B, 1997, 64: 573-578.

[20] Lu Y F, Song W D, Ang B W, et al. A theoretical model for laser removal of particles from solid surfaces[J]. Applied Physics A: Materials Science & Processing, 1997, 65 (1): 9.

[21] Lu Y F. Laser microprocessing and applications in microelectronics and electronics[M]//Guenther H. International trends in applied optics. Bellingham: SPIE Press, 2002.

[22] Ashfold M N R, Claeyssens F, Fuge G M, et al. Pulsed laser ablation and deposition of thin films[J]. Chemical Society Reviews, 2004, 33 (1): 23-31.

[23] Root R G. Laser-induced plasmas and applications[M]. New York: Marcel Dekker, 1989.

[24] Miller J C, Haglung R. Laser ablation: Mechanisms and applications[M]. Berlin: Sprnger-Verlag, 1991.

[25] Fogarassy E, Lazare S. Laser ablation of electronic materials: Basic mechanisms and applications[M]. Amsterdam: Elsevier, 1992.

[26] Srinivasan R, Braren B. Ultraviolet laser ablation of organic polymers[J]. Chemical Reviews, 1989, 89 (6):

1303-1316.

[27] Angleraud B, Girault C, Champeaux C, et al. Study of the expansion of the laser ablation plume above a boron nitride target[J]. Applied Surface Science, 1996, 96-98: 117-121.

[28] Balazs L, Gijbels R, Vertes A. Expansion of laser-generated plumes near the plasma ignition threshold[J]. Analytical Chemistry, 1991, 63 (4): 314-320.

[29] Sakka T, Takatani K, Ogate Y H, et al. Laser ablation at the solid-liquid interface: Transient absorption of continuous spectral emission by ablated aluminium atoms[J]. Journal of Physics D: Applied Physics, 2002, 35 (1): 65.

[30] Sakka T, Saito K, Ogata Y H. Emission spectra of the species ablated from a solid target submerged in liquid: Vibrational temperature of C_2 molecules in water-confined geometry[J]. Applied Surface Science, 2002, 197-198: 246-250.

[31] Saito K, Takatani K, Sakka T, et al. Observation of the light emitting region produced by pulsed laser irradiation to a solid-liquid interface[J]. Applied Surface Science, 2002, 197-198: 56-60.

[32] Saito K, Sakka T, Ogata Y H. Rotational spectra and temperature evaluation of C_2 molecules produced by pulsed laser irradiation to a graphite-water interface[J]. Journal of Applied Physics, 2003, 94 (9): 5530-5536.

[33] Park H K, Grigoropoulos C P, Poon C C, et al. Optical probing of the temperature transients during pulsed-laser induced boiling of liquids[J]. Applied Physics Letters, 1996, 68 (5): 596-598.

[34] Furusawa H, Sakka T, Ogata Y H. Characterization of laser-induced plasma plume: Comparison between Al and Al_2O_3 targets[J]. Applied Physics A, 2004, 79: 1291-1294.

[35] Wang J B, Hu Z S, Liu Q X, et al. Nanocrystallite preparation by pulsed-laser-induced reaction at liquid-solid interface[C]//Laser Processing of Materials and Industrial Applications II. SPIE, 1998, 3550: 65-71.

[36] El-Shall M S, Li S, Turkki T, et al. Synthesis and photoluminescence of weblike agglomeration of silica nanoparticles[J]. The Journal of Physical Chemistry, 1995, 99 (51): 17805-17809.

[37] Burr T A, Seraphin A A, Werwa E, et al. Carrier transport in thin films of silicon nanoparticles[J]. Physical Review B, 1997, 56 (8): 4818.

[38] Geohegan D B, Puretzky A A, Duscher G, et al. Time-resolved imaging of gas phase nanoparticle synthesis by laser ablation[J]. Applied Physics Letters, 1998, 72 (23): 2987-2989.

[39] Zhang C Y, Zhong X L, Wang J B, et al. Room-temperature growth of cubic nitride boron film by RF plasma enhanced pulsed laser deposition[J]. Chemical Physics Letters, 2003, 370 (3-4): 522-527.

第 3 章　液体环境中固体靶的激光烧蚀

固体材料的激光烧蚀已经有几十年的研究历史,并且已经在激光材料制备与加工领域显示出了巨大的应用前景,这些应用包括薄膜材料制备,纳米材料合成,激光切割、焊接、钻孔、表面清洁,以及器件制造等[1-6]。由于固体靶的激光烧蚀通常都是在真空或气体环境的传统沉积室中进行的,装置条件相对简单,并且激光烧蚀发生在气-固界面,因此,在激光与物质相互作用的研究方面,大多数研究人员将注意力集中在气体环境中的固体靶的激光烧蚀[7, 8]。所以,与真空或稀释气体环境中的激光烧蚀固体靶的研究相比,液体环境中的激光烧蚀固体靶的研究相对有限。然而,在国际上,LAL 技术已经广泛应用于纳米材料制备和纳米结构组装技术,许多研究小组专注于这一研究方向,并且采用 LAL 技术合成了丰富多彩且具有潜在技术应用的纳米材料如金属、金属合金、半导体和聚合物等。因此,为了将 LAL 技术发展成为功能更加强大的纳米制备技术,人们有必要对其基本物理和化学过程有一个全面、系统和深入的认识。本章,我们详细探讨 LAL 纳米晶合成和纳米结构组装中所涉及的基本物理和化学过程。

3.1　基本物理过程

根据 LAL 过程中所使用的激光脉宽的大小,我们可以将 LAL 分为 4 种类型:毫秒 LAL（ms-LAL）、纳秒 LAL（ns-LAL）、皮秒 LAL（ps-LAL）和飞秒 LAL（fs-LAL）。重要的是,这 4 种 LAL 的脉冲激光与固体靶的相互作用有着显著不同的物理过程。因此,我们分别阐述这些过程的基本原理。

3.1.1　液体环境中毫秒激光烧蚀固体靶及熔化过程

一般来说,与纳秒激光和飞秒激光相比,毫秒激光很少用于纳米材料合成,它们往往用于纳米材料加工,如焊接和切割金属[9]。因此,长期以来 ms-LAL 的作用机制和可能的应用领域不是十分清楚。在 ms-LAL 方面,Niu 等[10]率先使用低功率密度的毫秒激光可控地合成了多种金属纳米材料。他们发现,毫秒激光在 LAL 中的作用与传统的纳秒激光完全不同。例如,在 ns-LAL 中,短脉宽的纳秒激光的功率密度始终在 $10^8 \sim 10^{10}$ W/cm^2,这种高功率密度脉冲激光很容易导致固体靶电离产生等离子体羽。相比之下,在 ms-LAL 中,毫秒激光的功率密度较低

($10^6 \sim 10^7 \text{ W/cm}^2$),只能在 LAL 过程中产生纳米尺度液滴[11-13]。

ms-LAL 的物理机制主要是热效应,如图 3-1 所示。当毫秒激光开始烧蚀液体中的金属靶材时,首先它会在靶材表面产生熔融的金属液滴。金属液滴的温度很高,以至于它们可以加热周围的液体,使得这些液体以爆炸的方式剧烈蒸发。同时,受环境液体的束缚,液体蒸气及金属液滴被限制在了空泡中进而使得空泡内部产生高压。液体蒸气所产生的压强可提高靶材的沸点,这意味着即使温度超过靶材的标准沸点,被激光烧蚀的靶材也可以保持液态。由于高能蒸气诱导的强烈破碎效应,毫秒激光烧蚀靶材产生的大量毫米级金属液滴将爆炸性地喷射形成纳米液滴 [图 3-1(a)]。重要的是,高速摄影实验证实了该过程形成的是金属纳米液滴而非等离子体羽[10]。如图 3-1(c)所示,在 0~2.2 ms 的时间分辨图像中,我们可以观察到了许多微小液滴,但没有探测到等离子体羽。此外,由于这种金属纳米液滴的致密性,我们可以预测它们会从表面开始与环境液体发生化学反应 [图 3-1(b)]。因此,我们可以通过调控环境液体和激光参数,通过界面化学反应来合成不同形貌和尺寸的纳米材料[10, 11]。

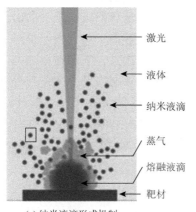

(a) 纳米液滴形成机制

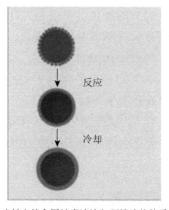

(b) 喷射出的金属纳米液滴与环境液体的反应

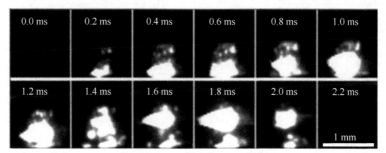

(c) 水中对 Ti 靶激光烧蚀的时间分辨图像

图 3-1 ms-LAL 的物理机制

3.1.2 液体环境中纳秒激光烧蚀固体靶及等离子体羽形成

Yang 在 2007 年发表了一篇关于 ns-LAL 的综述论文,这是国际 LAL 纳米制备技术研究的第一篇综述论文[14]。ns-LAL 主要包括等离子体羽的产生、相变和凝结,基本过程如下。首先,当脉冲激光辐照浸没在液体中的固体靶时,会在固体靶表面的液-固界面产生等离子体羽。等离子体羽产生的物理机制主要是由脉冲激光烧蚀固体靶表面所产生的气相粒子的多光子吸收、电离和韧致辐射逆过程。因此,等离子体羽包含大量来自固体靶的中性原子、离子和电子。我们称这种初始等离子体羽为激光诱导的等离子体羽,因为它是通过脉冲激光烧蚀固体靶直接形成的(图 3-2)。然而,与真空和气体环境中的等离子体羽不同,液体环境中激光诱导的等离子体羽的膨胀会受到环境液体的强烈束缚。在固体靶吸收脉冲激光后端能量而持续蒸发时,激光诱导的等离子体羽会发生绝热膨胀从而在环境液体中产生强大的冲击波,并且这种冲击波会反过来作用于激光诱导的等离子体羽,使其压强和温度进一步上升。我们把冲击波在等离子体羽中引起的附加压强称为等离子体羽诱导的附加压强。此外,由于等离子体羽诱导的附加压强的出现,等离子体羽的温度也会迅速升高。所以,冲击波通过在等离子体羽中产生附加温度和附加压强将等离子体羽推向在高温高压高密度的极端条件下的独特热力学状态。同时,在接下来的等离子体羽相变过程中,等离子体羽内部及液体与等离子体羽之间的界面处会发生 4 种化学反应(这些化学反应将在 3.2.1 节中详细讨论)。在环境液体的束缚下,等离子体羽会历经膨胀、冷却和凝结过程。但是,在这些过程中,等离子体羽会将它的能量传递给周围的环境液体,从而导致等离子体羽与环境液体

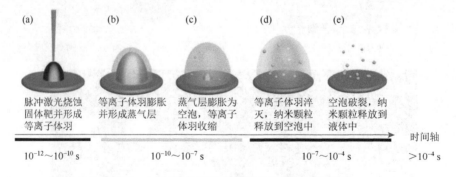

图 3-2 ns-LAL 的物理机制及各阶段的时间顺序

(a)当激光辐照固体靶表面时,入射脉冲激光的前端会烧蚀靶材,在表面产生来自固体靶的等离子体羽;(b)等离子体羽在环境液体中受束缚发生膨胀,将能量传递到周围的环境液体并在等离子体羽表面形成蒸气层;(c)蒸气层逐渐膨胀为空泡并压缩等离子体羽使其收缩;(d)等离子体羽淬灭,所有纳米颗粒释放到空泡中;(e)空泡破裂,纳米颗粒(粒子团簇)释放到液体中并形成胶体

之间产生薄薄的一层蒸气层。该蒸气层被认为是空泡产生的早期阶段。液体中等离子体羽演变的阶段还有冷却和凝结，受环境液体的束缚，伴随着压强和温度的降低，等离子体羽发生淬灭并释放出纳米颗粒。这一演化过程如图 3-3 所示[15]，它通过高速相机与快速阴影法实现可视化[16]，该图像是在水中利用激光烧蚀 Pt 丝时获得的，清楚地揭示了等离子体羽的膨胀和破裂过程。值得注意的是，等离子体羽具有极快的冷却速度，并且伴随着等离子体羽前端周围环境液体的蒸发和收缩。

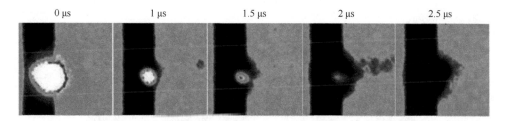

图 3-3 高速相机拍摄的水中 Pt 丝上激光诱导的等离子体羽和空泡图像的复合图像

（Pt 丝直径 = 545 μm，λ = 532 nm，辐射照度 = 12 GW/cm^2）

当蒸气层产生后，它逐渐膨胀并比周围环境液体更有效地将等离子体羽朝着靶材方向压缩［图 3-2（c）］。一般来说，空泡的膨胀和收缩过程遵从它的动力学。空泡的寿命时间不是固定值，它很大程度上取决于脉冲激光的特性，例如，激光能量密度[17]。空泡中的压强和温度可以使用范德耳斯模型及通过阴影图技术获得的空泡半径来估算[18]：

$$P(t) = \left(P_{\infty} + \frac{2\sigma}{R} \right) \left(\frac{R_{\infty}^3 - h^3}{R^3 - h^3} \right)^{\gamma} \tag{3-1}$$

$$T(t) = T_{\infty} \left(\frac{R_{\infty}^3 - h^3}{R^3 - h^3} \right)^{\gamma - 1} \tag{3-2}$$

式中，P_{∞} 和 T_{∞} 分别为液体的压强和温度；R 为空泡半径；R_{∞} 是空泡具有内部压强下的半径；σ 是液体的表面张力；$h = R_{\infty}/9.174$ 是水分子的半径；$\gamma = 1.22$。结合实验数据，我们估算出空泡内部的温度和压强分别为 1000 K 和 $10^7 \sim 10^8$ Pa[18]。值得注意的是，这种高温高压状态仅存在于初始阶段，空泡在随后的膨胀时会急剧冷却。

在实验中，我们可以通过快速阴影法实时观察空泡的产生、膨胀和破裂（溃灭）。图 3-4 显示了在不同的延迟时间下激光烧蚀水中的 Cu 靶所产生的空泡的形态。当激光脉宽分别为 19 ns、90 ns 和 150 ns 时，在 400 ns～400 μs 内皆可清楚地看到空泡的膨胀和溃灭[19]。在激光辐照 400 ns 后，从 90 ns 和 150 ns 激光产生的等离子体羽中可观察到明亮的光斑。显然，更长的纳秒激光会导致更亮的光斑，空泡大小与激光脉宽密切相关。这表明脉冲激光越长，空泡膨胀得越大。具体而言，空泡会在 1～100 μs 内不断增大，在 100 μs 时达到最大，然后收缩，直至最终溃灭并将其内部的成分释放到液体中，包括粒子团簇和微米/纳米颗粒 [图 3-2（e）]。需要注意的是，上述过程合成的粒子团簇和纳米颗粒初始状态不稳定且大概率会发生集聚。所以，大多数 LAL 合成的初始胶体溶液都会经历一个熟化的过程。

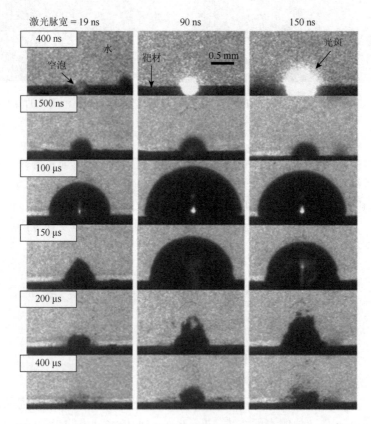

图 3-4　在 400 ns～400 μs 的不同延迟时间下激光烧蚀水中 Cu 靶，激光照射点产生的空泡的阴影图

图中分别是 19 ns、90 ns 和 150 ns 三种脉宽

如上所述，空泡起源于等离子体羽周围的薄蒸气层。由于等离子体羽将能量传递给了蒸气从而提高了蒸气温度。与等离子体羽类似，蒸气层也会膨胀并转变

为半径比蒸气层还要大的空泡。同时，环境液体的束缚增大了空泡内部的压强。此时，等离子体羽膨胀并将其能量转移到蒸气层后逐渐冷却，而空泡则继续向各个方向膨胀。在空泡膨胀的过程中，空泡不仅需要对抗环境液体的变化，还要对抗并压缩等离子体羽。等离子体羽和空泡的膨胀行为的差异在于，空泡分为膨胀阶段和收缩阶段，且该过程可以像阻尼振荡器一样重复多次[20]。在此阶段，空泡内部的压强会逐渐与周围环境液体的压强达到平衡。此时，空泡达到其最大半径，并在一定时间内保持这种准平衡状态。

当空泡溃灭时，等离子体羽冷凝时形成的纳米颗粒可以扩散到周围的环境液体中，形成胶体溶液[14]。换句话说，纳米颗粒主要是在等离子体羽淬灭过程中产生的，在 LAL 过程产生的激光诱导的等离子体羽的发射光谱中，人们可以观察到新生成小分子的信号[21]。因此，在等离子体羽凝结和破裂之后，由靶材粒子和环境液体分子化学反应产生的纳米颗粒在空泡中扩散，最后释放到环境溶液中。

关于 LAL 的物理机制，还有另一种理解，认为激光诱导的空泡在纳米颗粒的形成中发挥着重要作用[22-24]。在热力学上，空泡中的压强和温度可以用阴影法所定义的空泡半径来计算。例如，Giacomo 等根据空泡半径估算出空泡内的温度和压强分别可达 1000 K 和 10^8 Pa[18]。然而，由于仪器精度的限制，这些理论计算可能错过空泡从破裂到反弹的反转阶段，该阶段可能具有更高的温度和压强。Akhatov 等[25]的理论模型预测空泡内部温度和压强分别可达 10 000 K 和数千兆帕。在如此高温高压的环境下，纳米颗粒的相变和凝结似乎是合理的。从这个意义上说，空泡中的化学反应可能与等离子体羽中的化学反应相似。此外，空泡的持续时间比等离子体羽的持续时间长两个数量级[15]，这表明它存在的时间足以允许充分的化学反应进行和纳米晶的生长。

高时间分辨率小角度 X 射线散射已经证明了纳米颗粒可以在空泡中形成[26]，其可以穿透空泡并通过 X 射线散射来识别其中的物质，包括微小颗粒。例如，实验观察发现，直径为 8～10 nm 的 Au 纳米颗粒分布在整个空泡中，而较大的纳米颗粒（直径为 45 nm）仅存在于空泡的上部。激光散射技术也用于研究 LAL 的环境液体中纳米颗粒的生长。由于探测激光束被纳米颗粒散射，因此散射光的图像提供了纳米颗粒的位置。实验结果表明，纳米颗粒在脉冲激光后端 3 μm 处发生快速生长[27]。在空泡膨胀阶段，一小部分纳米颗粒会从空泡转移到水中，而大多数纳米颗粒仍被困在空泡中，直到空泡溃灭。

上述讨论表明，LAL 通过化学反应生成纳米晶的初始阶段应该存在多条动力学通道。实验证据表明，由于产生的高温和高压，等离子体羽和空泡可能都参与了化学过程，如初始成核过程可能发生在等离子体羽中，并且在空泡演化过程中发生相对较长的生长期。

3.1.3 液体环境中皮秒激光烧蚀固体靶及空泡产生

皮秒激光烧蚀固体靶的物理机制与纳秒激光的类似[28]，其反应过程可以描述为激光与固体靶相互作用后，从固体靶表面喷出高温高压致密的等离子体羽，并受环境液体束缚快速膨胀，产生强大的冲击波。通常，冲击波会于数十纳秒内形成，具体取决于激光辐照到固体靶上的能量。冲击波会以数百纳秒的时间尺度极速远离靶材，其后方的等离子体羽会与环境液体发生热交换，并产生包含液体蒸气、靶材气体和纳米颗粒的空泡。所产生的空泡会在数百微秒的时间尺度内首先膨胀然后破裂[28]。图3-5显示了通过快速阴影法得到的脉宽为25 ps、能量密度为6.3 J/cm^{2}的脉冲激光烧蚀水中Au靶时产生的冲击波和空泡时间演变的阴影图像。宏观上来说，由于靶材内部温度分布相同，所以纳秒激光和皮秒激光与材料的相互作用相似，这是由纳秒、皮秒激光较短的热扩散长度，以及较长的吸收长度导致的[29]。

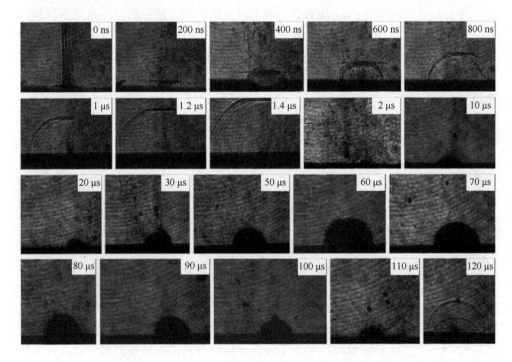

图3-5 脉冲激光烧蚀水中Au靶产生的冲击波和空泡时间演变的阴影图像
能量密度6.3 J/cm^2、脉宽25 ps、重复频率10 Hz、波长1064 nm

皮秒激光的物理机制与纳秒激光的相似，但是皮秒激光的高重复频率有利于提高LAL制备纳米颗粒的产量。Barcikowski等[30]发现，与相同激光能量密度下的纳秒激光相比,具有更高重复频率的 ps-LAL 产生的纳米颗粒产量高出三倍。而与传统的 fs-LAL 相比,ps-LAL 的产量提高了一个数量级。此外,fs-LAL 生成的无配体纳米颗粒和 Au-ssDNA 纳米缀合物,具有与 ns-LAL 生成的纳米颗粒相同的特性、表面组成、生物分子负载及相同水平的生物分子结构完整性[31]。许多研究表明,与纳秒激光和飞秒激光相比,皮秒激光在提高 LAL 合成纳米颗粒的产量方面具有优越性,甚至可以达到中试规模[32, 33]。

3.1.4 液体环境中飞秒激光烧蚀固体靶及多形态过程

飞秒激光与材料的相互作用,与纳秒激光和皮秒激光有很大区别[34, 35]。根据飞秒激光的功率密度,我们可将 fs-LAL 过程分为 4 种类型:相爆炸(phase explosion)、碎裂(fragmentation)、库仑爆炸(Coulomb explosion)和等离子体烧蚀(plasma ablation)。

在飞秒激光烧蚀固体靶过程中,光子能量会被靶材的载流子、金属的自由电子、半导体或绝缘体价带的电子及多光子电离吸收,并使电子具有一个初始动能,高能电子通过碰撞使靶材进一步电离并雪崩式地产生更多的额外自由载流子。图 3-6 显示了大气环境中飞秒激光烧蚀固体靶的计算机模拟结果。当入射功率密度低于 10^{13} W/cm^2 时,在飞秒激光辐照下(5×10^{11} W/cm^2),在脉冲激光开始约 0.7 ps 后,固体靶表面形成了具有高温(8000 K)和高压(10 GPa)的液体层[图 3-6(a)];随着激光烧蚀进一步浸入块体中,越来越多的固体靶转变为液态金属[图 3-6(b)];液体层内压强的弛豫会导致上层迅速膨胀,从而形成空洞[图 3-6(c)];空洞迅速增大[图 3-6(d)];最终喷射出大量物质[图 3-6(e)],这个过程称为相爆炸,其发生的时间尺度为 $10^{-12}\sim10^{-10}$ s;未被激光烧蚀的靶材随后在 $10^{-11}\sim10^{-9}$ s 内凝固[图 3-6(f)];如果功率密度增加到 1.1×10^{12} W/cm^2,则喷射出的物质具有弥散的团簇结构,这个过程为碎裂过程[图 3-6(g)][36, 37]。

目前,研究人员已经开发了一些基于 fs-LAL 的计算与模拟技术[120, 121]。Shih 等首次报告了液体环境中激光烧蚀金属靶材的动力学模拟结果,如图 3-7 所示,展示了激光与金属靶材相互作用的全原子描述,以及沉积在 SiO$_2$ 衬底上的激光烧蚀 Ag 薄膜的声阻抗匹配边界条件[120]。在相爆炸状态下的两个脉冲激光的计算与模拟结果描述了喷射诱导等离子体羽的扩散减速,以及环境液体与等离子体羽界面形成了密集过热金属熔融层。与金属熔融层接触的水进入了超临界状态,并形成不断膨胀的低密度金属-水混合区域,该区域是形成空泡的前体。同时,激光诱导等离子体羽膨胀的动力学受液体环境束缚的强烈影响。

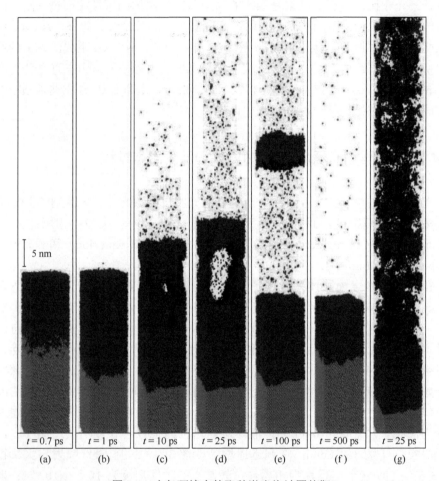

图 3-6　大气环境中的飞秒激光烧蚀固体靶

快照显示了在 266 nm、0.7~500 ps 下 Si 衬底上 Au 靶中引起的结构变化。(a)~(f) 0.225 J/cm² 下的脉冲激光；(g) 0.5 J/cm² 下的脉冲激光

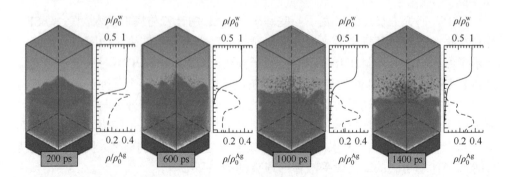

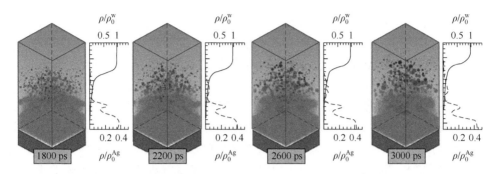

图 3-7 水中激光烧蚀 20 nm Ag 薄膜的分子动力学模拟的原子构型快照

该过程使用吸收通量为 400 J/m² 的 40 fs 激光辐照

Roeterdink 等[40]结合 Si 晶片的飞行时间（time-of-flight，TOF）光谱进行了飞秒激光实验。他们发现，在 TOF 光谱中，Si^+ 峰所对应的速度是观察到的 Si^{2+} 峰速度的一半，这表明发生了库仑爆炸。此时，如果功率密度继续增加，我们就会观察到等离子体羽烧蚀现象 [图 3-8（a）]。因此，当功率密度增加到非常接近激光烧蚀阈值时，就会发生库仑爆炸。库仑爆炸过程可以细分为如下几步：首先，材料吸收脉冲激光提供的高能量 [图 3-8（b）]；其次，电子通过光电和热电子发射从原子中剥离 [图 3-8（c）]，由于电子的剥离，辐照区域的表面上会产生非常高强度的电场，该电场将导致正离子之间产生非常强的排斥力 [图 3-8（d）]；最后，由于该排斥力大于键合强度，固体材料表面的正离子被剥离[41]。

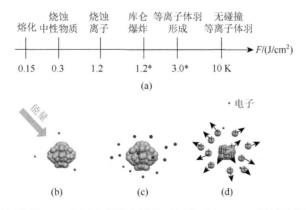

图 3-8 （a）不同能量下 Si 表面飞秒激光诱导不同的反应；（b）材料吸收脉冲激光提供的高能量；（c）电子通过光电和热电子发射从原子中剥离；（d）该电场将导致正离子之间产生非常强的排斥力

Werner 和 Hashimoto[42]发展了一种基于电子温度、晶格温度和颗粒周围的介质温度的双温度模型，并使用该模型进行了相关计算与模拟，解释了使用纳秒激光和飞秒激光诱导液态 Au 纳米颗粒尺寸缩小的机制。他们的结果表明，飞秒激光提供了足够的能量，可以将液态 Au 的电子温度提高到 7000 K，将固态 Au 的电子温度提高到 8000 K 以上，满足了库仑爆炸的要求。相比之下，纳秒激光的情况是由光热机制造成的。除了上述计算与模拟之外，他们还对飞秒激光诱导的液态 Au 纳米颗粒碎片进行了原位消光光谱和瞬态吸收光谱的测试[43]。这项研究首次实现直接通过光谱技术观察证实库仑爆炸会导致纳米晶体碎裂。如果功率密度增加到等离子体羽烧蚀阈值以上，那么固体颗粒直接转变为等离子体羽后会发生光学击穿[44, 45]，其中颗粒会被完全电离并蒸发，形成高温高密度的等离子体羽。

所以，虽然纳秒激光和飞秒激光都会产生等离子体羽，但这两种等离子体羽的特征并不相同。图 3-9（a）和图 3-9（b）分别显示了纳秒激光和飞秒激光诱导的等离子体羽温度和电子数密度与时间的关系[46]。与纳秒激光诱导的等离子体羽相比，飞秒激光诱导的等离子体羽温度和电子数密度较低且下降较快，飞秒激光没有表现出激光与等离子体羽的相互作用。图 3-9（c）和图 3-9（d）分别描绘了

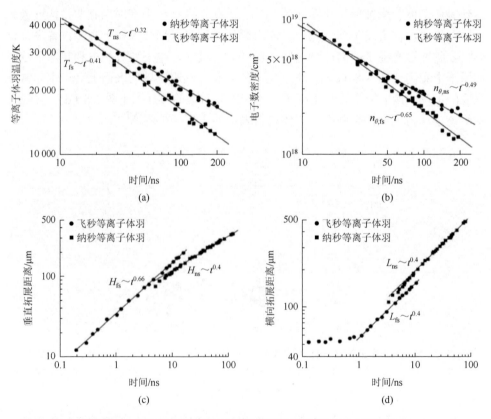

图 3-9　纳秒激光烧蚀和飞秒激光烧蚀的等离子体羽差异
（a）等离子体羽温度；（b）电子数密度；（c）垂直拓展距离；（d）横向扩展距离

纳秒激光和飞秒激光诱导的等离子体羽随时间变化的垂直拓展距离和横向拓展距离。纳秒激光产生的冲击波垂直拓展距离与 $t^{0.4}$ 成正比，类似球面传播；而飞秒激光产生的冲击波垂直拓展距离与 $t^{0.66}$ 成正比，符合一维拓展。在最初阶段内（<1 ns），飞秒激光诱导的等离子体羽主要在垂直于靶材表面的方向上拓展，几纳秒后，它在横向和垂直方向上膨胀，且垂直方向的膨胀比横向的膨胀快。相比之下，纳秒激光诱导的等离子体羽在两个方向上的膨胀速度基本相同。因此，纳秒激光和飞秒激光产生的等离子体羽存在显著差异。

3.2 基本化学过程

3.2.1 等离子体中的化学反应

在等离子体羽相变过程中，激光诱导的等离子体羽内部及等离子体羽与环境液体的界面处会发生 4 种典型的化学反应。重要的是，正如前文讨论的，由于液体的束缚效应，激光诱导的等离子体羽会被驱动并处于一个高温高压高密度的极端热力学状态，这是与在真空或气体环境中的激光烧蚀完全不同的热力学状态，同时，也是 4 种典型化学反应的起源。因为在 LAL 中，环境液体总是参与化学反应的，因此，这些化学反应不同于真空或气体环境中发生的化学反应[14]。Sakka 等[21]采用激光发射光谱研究 LAL 中的固-液界面。他们分别使用石墨靶和氮化硼靶，以及分别用水、苯、正己烷和四氯化碳作为环境液体进行 LAL 实验。在氮化硼-水系统的发射光谱中，他们发现在脉冲激光辐照 40 ns 后可明显观察到 BO 分子的存在。由于氮化硼中没有氧源，因此该分子中的氧原子必定来自水。此外，在苯、正己烷和四氯化碳等环境液体中氮化硼的发射光谱中检测到了 C_2（双原子碳）和 CN 的存在。同理，C_2 和 CN 中的碳原子应该来自环境液体。所以，BO 和 CN 等分子的存在证明了固体靶烧蚀的物质会与环境液体中的物质在 LAL 过程中发生化学反应。

激光诱导的等离子体羽内部及等离子体羽与环境液体的界面处发生的 4 种典型化学反应如表 3-1 和图 3-10 所示，包括它们的发生位置、反应前驱体和反应产物等。

表 3-1 激光诱导的等离子体羽内部及等离子体羽与环境液体界面处的 4 种典型化学反应

类型	发生位置	反应前驱体	反应产物	化学反应
1	激光诱导的等离子体羽内部	固体靶	亚稳相纳米结构	C(六方相)→C(立方相) Si(闪锌矿)→Si(面心立方) Ge(立方相)→Ge(四方相)

续表

类型	发生位置	反应前驱体	反应产物	化学反应
2	激光诱导的等离子体羽内部	固体靶和环境液体	氧化物 氮化物 碳化物	$2V + 5O(水) \rightarrow V_2O_5$ $3C + 4N(氢氧化铵) \rightarrow C_3N_4$ $Si + C(乙醇) \rightarrow SiC$
3	激光诱导的等离子体羽和环境液体的界面处	固体靶和环境液体	水合氧化物	$xW(簇) + (m+n)H_2O \rightarrow W_xO_m \cdot nH_2O + mH_2$ $6Sn(团簇) + 8H_2O \rightarrow 6SnO \cdot 2H_2O + 6H_2$
4	环境液体内部	固体靶和环境液体	氢氧化物	$Mn_3O_4 + 2H^+ \rightarrow 2MnOOH + Mn^{2+}(质子化作用)$ $Mn^{2+} + 1/6O_2(aq) + H_2O \rightarrow 1/3Mn_3O_4 + 2H^+(质子化作用)$

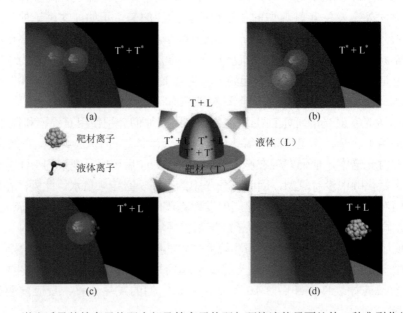

图 3-10 激光诱导的等离子体羽内部及等离子体羽与环境液体界面处的 4 种典型化学反应

(a) 化学反应发生在激光诱导的等离子体羽内部, 反应前驱体来自固体靶; (b) 化学反应发生在激光诱导的等离子体羽内部, 反应前驱体分别来自固体靶和环境液体; (c) 化学反应发生在激光诱导的等离子体羽和环境液体的界面处, 反应前驱体分别来自固体靶和环境液体; (d) 化学反应发生在环境液体内部, 反应前驱体分别来自固体靶和环境液体

(1) 第一种化学反应发生在激光诱导的等离子体羽内部。由于激光诱导的高密度等离子体羽处于高温高压状态, 固体靶烧蚀粒子之间存在高温化学反应, 并且极易产生亚稳相纳米结构 [图 3-10 (a)]。例如, 代表性的化学反应包括在 LAL

过程中六方碳转变为立方碳,以及闪锌矿硅转变为面心立方硅[47-50]等。

(2) 第二种化学反应也发生在激光诱导的等离子体羽内部。参与反应的前驱体不单是来自固体靶,还来自环境液体。正如我们知道的,激光诱导的等离子体羽前端的高温高压会导致等离子体羽与环境液体界面处的液体分子蒸发、激发及离化,从而在界面处生成新的等离子体羽。由于新等离子体羽是由激光诱导的等离子体羽在环境液体中激发产生的,因此,我们称它为等离子体羽诱导的等离子体羽。这个等离子体羽一旦产生,就会迅速融入激光诱导的等离子体羽中。所以,来自激光烧蚀固体靶的粒子与来自等离子体羽诱导的等离子体羽中的粒子会发生化学反应[图3-10(b)]。代表性的化学反应产物是各种氧化物、氮化物和碳化物[51-53]。

(3) 第三种化学反应发生在激光诱导的等离子体羽和环境液体的界面处。具有高温高压高密度的激光诱导的等离子体羽中的粒子可以直接与环境液体分子在界面处发生高温化学反应[图3-10(c)]。一个代表性的化学反应是在LAL过程中生成水合氧化物[54,55]。

(4) 第四种化学反应发生在环境液体内部。由于激光诱导的等离子体羽处于高温高压高密度状态,所以,等离子体羽中的粒子将会被极高的压强从等离子体羽中驱动到环境液体中,这样,来自等离子体羽中的粒子与环境液体分子的化学反应将会在环境液体内部发生[图3-10(d)]。一个代表性的化学反应就是LAL过程中氢氧化物的形成[56]。

由此可见,上述4种化学反应中有三种反应的前驱体都是来自固体靶和环境液体。因此,我们可以很方便地通过设计合适的固体靶的组分和环境液体的种类,在激光诱导的等离子体羽内部、等离子体羽-环境液体界面、环境液体内部设计合理的化学反应来合成我们所需要的纳米材料。例如,Stratakis等[57]选择了不同种类的环境液体如乙醇(饱和氢气的乙醇)和稀碱性溶液等,研究在LAL过程中这些环境液体与固体铝靶的化学反应。他们的研究发现,飞秒或皮秒LAL可以通过乙醇和铝靶制备Al纳米颗粒,这些纳米颗粒大多是具有单晶夹杂物和天然氧化物包裹层的无定形纳米颗粒。然而,如果环境液体是含有饱和氢气的乙醇,则产物会变成内部有空洞的Al纳米颗粒,这是由Al纳米颗粒在LAL凝固过程中氢气的释放所致[58]。注意,这项研究首次描述了LAL过程中溶解在环境液体中的轻质气体[58,59]。有趣的是,该研究组还报道了在稀碱性溶液中激光烧蚀铝靶时形成的新型自组织空泡[60,61],这种结构是在金属和氨水溶液发生化学反应过程中从环境液体表面释放出的氢空泡上升中形成的。我们知道,氨水溶液是弱碱性的,它与铝的化学相互作用会导致氢气的释放。在LAL过程中,这些空泡可以形成固定图案,该图案取决于激光烧蚀区域的几何形状。对形状为涡流的烧蚀区域进行特殊激光处理,无须额外施加力即可导致液体旋转。这种自组织空泡是激光烧蚀金属和铝靶相对缓慢的化学刻蚀的独特组合导致的结果[61]。

3.2.2 固体与液体界面的化学反应

与等离子体羽和空泡中的化学反应不同,在 ms-LAL 过程中会发生包括柯肯德尔效应(Kirkendall effect)在内的表面反应。柯肯德尔效应常常被用来在金属氧化物和硫化物纳米颗粒中制造空洞[62,63],其基本物理过程就是当金属颗粒氧化时金属核快速向外扩散,而氧化剂则缓慢向内扩散,这样就导致中空纳米颗粒的形成[64]。

如上所述,在 LAL 纳米制备技术中,长脉宽的激光[例如,具有低功率密度(10^6 W/cm^2)的毫秒激光]通常会产生纳米液滴而不是等离子体羽[11-13]。因此,我们可以推测,纳米液滴会发生表面反应,从而产生核-壳纳米结构的颗粒。在这里,我们以 ms-LAL 制备 ZnS 纳米颗粒为例阐述这一过程[11]。在毫秒激光烧蚀 Zn 靶过程中,固体靶会在环境液体中喷射出 Zn 纳米液滴。当这些"热"纳米液滴与富含硫的环境液体(如十二烷基硫醇)接触后,由于 Zn 纳米液滴与环境液体之间的界面温度较高,环境液体分子很容易与纳米液滴中的 Zn 发生化学反应。因此,Zn 纳米液滴的表面会被硫化并逐渐转变为 ZnS,形成 Zn/ZnS 核-壳纳米颗粒。由于在高温下 Zn 核的高扩散系数和 ZnS 壳的低扩散系数,Zn/ZnS 核-壳纳米颗粒逐渐转变为 ZnS 中空纳米颗粒。在这里,毫秒激光可以提供高功率密度能量以快速加热和烧蚀固体靶,随后低温的环境液体就会将高温相淬灭以形成纳米颗粒。这种相变是非常快的,以至于氧化和硫化不能在核的内部同时进行,从而产生了核-壳纳米颗粒。图 3-11(a)~图 3-11(e)显示了不同阶段观察到的 ZnS 中空纳米颗粒和相应的结构模型。基于同样的技术路线,Mg 纳米颗粒也可以通过毫秒激光烧蚀转化为 MgO 空心纳米球,这表明 Mg 原子向外扩散更快,且 O 原子向内扩散更慢[65]。因此,ms-LAL 诱导的柯肯德尔效应可以应用于金属氧化物和硫化物空心纳米结构的组装。

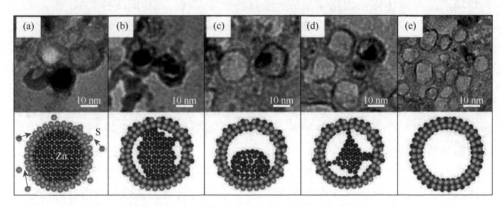

图 3-11 在液体中毫秒激光烧蚀产物的 TEM 图像
(a)~(e)不同阶段和结构模型中观察到的 ZnS 中空纳米颗粒

当然，空心纳米结构也可以采用其他类型脉冲激光的 LAL 进行组装。然而，它们的形成机制与毫秒激光完全不同。Yan 等[66]报道了纳秒准分子脉冲激光制备 MgO 纳米颗粒的方法和形成机制。MgO 纳米颗粒的形成机制与 ms-LAL 合成中空纳米颗粒的不同。正如我们上面讨论的，毫秒激光主要涉及表面反应。但是，纳秒激光合成的 MgO 纳米颗粒被认为是通过空泡表面钉扎作用在脉冲激光诱导的空泡上生成的。此过程有两种情况。第一种情况是，粒子团簇在空泡中形成，并且随着空泡振荡在空泡-液体界面处积聚，特别是在空泡溃灭期间。第二种情况是，粒子团簇分散在空泡周围的环境液体中，包括激光诱导的空泡和源自溃灭空泡的亚稳态超微空泡。对于上述两种情况，一旦空泡-液体界面吸收足够多的粒子团簇，它们就会开始彼此碰撞、合并并形成粒子团簇网络层。这些粒子团簇可以产生空心聚集体，甚至熔化形成光滑层，这可以为具有更厚壳的中空纳米颗粒的进一步成核和生长提供模板。

同样的机制也适用于通过纳秒激光 LAL 合成的 Al_2O_3 中空纳米颗粒[67]。纳秒准分子脉冲激光烧蚀固体靶首先在环境液体中产生纳米团簇，并且同时在环境液体中产生空泡，然后空泡-液体界面会捕获纳米团簇，导致中空纳米颗粒的形成。Viau 等[59]通过在饱和氢气的乙醇中激光烧蚀铝靶制备 Al 中空纳米颗粒。他们发现，由于在 LAL 凝固过程中释放出氢气，Al 纳米颗粒内部就会形成空腔。因此，只有毫秒激光合成的中空纳米结构才会受到热转导和柯肯德尔效应的影响。

在本节中，我们探讨了使用毫秒、纳秒和飞秒激光在 LAL 过程中发生的物理过程和化学反应，明确指出了不同脉宽的激光在 LAL 中的烧蚀机制存在显著差异。毫秒激光的熔融纳米液滴，纳秒激光的等离子体羽和空泡，以及飞秒激光的相爆炸、碎裂、库仑爆炸和等离子体烧蚀等多种过程。由于激光烧蚀机制的不同，每种类型的激光都有自己的优点和应用范围。因此，全面了解 LAL 的基本机制有利于设计和控制纳米材料合成和纳米结构组装。

3.3 热力学特征

显然，在 LAL 纳米材料制备中，我们通过对上述脉冲激光诱导的等离子体羽演变的理解发现，LAL 过程的热力学和动力学因素可以极大地影响纳米材料的制备。我们以石墨靶和铝靶在水中的 PLA 作为典型例子，通过对脉冲激光诱导的等离子体羽的三个重要热力学参数，即物质密度、温度和压强的详细表征，对 LAL 的热力学因素做出了基本描述。

通常，对于液体中来自 PLA 的激光诱导的等离子体羽中的物质密度，我们可以通过等离子体羽的膨胀体积和激光烧蚀后留在靶材表面上的孔的体积来计算烧蚀物质的量。首先，我们可以采用光谱技术，根据激光烧蚀固体靶在靶材表面上产生的发光区域的图像来测量激光诱导的等离子体羽的膨胀体积，图 3-12 显示了发光区域的图像和强度分布。考虑到等离子体羽是一个直径为半峰全宽（full with at half maximum，FWHM）强度的半球[68]，那么等离子体羽的体积可以估计为 $9.9×10^7$ cm^3。其次，考虑到留在靶材表面上的孔的体积随着脉冲激光数的增加而线性增加，确定单个脉冲激光的烧蚀体积为 $7.4×10^{-8}$ cm^3，图 3-13 显示了激光烧蚀后留在靶材表面的孔的垂直截面轮廓拟合[69]。这样，我们就可以计算出水中石墨靶的 Nd∶YAG PLA 产生的等离子体羽中物质密度为 $6.7×10^{-21}$ cm^3。

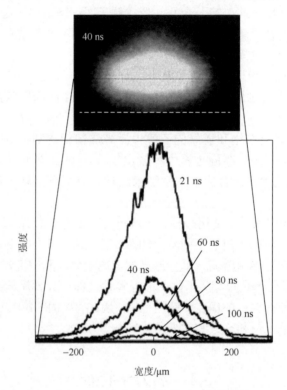

图 3-12　波长为 1064 nm、脉宽为 20 ns、能量密度为 10 J/cm^2 的 Nd∶YAG 脉冲激光烧蚀水中石墨靶产生的脉冲激光诱导的等离子体羽的图像和强度分布

在环境液体束缚下的脉冲激光诱导的等离子体羽的光学发射光谱是确定等离子体羽温度的有效方法[68-71]。例如，Sakka 等[69]根据对 C_2 分子发射光谱的测量，得到了脉冲激光诱导的等离子体羽的温度约为 5000 K，C_2 分子的产生是通过对水中石墨靶的 PLA（波长为 1064 nm、脉宽为 20 ns、能量密度为 10 J/cm^2 的 Nd∶YAG 脉冲激光）。

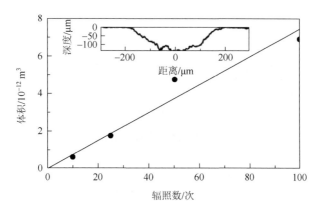

图 3-13 脉冲激光烧蚀石墨靶 100 次产生的孔的垂直截面轮廓拟合

内插图为距离与深度的关系

当 PLA 被环境液体束缚时,脉冲激光诱导的等离子体羽中的高压通常归因于两个贡献:一个是脉冲激光诱导的等离子体羽在环境液体约束下的绝热膨胀,另一个是由冲击波引起的额外压强增加。因此,Devaux 等开发了一系列实验技术[72-79],通过图 3-14 所示的冲击波特征来测量压强,并建立了水中脉冲激光诱导压强产生的理论模型。类似地,Lu 等通过记录等离子体羽在水中诱导的声波来测量脉冲激光诱导的等离子体羽中的压强[80, 81]。例如,Berther 等报道[74-77],当波长为 1064 nm、功率密度为 10 GW/cm²、脉宽为 20 ns 的脉冲激光辐照水中的铝靶时,脉冲激光诱导的等离子体羽中的最大压强高达 5.5 GPa,持续时间约为 50 ns。基于已经建立的理论模型[72-79],脉冲激光诱导的等离子体羽在水中产生的最大压强可以由式(3-3)得到:

$$P(\text{GPa}) = 0.01\sqrt{\frac{\alpha}{\alpha+3}}\sqrt{Z[\text{g}/(\text{cm}^2\cdot\text{s})]}\sqrt{I_0(\text{GW}/\text{cm}^2)} \quad (3\text{-}3)$$

式中,α 是用于热能的内能分数(通常约为 0.25);I_0 是入射功率强度;Z 是靶材和水之间减小的冲击阻抗,定义为

$$\frac{2}{Z} = \frac{1}{Z_{\text{water}}} + \frac{1}{Z_{\text{target}}} \quad (3\text{-}4)$$

式中,Z_{water} 和 Z_{target} 分别是水和靶材的冲击阻抗。例如,对于铝靶,$Z_{\text{water}} = 0.165\times 10^6$ g/(cm²·s),$Z_{\text{target}} = 1.5\times 10^6$ g/(cm²·s);对于硅靶,$Z_{\text{target}} = 2.1\times 10^6$ g/(cm²·s)。注意,激光诱导的等离子体羽中的温度和压强的关系与气体状态方程 $P = nN_AkT/V$ [n 为气体密度,N_A 为阿伏伽德罗(Avogadro)常数,k 为波尔兹曼(Boltzmann)常数,V 为气体体积]的预测并不完全一致,因为脉冲激光诱导的等离子体羽在环境液体束缚中是一个远离热力学平衡的过程,并且等离子体羽通常不被视为理想气体。

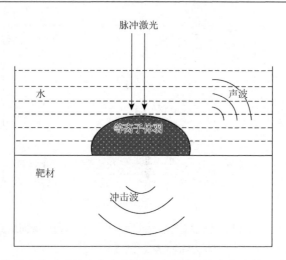

图 3-14　通过表征靶材中的冲击波和水中的声波对受限区域中脉冲激光诱导的等离子体羽的压强测量示意图

3.4　动力学特征

　　LAL 过程具有独特的动力学特性。与在真空和稀释气体环境中 PLA 相比，通过在环境液体束缚中 PLA 固体靶可以实现更高的烧蚀速率。显然，在液体中的 PLA，具有高温高压高密度的脉冲激光诱导的等离子体羽可以在等离子体羽-固体靶界面处连续烧蚀固体靶，持续不断地提高总烧蚀速率，如图 3-15（a）所示[80-83]。Zhu 等[80, 81]发现当使用脉宽为 23 ns 的 248 nm KrF 准分子脉冲激光在水中烧蚀 Si 靶时，Si 靶的激光烧蚀速率随着 Si 靶上方水层的厚度变化而变化，1.1 mm 的水层厚度使得激光烧蚀速率得到了最大的提高，如图 3-15（b）所示。因此，我们可以看出，似乎存在一个可以在脉冲激光诱导的等离子体羽中引起最强压强的水层厚度。具体而言，当水层覆盖 Si 靶时，激光烧蚀产生的冲击波首先发射到水层中，然后再引发水中等离子体羽的爆炸，随后冲击波会通过空气摩擦衰减为声波，这一现象被称为烧蚀活塞（ablative piston）效应，可以大大提高等离子体羽对固体靶的烧蚀速率。另外，水层也会吸收激光能量以削弱激光烧蚀。因此，当烧蚀活塞效应的等离子体羽烧蚀和水层的吸收达到平衡时，就会出现水层的最佳厚度。当水层小于最佳厚度时，激光烧蚀速率非常高，并且烧蚀活塞效应在该过程中占主导地位，激光烧蚀速率随着水层厚度的增加而增加。然而，当水层大于最佳厚度时，水层吸收了很多激光能量，激光烧蚀变得较弱，减少了烧蚀活塞效应，激光烧蚀速率迅速下降并随着水层厚度的增加而降低。类似地，Kim 等[83]揭示了液体层会导致烧蚀阈值的降低和烧蚀速率的提高。这些研究结果表明，PLA 在环境液体束缚中可以有较高的产额。

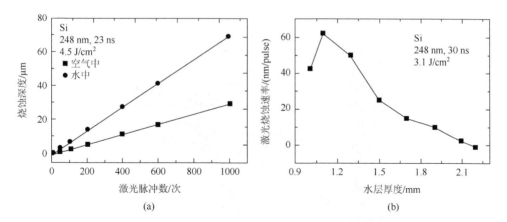

图 3-15 （a）在空气和水中 4.5 J/cm² 的激光能量下，留在靶材表面上的孔的烧蚀深度与激光脉冲数的关系；（b）在 3.1 J/cm² 的激光能量下，Si 靶的激光烧蚀速率与水层厚度的关系

在环境液体束缚下的脉冲激光诱导的等离子体羽可以实现更短的淬灭时间。激光诱导的等离子体羽在空气中和液体中持续时间的实验比较结果，如图 3-16 所示[84]。我们可以看到，固体靶在空气中的 PLA 的等离子体羽持续时间是其在液体中 PLA 的十倍。因此，环境液体的束缚作用加速了等离子体羽的淬灭时间。重要的是，LAL 这种动力学特征为纳米材料合成打开了一扇新的大门。根据前文讨论的 LAL 中脉冲激光诱导的等离子体羽演化的三个基本过程，我们知道，在等离子体羽的持续时间内，粒子团簇形成、成核和晶体生长是以特定的顺序发生的。所以，等离子体羽的淬灭时间实际上就是所生成相的生长时间，短的淬灭时间可以有效地限制生长粒子的大小。例如，当等离子体羽的淬灭时间在纳秒级时，所合成晶体的尺寸可以在纳米级。事实上，当使用脉宽小于 20 ns 的激光时，LAL 合成纳米晶的尺寸分布都是在纳米尺度范围[85-104]。

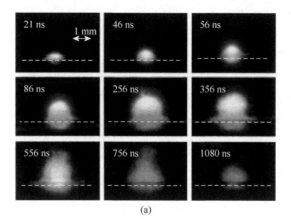

(a)

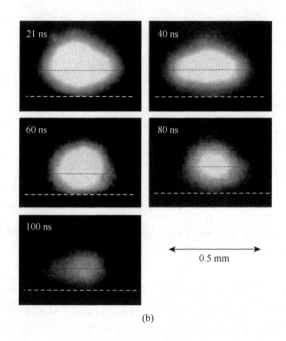

图 3-16 在空气和液体中脉冲激光烧蚀产生等离子体羽持续时间的比较

（a）空气和（b）水，由波长为 1064 nm、脉宽为 20 ns、能量密度为 10 J/cm² 的 Nd：YAG 脉冲激光烧蚀石墨靶产生的等离子体羽发射的系列图像

此外，环境液体对脉冲激光诱导的等离子体羽的冷却作用会促进等离子体羽相变过程中亚稳结构的形成。换言之，由于等离子体羽在环境液体中的冷却时间（淬灭时间）较短，一些亚稳相在从亚稳态向稳态转变的过程中会被冻结而保留在最后的产物中。例如，Yang 等[85]在水中石墨靶的 PLA 合成金刚石纳米晶过程中观察到了从六方石墨到立方金刚石转变的中间阶段。显然，LAL 这种独特的动力学行为为亚稳相的形成提供了动力学优势。

通过上面的讨论，我们可以看到，液体中的 PLA 与真空或气体环境中的 PLA 有许多的区别，可以概括为三个重要方面。第一，当激光烧蚀环境液体束缚固体靶时，在液-固界面处会产生等离子体羽，因为环境液体束缚等离子体羽膨胀，从而通过热力学效应（高温高压高密度状态）导致等离子体羽内部压强和温度的增加；而在真空或气体环境中 PLA 固体靶产生的等离子体羽不存在这种情况。第二，高温高压下的化学反应会发生在脉冲激光诱导的等离子体羽内部、等离子体羽与环境液体的界面处、环境液体中；而在真空或气体环境中的 PLA 是不存在这些高温化学反应的。第三，LAL 中脉冲激光诱导的等离子体羽的淬灭时间要比真空和稀释气体环境中 PLA 等离子体羽的淬灭时间短得多。

参 考 文 献

[1] Eason R. Pulsed laser deposition of thin films: Applications-led growth of functional materials[M]. Hoboken: John Wiley & Sons, 2007.

[2] White C W. Laser and electron beam processing of materials[M]. New York: Academic Press, 2012.

[3] Lv S J, Wang Y. An investigation of pulsed laser cutting of titanium alloy sheet[J]. Optics and Lasers in Engineering, 2006, 44 (10): 1067-1077.

[4] Ohl C D, Arora M, Dijkink R, et al. Surface cleaning from laser-induced cavitation bubbles[J]. Applied Physics Letters, 2006, 89 (7): 074102.

[5] Torkamany M J, Hamedi M J, Malek F, et al. The effect of process parameters on keyhole welding with a 400 W Nd∶YAG pulsed laser[J]. Journal of Physics D: Applied Physics, 2006, 39 (21): 4563.

[6] Lu Y F, Takai M, Komuro S, et al. Surface cleaning of metals by pulsed-laser irradiation in air[J]. Applied Physics A, 1994, 59: 281-288.

[7] Aneesh P M, Shijeesh M R, Aravind A, et al. Highly luminescent undoped and Mn-doped ZnS nanoparticles by liquid phase pulsed laser ablation[J]. Applied Physics A, 2014, 116: 1085-1089.

[8] Feng G Y, Yang C, Zhou S H. Nanocrystalline Cr^{2+}-doped ZnSe nanowires laser[J]. Nano Letters, 2013, 13 (1): 272-275.

[9] Oztoprak B G, Akman E, Hanon M M, et al. Laser welding of copper with stellite 6 powder and investigation using LIBS technique[J]. Optics & Laser Technology, 2013, 45: 748-755.

[10] Niu K Y, Yang J, Kulinich S A, et al. Morphology control of nanostructures via surface reaction of metal nanodroplets[J]. Journal of the American Chemical Society, 2010, 132 (28): 9814-9819.

[11] Niu K Y, Yang J, Kulinich S A, et al. Hollow nanoparticles of metal oxides and sulfides: Fast preparation via laser ablation in liquid[J]. Langmuir, 2010, 26 (22): 16652-16657.

[12] Sun M Y, Yang J, Lin T, et al. Facile synthesis of SnS hollow nanoparticles via laser ablation followed by chemical etching[J]. RSC advances, 2012, 2 (20): 7824-7828.

[13] Song S T, Cui L, Yang J, et al. Millisecond laser ablation of molybdenum target in reactive gas toward MoS_2 fullerene-like nanoparticles with thermally stable photoresponse[J]. ACS applied Materials & Interfaces, 2015, 7 (3): 1949-1954.

[14] Amendola V, Meneghetti M. What controls the composition and the structure of nanomaterials generated by laser ablation in liquid solution? [J]. Physical Chemistry Chemical Physics, 2013, 15 (9): 3027-3046.

[15] Dell'Aglio M, Gaudiuso R, De Pascale O, et al. Mechanisms and processes of pulsed laser ablation in liquids during nanoparticle production[J]. Applied Surface Science, 2015, 348: 4-9.

[16] Reich S, Schönfeld P, Letzel A, et al. Fluence threshold behaviour on ablation and bubble formation in pulsed laser ablation in liquids[J]. ChemPhysChem, 2017, 18 (9): 1084-1090.

[17] De Giacomo A, Dell'Aglio M, Santagata A, et al. Cavitation dynamics of laser ablation of bulk and wire-shaped metals in water during nanoparticles production[J]. Physical Chemistry Chemical Physics, 2013, 15 (9): 3083-3092.

[18] Sakka T, Masai S, Fukami K, et al. Spectral profile of atomic emission lines and effects of pulse duration on laser ablation in liquid[J]. Spectrochimica Acta Part B: Atomic Spectroscopy, 2009, 64 (10): 981-985.

[19] Petkovšek R, Gregorčič P. A laser probe measurement of cavitation bubble dynamics improved by shock wave

detection and compared to shadow photography[J]. Journal of Applied Physics, 2007, 102 (4): 044909.

[20] Yang G W. Laser ablation in liquids: Applications in the synthesis of nanocrystals[J]. Progress in Materials Science, 2007, 52 (4): 648-698.

[21] Sakka T, Iwanaga S, Ogata Y H, et al. Laser ablation at solid-liquid interfaces: An approach from optical emission spectra[J]. The Journal of Chemical Physics, 2000, 112 (19): 8645-8653.

[22] Tsuji T, Okazaki Y, Tsuboi Y, et al. Nanosecond time-resolved observations of laser ablation of silver in water[J]. Japanese Journal of Applied Physics, 2007, 46 (4R): 1533.

[23] Wagener P, Ibrahimkutty S, Menzel A, et al. Dynamics of silver nanoparticle formation and agglomeration inside the cavitation bubble after pulsed laser ablation in liquid[J]. Physical Chemistry Chemical Physics, 2013, 15 (9): 3068-3074.

[24] Tsuji T, Thang D H, Okazaki Y, et al. Preparation of silver nanoparticles by laser ablation in polyvinylpyrrolidone solutions[J]. Applied Surface Science, 2008, 254 (16): 5224-5230.

[25] Akhatov I, Lindau O, Topolnikov A, et al. Collapse and rebound of a laser-induced cavitation bubble[J]. Physics of Fluids, 2001, 13 (10): 2805-2819.

[26] Ibrahimkutty S, Wagener P, Menzel A, et al. Nanoparticle formation in a cavitation bubble after pulsed laser ablation in liquid studied with high time resolution small angle X-ray scattering[J]. Applied Physics Letters, 2012, 101 (10): 103104.

[27] Soliman W, Takada N, Sasaki K. Growth processes of nanoparticles in liquid-phase laser ablation studied by laser-light scattering[J]. Applied Physics Express, 2010, 3 (3): 035201.

[28] Tiberi M, Simonelli A, Cristoforetti G, et al. Effect of picosecond laser induced cavitation bubbles generated on Au targets in a nanoparticle production set-up[J]. Applied Physics A, 2013, 110: 857-861.

[29] Wackerow S, Abdolvand A. Generation of silver nanoparticles with controlled size and spatial distribution by pulsed laser irradiation of silver ion-doped glass[J]. Optics Express, 2014, 22 (5): 5076-5085.

[30] Barcikowski S, Menéndez-Manjón A, Chichkov B, et al. Generation of nanoparticle colloids by picosecond and femtosecond laser ablations in liquid flow[J]. Applied Physics Letters, 2007, 91 (8): 083113.

[31] Barchanski A, Funk D, Wittich O, et al. Picosecond laser fabrication of functional gold-antibody nanoconjugates for biomedical applications[J]. The Journal of Physical Chemistry C, 2015, 119 (17): 9524-9533.

[32] Streubel R, Bendt G, Gökce B. Pilot-scale synthesis of metal nanoparticles by high-speed pulsed laser ablation in liquids[J]. Nanotechnology, 2016, 27 (20): 205602.

[33] Streubel R, Barcikowski S, Gökce B. Continuous multigram nanoparticle synthesis by high-power, high-repetition-rate ultrafast laser ablation in liquids[J]. Optics Letters, 2016, 41 (7): 1486-1489.

[34] Tan D Z, Zhou S F, Qiu J R, et al. Preparation of functional nanomaterials with femtosecond laser ablation in solution[J]. Journal of Photochemistry and Photobiology C: Photochemistry Reviews, 2013, 17: 50-68.

[35] Tan D Z, Sharafudeen K N, Yue Y Z, et al. Femtosecond laser induced phenomena in transparent solid materials: Fundamentals and applications[J]. Progress in Materials Science, 2016, 76: 154-228.

[36] Lorazo P, Lewis L J, Meunier M. Thermodynamic pathways to melting, ablation, and solidification in absorbing solids under pulsed laser irradiation[J]. Physical Review B, 2006, 73 (13): 134108.

[37] Lorazo P, Lewis L J, Meunier M. Short-pulse laser ablation of solids: From phase explosion to fragmentation[J]. Physical Review Letters, 2003, 91 (22): 225502.

[38] Shih C Y, Wu C P, Shugaev M V, et al. Atomistic modeling of nanoparticle generation in short pulse laser ablation of thin metal films in water[J]. Journal of Colloid and Interface Science, 2017, 489: 3-17.

[39] Povarnitsyn M E, Itina T E, Levashov P R, et al. Mechanisms of nanoparticle formation by ultra-short laser ablation of metals in liquid environment[J]. Physical Chemistry Chemical Physics, 2013, 15 (9): 3108-3114.

[40] Roeterdink W G, Juurlink L B F, Vaughan O P H, et al. Coulomb explosion in femtosecond laser ablation of Si(111)[J]. Applied Physics Letters, 2003, 82 (23): 4190-4192.

[41] Annou R, Tripathi V K. Femtosecond laser pulse induced Coulomb explosion[J]. Plasma Physics, 2005, 0510014.

[42] Werner D, Hashimoto S. Improved working model for interpreting the excitation wavelength-and fluence-dependent response in pulsed laser-induced size reduction of aqueous gold nanoparticles[J]. The Journal of Physical Chemistry C, 2011, 115 (12): 5063-5072.

[43] Werner D, Furube A, Okamoto T, et al. Femtosecond laser-induced size reduction of aqueous gold nanoparticles: In situ and pump-probe spectroscopy investigations revealing Coulomb explosion[J]. The Journal of Physical Chemistry C, 2011, 115 (17): 8503-8512.

[44] Von der Linde D, Sokolowski-Tinten K, Bialkowski J. Laser-solid interaction in the femtosecond time regime[J]. Applied Surface Science, 1997, 109-110: 1-10.

[45] Von der Linde D, Schüler H. Breakdown threshold and plasma formation in femtosecond laser-solid interaction[J]. Journal of the Optical Society of America B, 1996, 13 (1): 216-222.

[46] Zeng X, Mao X L, Greif R, et al. Experimental investigation of ablation efficiency and plasma expansion during femtosecond and nanosecond laser ablation of silicon[J]. Applied Physics A, 2005, 80: 237-241.

[47] Wang J B, Yang G W. Phase transformation between diamond and graphite in preparation of diamonds by pulsed-laser induced liquid-solid interface reaction[J]. Journal of Physics: Condensed Matter, 1999, 11 (37): 7089.

[48] Yang L, May P W, Yin L, et al. Growth of diamond nanocrystals by pulsed laser ablation of graphite in liquid[J]. Diamond and Related Materials, 2007, 16 (4-7): 725-729.

[49] Du X W, Qin W J, Lu Y W, et al. Face-centered-cubic Si nanocrystals prepared by microsecond pulsed laser ablation[J]. Journal of Applied Physics, 2007, 102 (1): 013518.

[50] Liu P, Cao Y L, Chen X Y, et al. Trapping high-pressure nanophase of Ge upon laser ablation in liquid[J]. Crystal Growth & Design, 2009, 9 (3): 1390-1393.

[51] Liang Y, Liu P, Li H B, et al. Synthesis and characterization of copper vanadate nanostructures via electrochemistry assisted laser ablation in liquid and the optical multi-absorptions performance[J]. CrystEngComm, 2012, 14 (9): 3291-3296.

[52] Yang L, May P W, Yin L, et al. Ultra fine carbon nitride nanocrystals synthesized by laser ablation in liquid solution[J]. Journal of Nanoparticle Research, 2007, 9: 1181-1185.

[53] Yang S K, Kiraly B, Wang W Y, et al. Fabrication and characterization of beaded SiC quantum rings with anomalous red spectral shift[J]. Advanced Materials, 2012, 24 (41): 5598-5603.

[54] Xiao J, Wu Q L, Liu P, et al. Highly stable sub-5 nm $Sn_6O_4(OH)_4$ nanocrystals with ultrahigh activity as advanced photocatalytic materials for photodegradation of methyl orange[J]. Nanotechnology, 2014, 25 (13): 135702.

[55] Zhang H W, Duan G T, Li Y, et al. Leaf-like tungsten oxide nanoplatelets induced by laser ablation in liquid and subsequent aging[J]. Crystal Growth & Design, 2012, 12 (5): 2646-2652.

[56] Zhang H M, Liang C H, Tian Z F, et al. Organization of Mn_3O_4 nanoparticles into γ-MnOOH nanowires via hydrothermal treatment of the colloids induced by laser ablation in water[J]. CrystEngComm, 2011, 13 (4): 1063-1066.

[57] Stratakis E, Barberoglou M, Fotakis C, et al. Generation of Al nanoparticles via ablation of bulk Al in liquids with

short laser pulses[J]. Optics Express, 2009, 17 (15): 12650-12659.

[58] Kuzmin P G, Shafeev G A, Viau G, et al. Porous nanoparticles of Al and Ti generated by laser ablation in liquids[J]. Applied Surface Science, 2012, 258 (23): 9283-9287.

[59] Viau G, Collière V, Lacroix L M, et al. Internal structure of Al hollow nanoparticles generated by laser ablation in liquid ethanol[J]. Chemical Physics Letters, 2011, 501 (4-6): 419-422.

[60] Barmina E V, Kuzmin P G, Shafeev G A. Self-organization of hydrogen gas bubbles rising above laser-etched metallic aluminum in a weakly basic aqueous solution[J]. Physical Review E, 2011, 84 (4): 045302.

[61] Barmina E V, Kuzmin P G, Shafeev G A, et al. Self-organization of hydrogen gas bubbles rising from the surface of the laser-irradiated aluminum target under its etching in a dilute alkaline solution[J]. Physics of Wave Phenomena, 2012, 20: 159-165.

[62] El Mel A A, Buffière M, Tessier P Y, et al. Hollow nanostructures: Highly ordered hollow oxide nanostructures: The Kirkendall effect at the nanoscale (Small 17/2013) [J]. Small, 2013, 9 (17): 2837.

[63] Niu K Y, Park J W, Zheng H M, et al. Revealing bismuth oxide hollow nanoparticle formation by the Kirkendall effect[J]. Nano Letters, 2013, 13 (11): 5715-5719.

[64] Wang W H, Dahl M, Yin Y D. Hollow nanocrystals through the nanoscale Kirkendall effect[J]. Chemistry of Materials, 2013, 25 (8): 1179-1189.

[65] Niu K Y, Yang J, Sun J, et al. One-step synthesis of MgO hollow nanospheres with blue emission[J]. Nanotechnology, 2010, 21 (29): 295604.

[66] Yan Z J, Bao R Q, Busta C M, et al. Fabrication and formation mechanism of hollow MgO particles by pulsed excimer laser ablation of Mg in liquid[J]. Nanotechnology, 2011, 22 (26): 265610.

[67] Yan Z J, Bao R Q, Huang Y, et al. Hollow particles formed on laser-induced bubbles by excimer laser ablation of Al in liquid[J]. The Journal of Physical Chemistry C, 2010, 114 (26): 11370-11374.

[68] Sakka T, Takatani K, Ogata Y H, et al. Laser ablation at the solid-liquid interface: Transient absorption of continuous spectral emission by ablated aluminium atoms[J]. Journal of Physics D: Applied Physics, 2002, 35 (1): 65.

[69] Sakka T, Saito K, Ogata Y H. Emission spectra of the species ablated from a solid target submerged in liquid: Vibrational temperature of C_2 molecules in water-confined geometry[J]. Applied Surface Science, 2002, 197-198: 246-250.

[70] Saito K, Takatani K, Sakka T, et al. Observation of the light emitting region produced by pulsed laser irradiation to a solid-liquid interface[J]. Applied Surface Science, 2002, 197-198: 56-60.

[71] Saito K, Sakka T, Ogata Y H. Rotational spectra and temperature evaluation of C_2 molecules produced by pulsed laser irradiation to a graphite-water interface[J]. Journal of Applied Physics, 2003, 94 (9): 5530-5536.

[72] Devaux D, Fabbro R, Tollier L, et al. Generation of shock waves by laser-induced plasma in confined geometry[J]. Journal of Applied Physics, 1993, 74 (4): 2268-2273.

[73] Peyre P, Fabbro R. Laser shock processing: A review of the physics and applications[J]. Optical and Quantum Electronics, 1995, 27: 1213-1229.

[74] Berthe L, Fabbro R, Peyre P, et al. Shock waves from a water-confined laser-generated plasma[J]. Journal of Applied Physics, 1997, 82 (6): 2826-2832.

[75] Peyre P, Berthe L, Scherpereel X, et al. Laser-shock processing of aluminium-coated 55C1 steel in water-confinement regime, characterization and application to high-cycle fatigue behaviour[J]. Journal of Materials Science, 1998, 33 (6): 1421-1429.

[76] Berthe L, Fabbro R, Peyre P, et al. Wavelength dependent of laser shock-wave generation in the water-confinement regime[J]. Journal of Applied Physics, 1999, 85 (11): 7552-7555.

[77] Berthe L, Sollier A, Peyre P, et al. The generation of laser shock waves in a water-confinement regime with 50 ns and 150 ns XeCl excimer laser pulses[J]. Journal of Physics D: Applied Physics, 2000, 33 (17): 2142.

[78] Peyre P, Berthe L, Fabbro R, et al. Experimental determination by PVDF and EMV techniques of shock amplitudes induced by 0.6~3 ns laser pulses in a confined regime with water[J]. Journal of Physics D: Applied Physics, 2000, 33 (5): 498.

[79] Sollier A, Berthe L, Fabbro R. Numerical modeling of the transmission of breakdown plasma generated in water during laser shock processing[J]. The European Physical Journal: Applied Physics, 2001, 16 (2): 131-139.

[80] Zhu S, Lu Y F, Hong M H, et al. Laser ablation of solid substrates in water and ambient air[J]. Journal of Applied Physics, 2001, 89 (4): 2400-2403.

[81] Zhu S, Lu Y F, Hong M H. Laser ablation of solid substrates in a water-confined environment[J]. Applied Physics Letters, 2001, 79 (9): 1396-1398.

[82] Geiger M, Becker W, Rebhan T, et al. Increase of efficiency for the XeCl excimer laser ablation of ceramics[J]. Applied Surface Science, 1996, 96-98: 309-315.

[83] Kim D, Oh B, Lee H. Effect of liquid film on near-threshold laser ablation of a solid surface[J]. Applied Surface Science, 2004, 222 (1-4): 138-147.

[84] Saito K, Sakka T, Ogata Y H. Rotational spectra and temperature evaluation of C_2 molecules produced by pulsed laser irradiation to a graphite-water interface[J]. Journal of Applied Physics, 2003, 94 (9), 530-5536.

[85] Yang G W, Wang J B, Liu Q X. Preparation of nano-crystalline diamonds using pulsed laser induced reactive quenching[J]. Journal of Physics: Condensed Matter, 1998, 10 (35): 7923.

[86] Yeh M S, Yang Y S, Lee Y P, et al. Formation and characteristics of Cu colloids from CuO powder by laser irradiation in 2-propanol[J]. The Journal of Physical Chemistry B, 1999, 103 (33): 6851-6857.

[87] Poondi D, Singh J. Synthesis of metastable silver-nickel alloys by a novel laser-liquid-solid interaction technique[J]. Journal of Materials Science, 2000, 35: 2467-2476.

[88] Simakin A V, Voronov V V, Shafeev G A, et al. Nanodisks of Au and Ag produced by laser ablation in liquid environment[J]. Chemical Physics Letters, 2001, 348 (3-4): 182-186.

[89] Jiang Z Y, Huang R B, Xie S Y, et al. Synthesis of silver selenide bicomponent nanoparticles by a novel technique: Laser-solid-liquid ablation[J]. Journal of Solid State Chemistry, 2001, 160 (2): 430-434.

[90] Wang J B, Zhang C Y, Zhong X L, et al. Cubic and hexagonal structures of diamond nanocrystals formed upon pulsed laser induced liquid-solid interfacial reaction[J]. Chemical Physics Letters, 2002, 361 (1-2): 86-90.

[91] Compagnini G, Scalisi A A, Puglisi O. Ablation of noble metals in liquids: A method to obtain nanoparticles in a thin polymeric film[J]. Physical Chemistry Chemical Physics, 2002, 4 (12): 2787-2791.

[92] Liang C H, Shimizu Y, Sasaki T, et al. Synthesis of ultrafine SnO_{2-x} nanocrystals by pulsed laser-induced reactive quenching in liquid medium[J]. The Journal of Physical Chemistry B, 2003, 107 (35): 9220-9225.

[93] Liu Q X, Wang C X, Zhang W, et al. Immiscible silver-nickel alloying nanorods growth upon pulsed-laser induced liquid/solid interfacial reaction[J]. Chemical Physics Letters, 2003, 382 (1-2): 1-5.

[94] Zhang Y, Chen W Z, Zhang W G. Studies on nano-cobalt/ethanol sol prepared by pulsed laser ablation[J]. Chemical Journal of Chinese Universities, 2003, 24 (2): 337-339.

[95] Kabashin A V, Meunier M. Synthesis of colloidal nanoparticles during femtosecond laser ablation of gold in water[J]. Journal of Applied Physics, 2003, 94 (12): 7941-7943.

[96] Izgaliev A T, Simakin A V, Shafeev G A. Formation of the alloy of Au and Ag nanoparticles upon laser irradiation of the mixture of their colloidal solutions[J]. Quantum Electronics, 2004, 34 (1): 47.

[97] Zhang W G, Jin Z G. Research on successive preparation of nano-FeNi alloy and its ethanol sol by pulsed laser ablation[J]. Science in China Series B: Chemistry, 2004, 47: 159-165.

[98] Pearce S R J, Henley S J, Claeyssens F, et al. Production of nanocrystalline diamond by laser ablation at the solid/liquid interface[J]. Diamond and Related Materials, 2004, 13 (4-8): 661-665.

[99] Chen J W, Dong Q Z, Yang J, et al. The irradiation effect of a Nd-YAG pulsed laser on the CeO_2 target in the liquid[J]. Materials Letters, 2004, 58 (3-4): 337-341.

[100] Chen G X, Hong M H, Lan B, et al. A convenient way to prepare magnetic colloids by direct Nd:YAG laser ablation[J]. Applied Surface Science, 2004, 228 (1-4): 169-175.

[101] Pyatenko A, Shimokawa K, Yamaguchi M, et al. Synthesis of silver nanoparticles by laser ablation in pure water[J]. Applied Physics A, 2004, 79: 803-806.

[102] Zeng H B, Cai W P, Li Y, et al. Composition/structural evolution and optical properties of ZnO/Zn nanoparticles by laser ablation in liquid media[J]. The Journal of Physical Chemistry B, 2005, 109 (39): 18260-18266.

[103] Usui H, Sasaki T, Koshizaki N. Ultraviolet emission from layered nanocomposites of Zn(OH)$_2$ and sodium dodecyl sulfate prepared by laser ablation in liquid[J]. Applied Physics Letters, 2005, 87 (6): 063105.

[104] Kitazawa S, Abe H, Yamamoto S. Formation of nanostructured solid-state carbon particles by laser ablation of graphite in isopropyl alcohol[J]. Journal of Physics and Chemistry of Solids, 2005, 66 (2-4): 555-559.

第 4 章　LAL 中纳米晶形成的物理化学过程

由于 LAL 会产生局域高温高压高密度的热力学极端环境，所以在亚稳相尤其是高温高压相的合成中具有显著的优势。然而，LAL 纳米制备技术是一种相对新颖的激光材料制备与加工技术，纳米材料在 LAL 中所涉及的成核、相变及生长的机制尚不明晰。因此，我们需要发展有效的理论工具来处理 LAL 纳米制备中纳米相的成核、相变和生长。幸运的是，我们已经发展了纳米尺度下材料生长和相变的热力学理论即纳米热力学理论[1]，迄今为止，我们的纳米热力学方法已经广泛用于纳米材料的生长和设计[2-15]。本章，我们基于已建立的纳米热力学理论，讨论 LAL 中纳米相成核、相变、生长的基本物理和化学过程。

4.1　成核热力学

一般来说，在合适的热力学环境中，母相中粒子团簇的成核是一种常见的热力学行为，例如，成核会在气体冷凝、液体蒸发和晶体生长等基本的物态转化过程中发生。LAL 纳米制备中纳米晶的形成，包括成核和生长，通常发生在激光诱导的等离子体羽的相变和凝结阶段，这方面我们将在 4.2 节中详细讨论。我们以金刚石成核为例，讨论 LAL 中纳米晶的成核热力学。需要注意的是，成核热力学是基于热力学平衡相图，并且该理论是基于以下假设[1]：①晶核是完美的球形，晶体结构与相应体材料一样；②晶核与晶核之间没有相互作用。

众所周知，我们可以用吉布斯自由能来描述在竞争相之间发生相变所需要的能量。在给定的热力学条件下，金刚石相可以和石墨相共存。然而，两个相中只有自由能较小的那个相是稳定的，另一个相则是亚稳的并且可以转变为稳定的。从热力学角度来看，两个相自由能的差异可以驱动相变发生。具体来说，热力学相的吉布斯自由能是压强和温度的函数，可用通用坐标或反应坐标表示[16]。

我们以脉冲激光烧蚀水中的石墨靶合成金刚石纳米晶为例来描述 LAL 中金刚石的成核。在碳的热力学平衡相图（简称为碳相图）中，金刚石成核的压强和温度区域分别限制在 10~15 GPa 和 4000~5000 K 内[17]。因此，我们的理论和计算都集中在这个热力学相区。在假设 LAL 成核为各向同性的球形晶核的前提下，由金刚石晶核的纳米尺寸效应所引起的附加压强 ΔP 可由拉普拉斯-杨方程

$\Delta P = 2\gamma/r$ 给出[3]。另外,碳的热力学平衡相图中石墨和金刚石的平衡相界可表示为 $P^e = 2.01\times10^6 T + 2.02\times10^{9}$ [9]。由于附加压强 ΔP 的作用,从石墨相转变为金刚石相所需的外部压强将减少相同的量。由此,我们可以得到纳米尺寸相关的平衡相界 $P^e = 2.01\times10^6 T + 2.02\times10^9 - 2\gamma/r$。当压强-温度处于相平衡线上时,可以得到从石墨相转变为金刚石相的摩尔体积吉布斯自由能差 $\Delta g^d_{T,P} = \Delta V \times (P - 2.01\times10^6 T - 2.02\times10^9 + 2\gamma/r)$,其中,$\Delta V$ 是金刚石和石墨的摩尔体积差。那么,考虑纳米尺寸效应后,石墨到金刚石相变的吉布斯自由能差表示为[9]

$$\Delta G(r) = \frac{4}{3}\pi r^3 \Delta V \times (P - 2.01\times10^6 T - 2.02\times10^9 + 2\gamma/r)/V_m + 4\pi r^2 \gamma \quad (4-1)$$

式中,V_m 是金刚石的摩尔体积。当 $\frac{\partial \Delta G(r)}{\partial r} = 0$ 时,金刚石晶核的临界半径可以表示为

$$r^* = 2\gamma\left(\frac{2}{3} + \frac{V_m}{\Delta V}\right)/(2.73\times10^6 T + 7.23\times10^8 - P) \quad (4-2)$$

将方程(4-2)代入式方程(4-1),我们可以得到金刚石晶核的临界吉布斯自由能(临界形成能)

$$\Delta G(r^*) = \frac{4}{3}\pi r^{*3}\Delta V \times (P - 2.73\times10^6 T - 7.23\times10^8 + 2\gamma/r^*)/V_m + 4\pi r^{*2}\gamma \quad (4-3)$$

这样,在温度为 4000~5000 K、压强为 10~15 GPa 的条件下,金刚石晶核临界半径 r^* 与温度的关系,如图 4-1(a)所示。在相同压强和温度的条件下,临界半径为 r^* 的金刚石晶核的 $\Delta G(r^*)$-T 曲线,如图 4-1(b)所示。我们可以清楚地看到,这些具有临界半径 r^* 和低形成能 $\Delta G(r^*)$ 的金刚石晶核可以通过 LAL 来实现。例如,当 $P = 10$ GPa、$T = 5000$ K 时,金刚石晶核的临界半径 r^* 可以达到 3 nm,

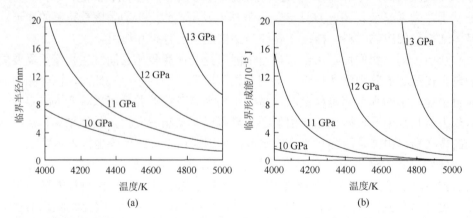

图 4-1 在不同压强条件下,金刚石晶核临界半径 r^* 和临界形成能 $\Delta G(r^*)$ 与温度的依赖关系
(a) r^*;(b) $\Delta G(r^*)$

而形成能小于 1×10^{-15} J。因此,这些结果表明,LAL 所创造的热力学环境是有利于金刚石成核的。此外,我们看到,金刚石晶核临界半径 r^* 和临界形成能 $\Delta G(r^*)$ 都随着温度的升高而减小,随着压强的升高而增大。因此,这些理论结果可以用来指导 LAL 合成金刚石纳米晶的实验参数选择,以获得所需要的纳米晶尺寸。事实上,目前 LAL 合成金刚石纳米晶的尺寸分布恰好落在上述理论和计算所预期的范围内[18,19],这表明,我们建立的 LAL 的热力学成核理论可以作为处理 LAL 纳米晶合成的普适理论。

4.2 相变热力学

在 LAL 纳米制备中,环境液体的束缚作用对于脉冲激光诱导的等离子体羽的相变和凝结过程具有重要意义,尤其是对于相变,因为相变决定了 LAL 的生成相。因此,我们有必要为此建立一个普适的相变热力学理论来处理 LAL 中发生的相变,为纳米材料的设计和生长提供理论参考。本节,我们基于纳米热力学理论建立了 LAL 过程中的相变模型,以阐明脉冲激光诱导的等离子体羽相变和凝结中的生成相演化。我们以纳秒激光烧蚀水中六方氮化硼(h-BN)靶合成立方氮化硼(c-BN)纳米晶为例介绍我们的相变热力学。

从热力学角度来看,两相自由能的差异促进了相变的发生,因为相变是由 h-BN 结构越过中间相势垒的相变几率定量确定的[20]。因此,热力学相的吉布斯自由能 $G_{T,P}$ 可以表示为压强和温度的函数。从亚稳相到稳定相的相变几率 f 不仅取决于吉布斯自由能差 $\Delta G_{T,P}$,还取决于相变所需的活化能 $E_a - \Delta G_{T,P}$。当两相处于热力学平衡状态时,即 $\Delta G_{T,P}=0$ 时,E_a 为两侧相对于总坐标的最大势能。对于从 h-BN 到 c-BN 的相变,$f=f_c$ 且 $\Delta G_{T,P}^c = G_{T,P}^h(\text{h-BN}) - G_{T,P}^c(\text{c-BN})$,其中,$E_a - \Delta G_{T,P}^c$ 是 h-BN 到 c-BN 的相变所需的活化能。对于 c-BN 到 h-BN 的相变,$f=f_h$ 且 $\Delta G_{T,P}^h = G_{T,P}^c(\text{c-BN}) - G_{T,P}^h(\text{h-BN})$,其中[11],

$$f_h = \exp[-(E_a/RT)] - \exp\{-[(E_a - \Delta G_{T,P}^h)/RT]\} \tag{4-4}$$

Berman 和 Simon 曾推断,当相变条件处在 c-BN 和 h-BN 的相平衡线上时(即 $\Delta G_{T,P}=0$ 时),P^e 和 T 满足关系 $P^e = 2.985\times10^6 T - 4.615\times10^9$ [21]。如果考虑纳米尺寸引起的附加压强 ΔP,则可以得到纳米尺寸相关的平衡相界 $P^e = 2.985\times10^6 T - 4.615\times10^9 - 2\gamma/r$。因此,单位体积内 h-BN 到 c-BN 转变的 $\Delta G_{T,P}^c$ 可以由式(4-5)给出[22]:

$$\Delta G_{T,P}^c = 3.79\times10^{-6}(P - 2.985\times10^6 T + 4.615\times10^9 + 2\gamma/r) \tag{4-5}$$

值得注意的是,单位体积内 c-BN 向 h-BN 转变的 $\Delta G_{T,P}^h$ 可以表示为

$$\Delta G_{T,P}^h = 3.79\times10^{-6}(2.985\times10^6 T - 4.615\times10^9 - 2\gamma/r - P) \tag{4-6}$$

这样，我们就可以计算出相变几率的分布，如图 4-2 所示，其中，$r = 20$ nm。f_c 常数呈 "V" 形，一侧接近 Berman-Simon（B-S）线，另一侧几乎垂直。f_c 随着温度的升高而迅速增加。当压强为 2.5～3 GPa、温度为 1800～2100 K 时，即高温高压条件下，f_c 大约为 10^{-6}[21]。然而，当压强为 4～6 GPa、温度为 3000～4000 K 时，即 LAL 合成 c-BN 纳米晶的条件，f_c 高达 $10^{-5} \sim 10^{-4}$。显然，LAL 的 f_c 比高温高压法中的 f_c 高出两个数量级。因此，这些理论结果表明，LAL 是一种高效合成 c-BN 纳米晶的方法。另外，基于上述热力学模型，我们可以计算 h-BN 向 c-BN 的相变几率与 c-BN 晶粒临界半径的关系，如图 4-3 所示，其中，$T = 3200$ K，$P = 5$ GPa。从图 4-3 可以看出，随着晶粒临界半径的增加，f_c 减小得非常快。特别是当晶粒临界半径大于 10 nm 时，f_c 变得非常小。这些结果说明较小半径的晶核更容易形成。

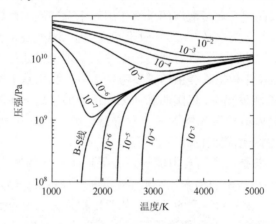

图 4-2　LAL 中 h-BN 和 c-BN 的相变几率 f 曲线

h-BN 到 c-BN 的相变几率在 B-S 线上方，c-BN 到 h-BN 的相变几率在 B-S 线下方

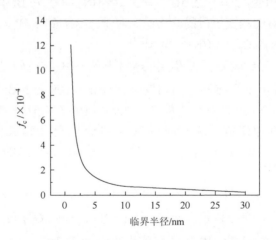

图 4-3　LAL 中 h-BN 到 c-BN 的相变几率与临界半径 r^* 的关系

因此，我们所建立的相变热力学理论不仅可以让人们更深入地理解 LAL 中脉冲激光诱导的等离子体羽相变和凝结中的基本物理过程，而且还为 LAL 纳米材料的可控制备提供了重要的理论指导。

4.3 生长动力学

通常，LAL 中脉冲激光诱导的等离子体羽演化的三个阶段中，生成相的生长发生在等离子体羽的凝结阶段。一旦成核发生，那么这些晶核就会开始生长。因此，建立晶核生长动力学理论是非常重要的，不仅有利于我们清晰地理解晶核的生长过程，而且为我们控制纳米晶的生长提供了动力学理论参考。本节，以纳秒激光烧蚀水中石墨靶合成金刚石纳米晶为例，我们发展了一种普适的生长动力学方法来阐明 LAL 中纳米晶的成核与生长[15]。重要的是，我们的动力学理论可以预测 LAL 中纳米晶的成核时间、生长速率和生长尺寸。

LAL 中金刚石纳米晶成核与生长的基本物理和化学过程可分为三个连续阶段：①等离子体羽形成。激光烧蚀固体靶后，固-液界面处会立即产生致密的等离子体羽，随后等离子体羽中的压强和温度会分别达到峰值。②稳态成核发生。当等离子体羽的压强和温度从峰值开始下降，等离子体羽的凝结会导致粒子团簇的形成，此时粒子间的相互作用比等离子体羽中的粒子的相互作用要强得多。来自固体靶的粒子可以通过扩散和碰撞聚集形成团簇，直到这些团簇达到临界成核尺寸时，成核发生。③晶体的稳态生长。临界晶核形成后，晶核开始长大。晶核的生长周期约为该阶段脉冲激光持续时间的 2 倍[23]。随着时间的延长，晶核的生长将因压强和温度降低而停止。

因此，假设粒子团簇和周围等离子体羽具有相同的温度 T，那么等温成核时间 τ 由式（4-7）给出[24]：

$$\tau = \sqrt{2\pi mkT} \frac{kT\gamma}{p_S(T)(\Delta\mu)^2} \tag{4-7}$$

式中，m、k、T 和 γ 分别表示单个粒子的质量、玻尔兹曼常数、绝对温度和金刚石的表面能密度；$p_S(T)$ 是在温度 T 下原子团簇的饱和蒸气压，由凝聚相的 P-T 相图计算得出，例如，在碳的热力学平衡相图中，石墨和金刚石的 $p_S(T) = 2.01\times10^6 T + 2.02\times10^9$ [25]；$\Delta\mu$ 是原子化学势差，可以表示为[1]

$$\Delta\mu = \frac{\Delta V(p - 2.01\times10^6 T - 2.02\times10^9)}{V_m N_A} \tag{4-8}$$

式中，ΔV、p、V_m 和 N_A 分别表示石墨和金刚石的摩尔体积差、实验压强、金刚石摩尔体积和阿伏加德罗常数。

根据式（4-7）和式（4-8），我们可以计算出不同温度下 LAL 中压强与成核时间的关系，如图 4-4 所示[15]。显然，我们可以看到，在合成金刚石的情况下，相界（P-T）区域附近的成核时间约为 $10^{-10} \sim 10^{-9}$ s。换言之，在该压强-温度区域内，随着等离子体羽压强的变化，成核时间会发生显著的变化。请注意，在碳的热力学平衡相图中，合成金刚石的压强-温度区域（图 4-4 中的插图）位于金刚石和石墨的边界线上方且靠近边界线。这些理论结果意味着 LAL 的成核是一个瞬态过程。此外，图 4-4 表明，在一定温度下，成核时间随着压强的增加而减少。同样，在一定压强下，成核时间随着温度的升高而增加。这可能是由于随着压强-温度区域逐渐接近边界线，原子化学势变小所导致的。

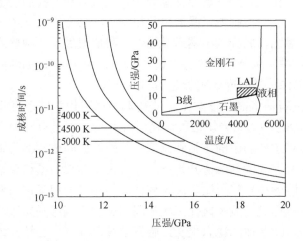

图 4-4　不同温度下压强与成核时间的关系曲线

插图为基于压强和温度的碳热力学平衡相图

下面，我们讨论晶核的生长动力学理论。根据 Wilson-Frenkel 生长定律，一般晶核的生长速率 V 可以表示为[26]

$$V = h\nu \exp(-E_a / RT)[1 - \exp(-|\Delta G_m| / RT)] \quad (4\text{-}9)$$

式中，h、ν、E_a、R 和 T 分别为 LAL 合成金刚石纳米晶过程中金刚石晶核生长方向的晶格常数、热振动频率、表面吸附原子的摩尔吸附能、气体常数和温度[27-29]；ΔG_m 是每摩尔体积的吉布斯自由能差，可表示为 $\Delta G_m = -RT \ln\left(\dfrac{P}{P_s}\right)$，其中，$P$ 和 P_s 分别是等离子体羽中金刚石的有效压强和饱和蒸气压。

根据方程（4-9），我们可以计算出金刚石晶核的生长速率与温度的关系曲线，如图 4-5 所示[15]。金刚石晶核的直径可以表示为 $d = V(2\tau_d - \tau) + 2r^*$，其中，$\tau_d$ 和 r^* 分别是脉冲激光持续时间和金刚石晶核的临界半径。图 4-6 显示了在不同温度下，压强与金刚石晶核尺寸的关系曲线。由此可见，我们预测纳秒激光烧蚀水中石墨靶合成的金刚石纳米晶的直径范围为 25~250 nm。重要的是，实验通过 LAL 合成的金刚石的尺寸在 40~200 nm 变化[18]，理论结果与实验数据非常一致，这表明本节所建立的生长动力学模型可以作为研究 LAL 中纳米晶生长所涉及的基础物理和化学的理论方法，同时为 LAL 纳米材料的可控制备提供理论参考。

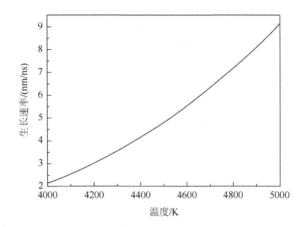

图 4-5　LAL 合成的金刚石晶核的生长速率对温度的依赖

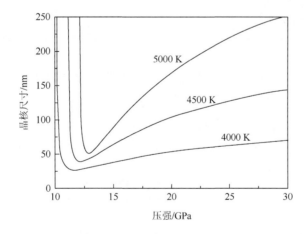

图 4-6　在不同温度下，LAL 合成的金刚石晶核尺寸对压强的依赖

参 考 文 献

[1] Wang C X, Yang G W. Thermodynamics of metastable phase nucleation at the nanoscale[J]. Materials Science and Engineering: R: Reports, 2005, 49 (6): 157-202.

[2] Wang C X, Yang Y H, Liu Q X, et al. Nucleation thermodynamics of cubic boron nitride upon high-pressure and high-temperature supercritical fluid system in nanoscale[J]. The Journal of Physical Chemistry B, 2004, 108 (2):

728-731.

[3] Zhang C Y, Wang C X, Yang Y H, et al. A nanoscaled thermodynamic approach in nucleation of CVD diamond on nondiamond surfaces[J]. The Journal of Physical Chemistry B, 2004, 108（8）: 2589-2593.

[4] Liu Q X, Wang C X, Li S W, et al. Nucleation stability of diamond nanowires inside carbon nanotubes: A thermodynamic approach[J]. Carbon, 2004, 42（3）: 629-633.

[5] Wang C X, Yang Y H, Yang G W. Nanothermodynamic analysis of the low-threshold-pressure-synthesized cubic boron nitride in supercritical-fluid systems[J]. Applied Physics Letters, 2004, 84（16）: 3034-3036.

[6] Liu Q X, Wang C X, Yang Y H, et al. One-dimensional nanostructures grown inside carbon nanotubes upon vapor deposition: A growth kinetic approach[J]. Applied Physics Letters, 2004, 84（22）: 4568-4570.

[7] Wang C X, Yang Y H, Xu N S, et al. Thermodynamics of diamond nucleation on the nanoscale[J]. Journal of the American Chemical Society, 2004, 126（36）: 11303-11306.

[8] Wang C X, Liu Q X, Yang G W. A nanothermodynamic analysis of cubic boron nitride nucleation upon chemical vapor deposition[J]. Chemical Vapor Deposition, 2004, 10（5）: 280-283.

[9] Wang B, Yang Y H, Wang C X, et al. Nanostructures and self-catalyzed growth of SnO_2[J]. Journal of Applied Physics, 2005, 98（7）: 073520.

[10] Liu Q X, Wang C X, Yang G W. Nucleation thermodynamics of cubic boron nitride in pulsed-laser ablation in liquid[J]. Physical Review B, 2005, 71（15）: 155422.

[11] Liu Q X, Wang C X, Xu N S, et al. Nanowire formation during catalyst assisted chemical vapor deposition[J]. Physical Review B, 2005, 72（8）: 085417.

[12] Wang C X, Wang B, Yang Y H, et al. Thermodynamic and kinetic size limit of nanowire growth[J]. The Journal of Physical Chemistry B, 2005, 109（20）: 9966-9969.

[13] Liang L H, Yang G W, Li B W. Size-dependent formation enthalpy of nanocompounds[J]. The Journal of Physical Chemistry B, 2005, 109（33）: 16081-16083.

[14] Wang C X, Chen J, Yang G W, et al. Thermodynamic stability and ultrasmall-size effect of nanodiamonds[J]. Angewandte Chemie International Edition, 2005, 44（45）: 7414-7418.

[15] Wang C X, Liu P, Cui H, et al. Nucleation and growth kinetics of nanocrystals formed upon pulsed-laser ablation in liquid[J]. Applied Physics Letters, 2005, 87（20）: 201913.

[16] Yang G W, Liu B X. Nucleation thermodynamics of quantum-dot formation in V-groove structures[J]. Physical Review B, 2000, 61（7）: 4500.

[17] Wang J B, Yang G W. Phase transformation between diamond and graphite in preparation of diamonds by pulsed-laser induced liquid-solid interface reaction[J]. Journal of Physics: Condensed Matter, 1999, 11（37）: 7089.

[18] Yang G W, Wang J B, Liu Q X. Preparation of nano-crystalline diamonds using pulsed laser induced reactive quenching[J]. Journal of Physics: Condensed Matter, 1998, 10（35）: 7923.

[19] Wang J B, Zhang C Y, Zhong X L, et al. Cubic and hexagonal structures of diamond nanocrystals formed upon pulsed laser induced liquid-solid interfacial reaction[J]. Chemical Physics Letters, 2002, 361（1-2）: 86-90.

[20] Liu Q X, Yang G W, Zhang J X. Phase transition between cubic-BN and hexagonal-BN upon pulsed laser induced liquid-solid interfacial reaction[J]. Chemical Physics Letters, 2003, 373（1-2）: 57-61.

[21] Solozhenko V L. Synchrotron radiation studies of the kinetics of c-BN crystallization in the NH_4F-BN system[J]. Physical Chemistry Chemical Physics, 2002, 4（6）: 1033-1035.

[22] Compagnini G, Scalisi A A, Puglisi O. Production of gold nanoparticles by laser ablation in liquid alkanes[J].

Journal of Applied Physics,2003,94(12):7874-7877.

[23] Fabbro R,Fournier J,Ballard P,et al. Physical study of laser-produced plasma in confined geometry[J]. Journal of Applied Physics,1990,68(2):775-784.

[24] Feder J,Russell K C,Lothe J,et al. Homogeneous nucleation and growth of droplets in vapours[J]. Advances in Physics,1966,15(57):111-178.

[25] Bundy F P,Bassett W A,Weathers M S,et al. The pressure-temperature phase and transformation diagram for carbon: Updated through 1994[J]. Carbon,1996,34(2):141-153.

[26] Wilson A. Philosophical Magazine,1909,50:609.

[27] Gogotsi Y,Welz S,Ersoy D A,et al. Conversion of silicon carbide to crystalline diamond-structured carbon at ambient pressure[J]. Nature,2001,411:283-287.

[28] Xie J J,Chen S P,Tse J S,et al. High-pressure thermal expansion, bulk modulus, and phonon structure of diamond[J]. Physical Review B,1999,60(13):9444.

[29] Mehandru S P,Anderson A B. Adsorption and bonding of C_1H_x and C_2H_y on unreconstructed diamond (111). Dependence on coverage and coadsorbed hydrogen[J]. Journal of Materials Research,1990,5(11):2286-2295.

第 5 章　LAL 合成纳米金刚石及新碳相纳米材料

众所周知，纳米晶合成的挑战之一是在中等温度和压强的条件下合成具有高温高压亚稳相的纳米晶。事实上，在过去的数十年里，研究人员发现越来越多的化学和物理途径可以在中等温度和压强下合成在相应的热力学平衡相图中具有较稳定结构的高温高压亚稳相[1-7]。正如前文所阐述的那样，LAL 在液体环境中产生的局域高温高压高密度的极端热力学状态是有利于亚稳相形成的。因此，与其他制备亚稳相纳米材料的方法如高温高压法等相比，在液体环境中激光烧蚀固体靶方法具有三个显著优点：化学方法"简单、清洁"的合成（减少了副产物的形成、更简单的初始材料、不需要催化剂等）、独特的合成环境、可以获得常规温和条件（中等温度和压强）的制备方法无法制备的亚稳相。金刚石是一种典型的亚稳相材料，与石墨相比，它具有亚稳结构。因此，当 Ogale 等[8]发现通过红宝石脉冲激光辐照沉浸在液态苯中的石墨靶合成了许多碳颗粒并观察到了金刚石相时，金刚石纳米晶及相关新碳相纳米材料的 LAL 合成成为应用 LAL 方法探索常态下亚稳相纳米材料合成研究的焦点。本章，我们介绍纳米金刚石及相关新碳相纳米材料的 LAL 合成。

5.1　纳米金刚石

近年来，纳米金刚石（nanodiamond，ND）引起了学术界和工业界的极大关注[47-54]，因为它们独特的物性使其适合用作生物活性基底[51]、生物传感器[52]、诊断和治疗生物医学成像探针[53]等。金刚石的物理硬度、高热导率和光学透明性等独特性质激发了人们对其合成的兴趣。自 20 世纪 50 年代首次采用高温高压法合成金刚石以来，已经发展了许多制备金刚石材料的方法[9]。在过去的几十年里，PLD 技术已经被证明是制备功能薄膜材料的有效方法[10]。特别是，研究人员已经广泛报道了在真空中激光烧蚀石墨靶制备具有类金刚石特性的非晶碳膜[11-24]的研究，其中一些研究人员观察到非晶碳膜中有晶体颗粒的生长，但是在所有报道的拉曼光谱中缺乏金刚石特征峰。直到 1995 年，Polo 等[25]首次发现了 PLD 样品的拉曼光谱中存在尖锐的 $1332\ cm^{-1}$ 峰，从而证实了金刚石立方结构的存在。1998 年，Yang 等首次发表应用 LAL 方法，分别以水、丙酮和乙醇等为环境液体，以固体石墨为靶材，使用 Nd：YAG 纳秒激光器，合成了具有良好单晶形态的纳米金刚石并给出了相对完整的金刚石拉曼光谱[26, 27]。随后，Pearce 等[28]成功地复制了 Yang 等的研究。现在，LAL 方法已经被公认为是合成纳米金刚石的重要方法之一[29]。

5.1.1 Bottom-up 合成纳米金刚石

Yang 等[26]在水、丙酮和乙醇等液体中使用脉冲激光烧蚀固体石墨靶合成出纳米金刚石，属于纳米材料的自下而上（bottom-up）制备方式，图 5-1 展示了 LAL 合成的纳米金刚石的形貌和相应的选区电子衍射（SAED）图像。我们可以看到合成的纳米晶有着清晰的晶面，而选区电子衍射图像 [图 5-1（b）] 表明这里合成的纳米金刚石具有立方（cubic）和六方（hexagonal）两种结构，即混合相结构，并且两种结构的取向关系为$[110]_c // [1\bar{2}10]_h$、$[111]_c // [0001]_h$ 和 $[112]_c // [10\bar{1}0]_h$（图 5-2）。

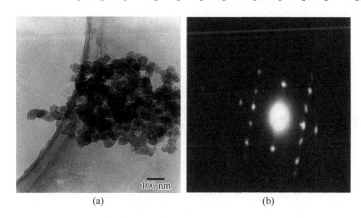

图 5-1 LAL 合成纳米金刚石的形貌和相应的选区电子衍射（SAED）图像

（a）纳米金刚石约为 30 nm 的球形颗粒，是一种具有六方和立方两种结构的晶体；（b）相应的 SAED 图像

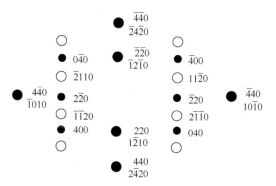

图 5-2 根据 SAED 图像标定的立方结构和六方结构的取向关系

图中基轴分别是 B = [112]和 B = [10$\bar{1}$0]

正如我们所知，拉曼光谱通常是表征各种晶型碳化学结构的重要方法[30, 31]。石墨和金刚石的拉曼光谱是众所周知的。例如，金刚石的一阶拉曼光谱是 $1332\ cm^{-1}$ 的单条线[32]，而石墨的拉曼光谱为 $1580\ cm^{-1}$[33]。此外，在碳材料的拉曼光谱分析中，所谓的"G"线被认为来自拉曼散射允许的 E_{2g} 模[34]，而 $1355\ cm^{-1}$ 是对应于多晶石墨的。然而，所谓的"D"线是来自无序晶格的 k 矢量守恒规则的破坏[35]。所以，通常在碳材料的拉曼光谱分析中，大家都使用"G"和"D"两条线来表征碳结构。LAL 合成纳米金刚石的拉曼光谱如图 5-3 所示，我们可以看到，三个宽带的中心分别为 $1352\ cm^{-1}$、$1100\ cm^{-1}$ 和 $623\ cm^{-1}$。由于在样品中纳米金刚石所占比例相对较低，所以金刚石的特征峰不能够明显观察到。另外，纳米金刚石的特征峰通常会变得更宽、更弱，并且向更低的频率偏移[36]。因此，需要对拉曼光谱进行分峰处理。

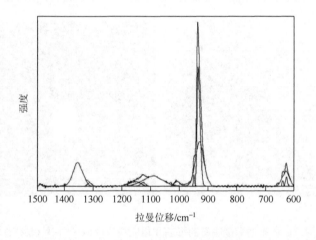

图 5-3　LAL 合成纳米金刚石的 $600\sim1500\ cm^{-1}$ 拉曼光谱及去卷积方法分析

三个宽带的中心分别为 $1352\ cm^{-1}$、$1100\ cm^{-1}$ 和 $623\ cm^{-1}$，图中 $930\ cm^{-1}$ 最强的特征峰的来源还不是很清楚

虽然，在图 5-3 中我们没有看到 $1332\ cm^{-1}$ 的特征峰，但是，从以 $1352\ cm^{-1}$ 为中心的宽带中分出来的 $1307\ cm^{-1}$ 特征峰通常被认为来自立方金刚石[37]。进一步地，我们分析制备样品在 $1000\sim1200\ cm^{-1}$ 的拉曼光谱（图 5-4），高斯线形的 4 个峰值与 $1152.5\ cm^{-1}$、$1125.5\ cm^{-1}$、$1090.9\ cm^{-1}$ 和 $1005.8\ cm^{-1}$ 完全匹配。Nemanich 等[38]认为微晶六方金刚石应该在 $1175\ cm^{-1}$ 产生最强的特征峰，而 Maruyama 等[39]报道六方金刚石的特征峰位于 $1150\ cm^{-1}$。因此，综合考虑，我们认为观察到的 $1152.5\ cm^{-1}$ 的特征峰应该来自微晶六方金刚石；中心位于 $1090.9\ cm^{-1}$ 的特征峰与已报道的纳米金刚石的特征峰是一致的[40]，并且其位置接近 Beeman 等[41]理论计算的金刚石团簇的振动态密度（VDOS）的主峰；对于

623 cm^{-1} 的特征峰，理论计算和实验结果都证明它是由金刚石引起的。Mao 和 Hemerly[42]在超高压下测量了金刚石的拉曼光谱并且发现了中心位于 590 cm^{-1} 的宽谱，Yoshikawa 等[36]也观察到了金刚石的中心位于大约 600 cm^{-1} 的宽谱。此外，根据文献报道[38-40]，我们推测 926 cm^{-1} 的特征峰应该是纳米金刚石的特征峰之一。需要指出的是，图 5-3 是首次给出的纳米金刚石的完整拉曼光谱。

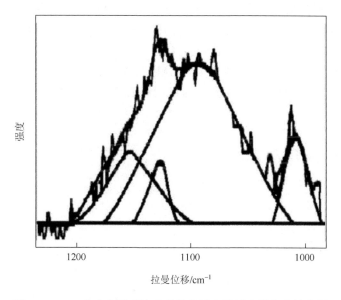

图 5-4　LAL 合成金刚石纳米晶的拉曼光谱及去卷积方法分析

如上所述，与气体或真空-固体界面的激光烧蚀相比，当脉冲激光辐照浸没在液体中的固体靶时，会在液-固体界面产生一个高温高压高密度等离子体羽。因此，纳米晶是在等离子体羽在液体束缚下快速淬灭的冷凝过程中产生的。通常，在 LAL 合成纳米晶过程中[38]，激光诱导的等离子体羽包含许多来自固体靶的具有 sp^2 键合的原子团簇及其离子。同时，激光诱导的额外压强会将激光诱导的等离子体羽驱动进入一个高温高压高密度状态。例如，在石墨-水系统 LAL 合成金刚石的情况下，压力和温度区域在 10～15 GPa 和 4000～5000 K[26]，这属于碳的热力学平衡相图中金刚石的稳定区域（图 5-5）。所以，在随后的等离子体羽淬灭过程中，可能发生金刚石成核和从石墨到金刚石的相变，因为 LAL 产生的热力学环境中，具有 sp^3 键合的金刚石相是稳定相，而具有 sp^2 键合的石墨相则是亚稳相。因此，在等离子体羽中金刚石的核的形成比石墨的更优先。另外，由于等离子体羽的高密度，有足够的离子供应，包括 OH$^-$、H$^+$等，以保持扩散机制的进行，并且这些离子可以通过抑制石墨 sp^2 键的形成来促进金刚石核的生长[43, 44]。此外，由于金刚石核的生长时间（等离子体羽淬灭时间）非常短，所以生长晶体的直径通常在纳米级。

在激光辐照下石墨直接转变成金刚石。众所周知，激光诱导的等离子体羽中包括来自固体靶本身的团簇、颗粒、液滴等[45]，例如，石墨-水系统 LAL 合成金刚石中存在石墨结构团簇、颗粒、液滴等。因此，由于高温高压高密度的热力学状态，金刚石可以在等离子体羽中从这些石墨碎片中成核[46]。例如，Pearce 等[28]通过在没有 H 和 O 元素的环境液体中激光烧蚀石墨靶合成了金刚石纳米颗粒。因此，他们得出结论，在 LAL 合成金刚石中，H^+ 和 OH^- 似乎不是形成金刚石相的必要离子，高温高压机制比化学反应更合理。

综上所述，我们可以清楚地看到，LAL 独特的物理和化学过程在合成纳米金刚石中得到了充分体现，这说明我们可以应用 LAL 方法去探索更多、更新颖的碳纳米材料的合成。

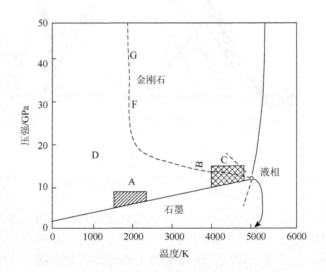

图 5-5　Bundy 建立的碳的热力学平衡相图

实线表示平衡相界。A 区表示高温高压催化石墨合成金刚石区；B 线表示石墨向金刚石快速（小于 1 ms）固-固转变的 P/T 阈值线；C 区表示 LAL 合成金刚石区；D 区表示六方石墨转变为六方金刚石区；F 线和 G 线表示无论是石墨或六方金刚石转化为立方金刚石的快速 P/T 阈值

5.1.2　Top-down 合成纳米金刚石

纳米颗粒尺寸与细胞生物学的相关性很大，例如，荧光 ND 可以作为量子点的无毒和稳定荧光替代品[54]。但是，合成尺寸小于 4 nm 而不团聚的 ND 和组装个位数（single-digit）ND 胶体溶液仍然具有挑战性[49, 55]。Chen 等[56]发展了一种简单的自上而下（top-down）的策略，应用 LAL 方法从商业高压高温金刚石微晶（工业原料）的悬浮液中合成平均尺寸为 3.6 nm 的单分散荧光 ND，并且通

过 ND 表面酯基和酮基的原位共价连接，实现可调谐的高性能荧光生物成像。简单来说，他们将 2 mg 金刚石微晶放入 10 ml 无水酒精的烧杯中，然后使用波长 532 nm、脉宽 10 ns、频率 19 Hz 和能量 180 mJ 的脉冲激光聚焦在烧杯的中间进行辐照（图 5-6），持续 30 min 后，组装出近乎完美的单分散荧光 ND 胶体溶液，如图 5-7 所示。

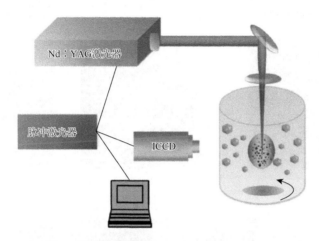

图 5-6　LAL 从金刚石微晶悬浮液中合成单分散荧光 ND 胶体溶液示意图

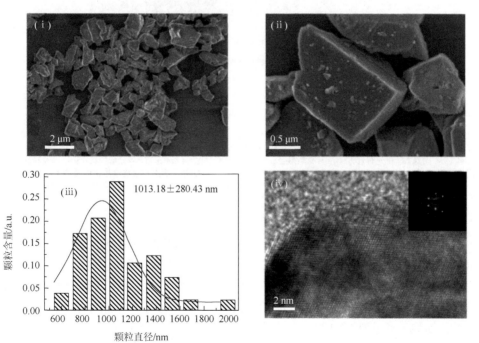

(a)

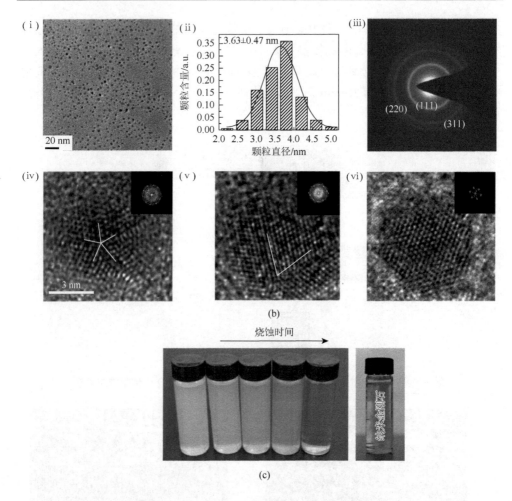

图 5-7 LAL 转化金刚石微晶为 ND（后附彩图）

(a) 初始金刚石微晶 SEM 图像（i）和（ii），分布直方图及其高斯拟合曲线（iii）显示颗粒尺寸为 1000 nm 左右，多分散度约 28%，以及相应高分辨透射电镜（HRTEM）（iv）和 SAED 图像（插图）；(b) 个位数 ND 的低放大率 TEM 图像（i）和分布直方图及其高斯拟合曲线（ii）表明尺寸为 4 nm 左右，多分散度约 13%，以及相应 SAED 图像（iii），（iv）～（vi）为 ND 各种结构的 HRTEM 图像，包括五重孪晶、三重孪晶和单晶结构，插图显示了相应的傅里叶变换衍射图，表明金刚石结晶良好；(c) 随着 LAL 过程的持续，合成胶体溶液颜色发生变化（左）和第五个瓶子的侧视图（右）

从图 5-7 我们可以清楚地看到，金刚石微晶是形状不规则的微米颗粒，而合成的 ND 则是球状尺寸为 4 nm 左右的纳米晶。需要关注的是，在 LAL 过程中，合成胶体溶液的颜色会从不透明的白色变化为透明的浅黄色（我们可以看到溶液小瓶后面的标签上写着"纳米金刚石"），表明形成了尺寸较小的 ND［图 5-7(c)］。所以，这些研究结果充分表明，通过 LAL，初始的金刚石微晶可以转化为平均尺寸为 3.6 nm 的单分散荧光 ND。

Chen 等使用高速摄影技术记录了脉冲激光烧蚀初始金刚石微晶悬浮液过程中等离子体羽的产生、演化和淬灭,如图 5-8(a)所示。我们可以看到,发光区域(最亮的部分)是等离子体羽的图像。显然,这些图像揭示了在单脉冲(脉宽 10 ns)激光烧蚀中等离子体羽的演化,如等离子体羽的产生、膨胀和消失,从等离子体羽产生到淬灭,时间约为 40 ns。如此短的淬灭时间使得在环境液体束缚中的激光诱导等离子体羽冷凝形成的产物为纳米尺度。

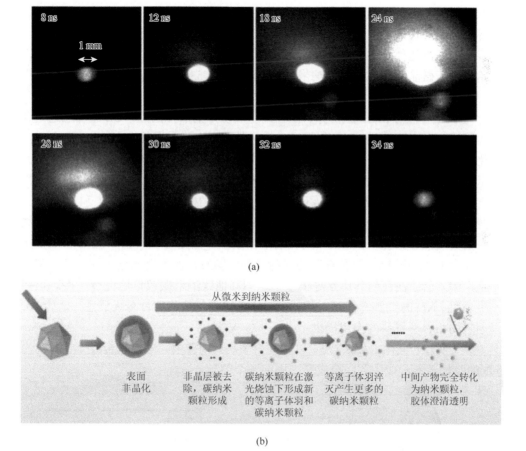

图 5-8 LAL 过程 ND 形成机制

(a) LAL 过程中等离子体羽演化图像,每个图中都提供了脉冲激光烧蚀时间;(b) 从金刚石微晶到个位数 ND 转变过程

尽管大块金刚石显示出最大的硬度,但是由于高能等离子体羽的产生,金刚石在 LAL 过程中会被刻蚀。金刚石的 LAL 会伴随着金刚石表面结构的重构,这

意味着在金刚石表面会发生石墨化或形成一层薄的非晶层。有趣的是，这种石墨化或非晶层是可以通过"激光清洁"去除的[57]。所以，基于实验观察和上述基本物理和化学过程，图 5-8（b）提出了 LAL "top-down" 合成制造个位数 ND 的机制。在第一阶段，首先，通过激光烧蚀金刚石微晶产生等离子体羽；然后，等离子体羽的淬灭导致金刚石微晶表面的 sp^3 碳转化为 sp^2 碳。在这里，金刚石在表面非晶化的过程中变成了无定形碳。在第二阶段，通过"激光清洁"去除非晶层，并且在去非晶化的过程中会产生小的碳纳米颗粒。在第三阶段，首先，通过持续的激光辐照，产生的小碳纳米颗粒会被激光烧蚀以产生新的等离子体羽；然后，新等离子体羽在高温高压高密度环境中淬灭会产生 ND，同时，发生非晶化。上述步骤的多次迭代发生，直到实现所有中间产物完全转化为 ND。在最后阶段，初始金刚石微晶的尺寸减小到纳米级，并形成了对 532 nm 激光辐照呈透明的稳定 ND 胶体溶液。

为了研究这些 LAL 合成的 ND 在细胞成像中的潜在用途，Chen 等通过向合成的胶体溶液中添加 PEG_{200N}，将 ND 重新分散到生物相容性溶剂中。用 PEG_{200N} 钝化的 ND 的荧光光谱与直接合成的分散在乙醇中的 ND 的相同，如图 5-9（c）所示。值得注意的是，PEG_{200N} 与水溶液相容，并且可以容易地与抗体或其他生物活性分子缀合[58]。生物光学成像研究采用了两种类型的细胞：A549 和 CNE-2。当荧光 ND 被细胞吸收后，在 405 nm、488 nm 和 552 nm 的激光辐照下，使用激光扫描共聚焦显微镜（CLSM）及亮场进行成像，如图 5-9（e）和图 5-9（f）所示，荧光 ND 能够标记两种细胞的细胞膜和细胞质，而不会到进入细胞核。重要的是，ND 已经被证明其比半导体量子点和其他碳纳米颗粒毒性更小[59,60]。这些结果表明，ND 的细胞毒性可以忽略不计，其体外生物相容性令人满意，这为 ND 作为生物相容性材料的应用提供了进一步的支持。

综上所述，应用 LAL 方法，以金刚石微晶为前驱物，宏观量合成个位数荧光 ND 的胶体溶液，成为 ND 合成和应用的一个突破。

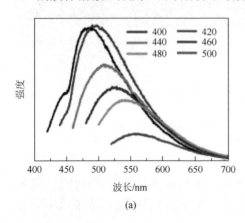

(a)

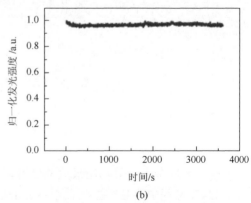

(b)

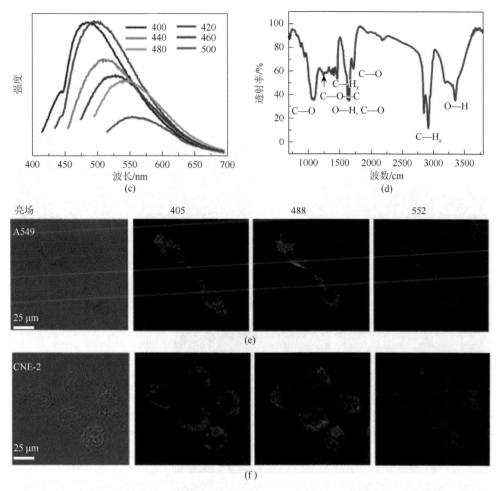

图 5-9　用于生物成像的个位数 ND 的各种光谱（后附彩图）

（a）不同波长下的发光光谱，从 400 nm 开始，以 20 nm 为增量；（b）450 W 氙灯照射的 ND 发光稳定性；（c）用 PEG$_{200N}$ 钝化后的 ND 的荧光光谱；（d）ND 的傅里叶变换红外光谱仪（FTIR）光谱，显示出各种表面基团的信号；（e）和（f）在 405 nm、488 nm 和 552 nm 的激光辐照下，用荧光 ND 标记的 A549 和 CNE-2 细胞的激光扫描共聚焦显微镜图像，以及相应的亮场图像

5.1.3　来自煤的纳米金刚石

煤是地球上最丰富的能源之一，一般通过燃烧来产生能量[61]。煤是一种分子固体，它具有非常复杂的结构[62-64]。煤的结构包含大量的无规则的、聚合的碳氢化合物单元，而一些非常小的结晶碳镶嵌在这些非晶网络里面[65, 66]。与其他结晶碳如石墨和金刚石相比，煤的应用只限于燃烧，而在其他领域还没发现其重要价值。煤价格便宜而且存量丰富，因此可以考虑将它作为原料合成金刚石。Xiao 等[66]以工业煤为原料，应用 LAL 方法合成出直径约为 3 nm 的立方纳米金刚石，而且合成

的产物可溶于溶剂中并有良好的分散性，同时有着强烈而稳定的荧光发射，在生物医学成像、光伏、光电子学等方面有着潜在的应用价值。

Xiao 等首先将 5 mg 的工业煤放入 10 mL 无水乙醇的烧杯中，然后应用脉冲激光（波长 532 nm，脉宽 10 ns，频率 10 Hz，能量 200 mJ）聚焦在烧杯的中部，激光烧蚀 2 h 后，合成出的具有良好分散性的纳米金刚石胶体溶液，如图 5-10

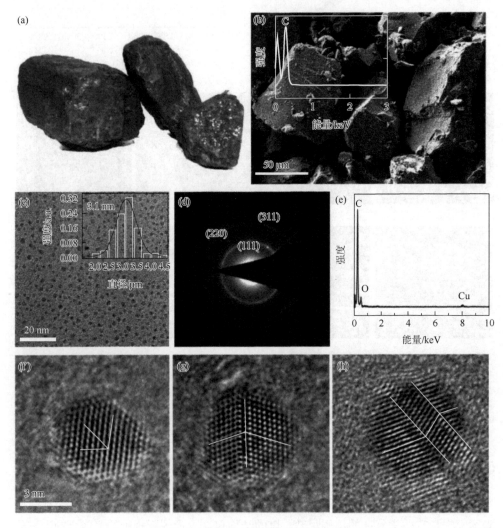

图 5-10 来自于无烟煤中的纳米金刚石

（a）无烟煤的宏观图像；（b）研磨后无烟煤的 SEM 图像，颗粒直径从几微米到数百微米，插图显示了相应的能量色散 X 射线谱（EDS），表明无烟煤含有纯碳，没有其他杂质；（c）LAL 合成的纳米金刚石的低放大率 TEM 图像及其尺寸分布（插图），可以看到纳米金刚石分散性好、尺寸均匀；（d）SAED 图显示了金刚石典型的三个强衍射环；（e）EDS 分析，其中 O 和 Cu 分别源于 C=O 官能团和铜栅格；（f）～（h）金刚石各种结构的 HRTEM 图像，包括单晶、孪晶结构等

所示。需要说明的是，Xiao 等通过 XRD、拉曼光谱和 X 射线光电子能谱法（XPS）等技术对合成产物进行了系统表征，证明它们是金刚石单晶，如图 5-11 所示。

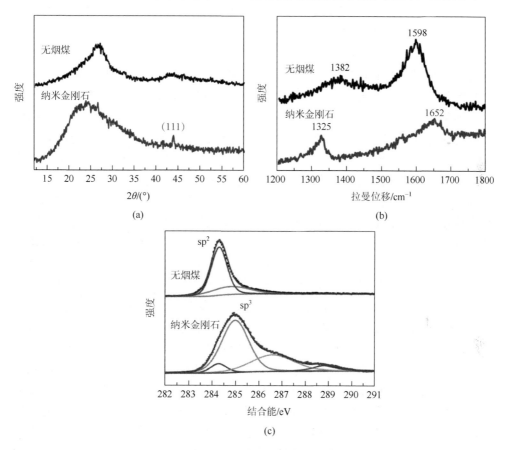

图 5-11　无烟煤和纳米金刚石的结构

（a）无烟煤的 XRD 谱（上方线）表明材料为非晶态，纳米金刚石的 XRD 谱（下方线）在 44°显示出衍射峰，与立方金刚石的（111）面一致；（b）无烟煤（上方线）和纳米金刚石（下方线）的拉曼光谱；（c）无烟煤（上方方块）和纳米金刚石（下方方块）的高分辨 C 1s XPS

我们可以从图 5-11 中清楚看到，XRD 显示原料煤在 26°附件有一个宽峰，应该是来源于非晶碳；合成产物在 44°有一个小峰，可以归因于金刚石的（111）面。原料煤在 1382 cm^{-1} 和 1598 cm^{-1} 有两个峰，分别对应 D 峰和 G 峰；而合成产物的峰出现在 1325 cm^{-1}[67]。原料煤中的 XPS 的 C 的结合能的峰位是 284.3 eV，对应于 sp^2 碳；而合成产物的峰形不对称，可以断定是由几种峰叠加而成。通过分峰处理后，我们可以发现合成产物的 XPS 的信号是由 4 种信号组合而成，分别为 284.3 eV 来源于 sp^2 碳原子、285 eV 来源于 sp^3 碳原子[68]、286.5 eV 来源于 C—O 键、288.7 eV 来源于 C=O 键[69]。另外，我们可以通过计算合成产物中 sp^3 的含量的来估计纳米金刚石的产率。煤原料的 sp^3 含量为 26.6%，意味着煤原料中混合着 sp^2 和 sp^3 的成分。在 LAL 作用后，sp^3 的成分增加为 54.4%。显然，sp^3 成分的增加主要源于纳米金刚石的生成。考虑到产物中的非晶碳也含有 sp^3 的成分，因此，LAL 合成纳米金刚石的产率约为 5%～10%。

LAL 过程中煤转变为纳米金刚石的机制包括成核与相变。如果注意观察，我们可以发现，在 LAL 作用过程中，合成胶体溶液的颜色从不透明的灰色变到深红棕色，再变到透明黄色。这表明在不同作用阶段有不同的产物生成。所以，我们特别关注从煤变成纳米金刚石过程中的中间产物。TEM 图像显示出中间产物大部分都是交叉重叠的网络结构，同时 SAED 图像确认这些结构是非晶碳［图 5-12（a）～图 5-12（f）］。接着，这些非晶碳网络破裂成一些小链段，最后这些链段中间的连接消失，破碎成非晶碳小球。因此，非晶碳小球是煤到纳米金刚石相变的中间产物。另外，从结构上，煤属于分子固体，包含了大量的无规则的、聚合的芳香烃的单元。这些芳香烃的单元在激光的作用下会分裂成非晶碳颗粒。在芳香碳分解的时候，会产生大量的 C_2 单元，我们用高速光谱仪捕捉到了这些 C_2 分子的信号，如图 5-12（g）和图 5-12（h）所示。事实上，C_2 分子确实在合成纳米金刚石上有很重要的作用。Zhou 等发表了纳米金刚石生长速率与等离子体羽中的 C_2 浓度关系，确定 C_2 是纳米金刚石的生长结构单元。

这样，LAL 中煤转变为纳米金刚石的过程如下［图 5-12（i）］：在激光的作用下，煤中的有机碳，如芳香烃的有机分子会分解成 C_2 分子，这些 C_2 分子形成在激光诱导的高温高压高密度等离子体羽的区域。基于金刚石的热力学平衡相图，金刚石成核的压强和温度的区域分别为 10～15 GPa 和 4000～5000 K，与 LAL 提供的热力学环境几乎完全吻合。因此，这个 C_2 分子就成了形成纳米金刚石的结构单元，是非晶碳小球的主要结构。当等离子体羽淬灭的时候，中间产物非晶碳小球会转变成金刚石晶核。同时，这些金刚石晶核的表面活性很大，环境液体如酒精中的活性基团会吸附在它的表面，这些表面基团不仅使得纳米金刚石结构变得稳定，而且能赋予纳米金刚石一些特异的性能如荧光性质等。

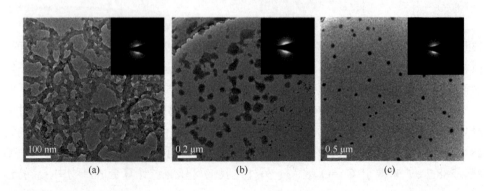

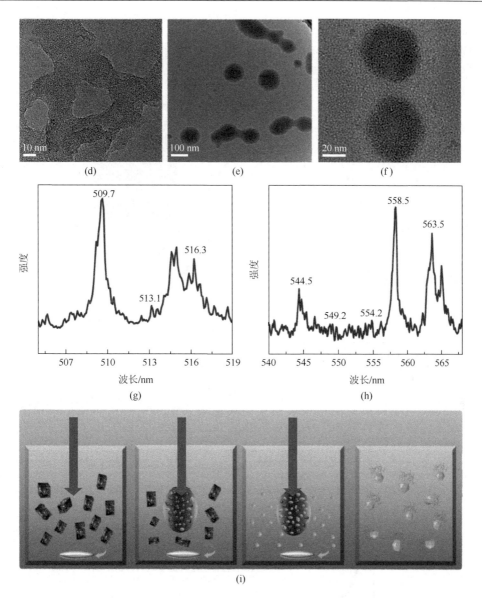

图 5-12 （a）~（f）LAL 过程中煤转变为纳米金刚石的中间产物；（g）和（h）C$_2$ 分子的发射光谱；（i）煤转变成纳米金刚石的机理示意图

LAL 合成的纳米金刚石的荧光光谱如图 5-13（a）所示。当我们用 420 nm 的激光激发时，纳米金刚石荧光峰位于 520 nm。分别以无烟煤、烟煤和焦炭等原料煤为初始材料合成的纳米金刚石的荧光寿命分别是 1.84 ns、0.96 ns 和 1.11 ns。值得注意的是，使用氙灯辐照 3 h，纳米金刚石荧光强度没有发生明显的衰减[图 5-13（b）]。有趣的是，水作为环境液体合成的纳米金刚石发出的是蓝光，而酒精作为环境液体时样品发出的是绿光。此外，在水中和酒精中制备纳米金刚石的荧光量子产率分别为 0.06 和 0.035。不同环境液体的 LAL 合成的纳米金刚石的荧光峰不一致主要是由表面吸附的活性基团不一样而导致的。

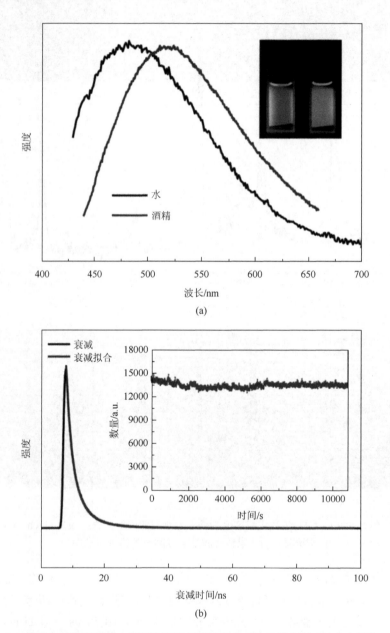

图 5-13 （a）波长与强度的关系，插图为不同环境液体合成的纳米金刚石的荧光颜色；（b）衰减时间与强度的关系，插图为荧光稳定性（后附彩图）

5.1.4 纳米金刚石的荧光起源

荧光纳米金刚石胶体的发现始于 ms-LAL 在有机溶剂中激光烧蚀商业碳粉末[70]，并发现用短脉冲的激光也能合成荧光纳米金刚石胶体，如 fs-LAL[71]。荧光纳米金刚石的发现为荧光碳纳米材料家族添加了一名新的成员。尽管已经有大量的关于荧光石墨烯量子点和碳点荧光的研究工作[58, 72–80]，但是关于荧光纳米金刚石的研究则有限，对于荧光纳米金刚石的起源知之甚少。众所周知，金刚石的带隙非常大，约 5.5 eV，因此不可能发出可见光。那么，是什么原因导致它有荧光发射呢？有研究指出可能是纳米金刚石的结构影响了它的发光，它的发光带隙取决于 sp^2 键和 sp^3 键的比例[60]。但与此同时，也存在不同的看法，有些研究人员认为是缺陷能级的辐射复合引起发光。然而，纳米金刚石胶体的荧光机制仍然存在比较大的争议，例如，为什么荧光发射峰位取决于激光波长，是否纳米金刚石的不同粒径影响发光，这些问题仍然没有得到解决。

有关纳米颗粒毒性的试验证实纳米金刚石比非金刚石相的碳纳米颗粒毒性要小。例如，当纳米尺度的碳黑与细胞共同培养时，与相同浓度的纳米金刚石相比，细胞死亡率高 10%，而且，纳米金刚石产生较少的活性氧物质[60]。因此，基于这些毒性分析数据，纳米金刚石比碳量子点似乎更适合用于生物体内成像。

所以，应用 LAL 方法合成"清洁"的荧光纳米金刚石，系统研究它们的荧光机制不仅仅是为了理解它们的荧光起源，更重要的是为了它们的实际应用。我们关注的焦点在于荧光起源及如何调控它们的荧光发射。

在 LAL 合成纳米金刚石中，我们使用了两种原料即单晶微米金刚石（1 μm）和爆轰法纳米金刚石（5 nm），酒精和水作为环境液体，合成的荧光纳米金刚石样品分别用 MD 和 DND 命名。图 5-14（a）和图 5-14（b）分别显示 MD 和 DND 的特征形貌，两种样品都显示出很好的分散性和均匀的粒径分布，DND 的粒径较大，为 5.4 nm；而 MD 的粒径比较小，只有 3.5 nm。此结果说明不同原料对 LAL 合成纳米金刚石的粒径有影响。MD 和 DND 的 HRTEM 图像［图 5-14（d）和图 5-14（e）］的晶面间距为 0.206 nm，与金刚石的（111）晶面一致。SAED 图像的三个强衍射环分别对应金刚石（111）、（220）和（311）晶面，相关的拉曼表征也证实合成的是纳米金刚石。

接着我们用紫外可见吸收光谱（UV-visible absorption spectrum）来表征 LAL 合成纳米金刚石胶体溶液的光吸收，如图 5-14（g）所示。我们可以观察到，在可见光区域没有明显的吸收峰，大量的吸收峰出现在 200～300 nm 的区域。在 218 nm（5.8 eV）的吸收峰可以归结于纳米金刚石的本征吸收，它比块状金刚石的带隙（5.5 eV）要大，这是因为小尺寸导致了蓝移；在 226 nm 和

239 nm 的吸收峰应该属于相对小的方向烃结构和多芳香烃生色团；将 252 nm、275 nm、297 nm 的信号归结于 C═C 双键的 π-π^*、C═O 的 π-π^* 和 n-π^* 的跃迁；261 nm 的吸收峰的起源不是很清楚，很有可能源于多芳香烃生色团的衍生物。我们注意到，两个样品吸收光谱的区别很小。因此，在酒精中合成的纳米金刚石可以代表多芳香烃生色团跃迁的集合。

同时，我们还注意到，两个样品的发射光谱强烈依赖激发光谱，这个现象与石墨烯量子点的荧光现象类似[81]。我们知道，斯托克斯位移（Stokes shift）是发射光谱和激发光谱的峰值波长（最强峰）的能差，它被广泛地用于纳米颗粒的量子限域效应[82]。如果荧光机制是由纳米颗粒粒径分布所决定的，那么当激发波长增加时，斯托克斯位移会逐渐减小，最后趋近零，说明存在块体材料的带隙。然而，这里的纳米金刚石不满足这个条件，如图 5-14（i）所示，当激发波长变大时，斯托克斯位移先减小，逐渐趋于稳定。这些结果说明纳米金刚石的荧光不是由尺寸导致的量子限域效应产生的，而是由其他因素控制的。

在排除了尺寸因素导致的纳米金刚石荧光机制后，我们继续找寻其他可能的因素。我们通过研究合成纳米金刚石胶体溶液荧光发射的温度依赖性来考察表面活性基团的影响。图 5-15（a）是 LAL 合成纳米金刚石的荧光光谱，发射波长主要集中于蓝光区域，当激发波长为 380 nm 时，发光最强峰 480 nm；当

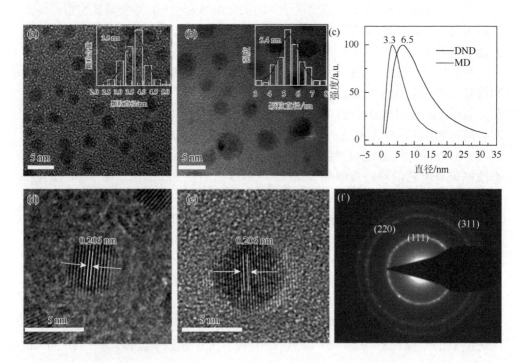

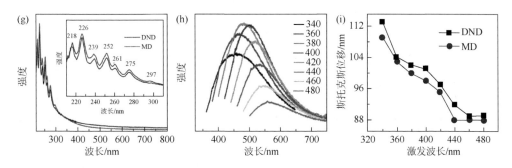

图 5-14 LAL 纳米金刚石的尺寸无关发光（后附彩图）

（a）MD 尺寸 3.5 nm 和（b）DND 尺寸 5.4 nm 的良好分散纳米金刚石的典型形态；（c）测量的纳米金刚石的有效直径；（d）MD 和（e）DND 的 HRTEM 图像；（f）相应 SAED 图像；（g）纳米金刚石胶体紫外可见吸收光谱，插图是 210~310 nm 的放大图；（h）不同激发波长的发光光谱，两种类型的纳米金刚石表现出相似发光行为；（i）DND 和 MD 的斯托克斯位移与激发波长的关系

增加激发波长时，荧光红移到绿光区域（420 nm 激发波长，518 nm 发射波长）和黄光区域（480 nm 激发波长，568 nm 发射波长）。合成的纳米金刚石胶体在 65℃的环境下加热不同的时间（1~5 h，间隔 1 h），相应的荧光光谱如图 5-15（b）~图 5-15（f）所示。显然，在加热不同时间后，光谱的形状并没有发生明显变化。但是，光谱最强峰的位置发生了变化。随着加热时间的延长，最佳激发波长从 380 nm 红移到 480 nm，相应的发射波长也从 480 nm 红移到 580 nm，如图 5-15（g）所示。Tan 等[83]也发表了金刚石结构的碳点在紫外光的辐照下发生了红移，他们认为这是光化学氧化的结果。当我们固定发射波长，观察在不同加热条件下的发射峰的变化，发现与最优化的发射峰相比，它们红移量变小了，分别为 40 nm、32 nm、30 nm、27 nm、23 nm 和 19 nm。可以看到，当激发波长增加时，红移量逐渐减少，而且远远小于最大峰的红移量（100 nm）。我们假设发射光谱整体红移，那么红移量应该是一样。但是，我们的结果已经证实了用特定发射波长激发时的红移与最强峰的红移不同步。因此，我们相信红移量是由纳米金刚石表面不同官能团的量来决定的。

为了判断氧化性基团在纳米金刚石发光中所起的作用，我们使用了微区傅里叶变换红外光谱仪（MFTIR）开展研究。我们把一滴合成胶体液体滴在衬底上，然后

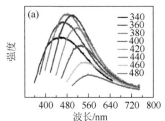
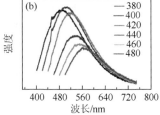
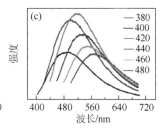

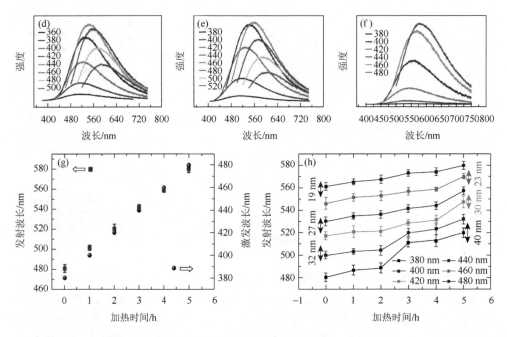

图 5-15 荧光纳米金刚石温度引起的发光红移（后附彩图）

（a）LAL 合成纳米金刚石荧光光谱；（b）～（f）合成的纳米金刚石胶体在 65℃下加热 1～5 h 后的荧光光谱；（g）由最佳激发波长激发的最强峰的变化；（h）在加热不同时间后，在特定发射波长激发下每个发射峰的红移

在红外灯的烘烤下让酒精完全挥发。整个实验装置示意图如图 5-16（a）所示，图 5-16（b）是样品在干燥后的照片。注意，纯酒精的 MFTIR 信号在 1000～1200 cm^{-1} 的吸收峰是 C—O 键相关的非对称和对称振动，在 3350 cm^{-1} 的吸收峰是 O—H 键的伸缩振动。LAL 合成样品在 1600～1800 cm^{-1} 出现了一个新的吸收峰，属于羰基（C=O）的信号。通过仔细地分析后，我们发现 C=O 基团包括两个类别：酮基 C=O 和酯基 C=O。在酯基 C=O 键的红外信号一般都比酮基 C=O 键的要高；在 1665～1685 cm^{-1} 的振动信号属于 α, β-未饱和酮基 C=O 键的信号，而 1730～1750 cm^{-1} 属于酯基 C=O 键的信号[84-86]。为了更清楚地显示基团的特征峰，我们把 1500～2000 cm^{-1} 的红外峰放大，如图 5-16（d）所示。在加热 1 h 后，羰基信号的强度增加，意味着氧化程度变强 [图 5-16（c）的曲线（iii）]，注意，酮基 C=O 键吸收峰的信号仍然大于酯基 C=O 键吸收峰的信号；随着加热时间延长，这两个基团信号强度变得一致 [图 5-16（d）的曲线（vi）]；在加热 4 h 后，酯基 C=O 键的信号强度超过了酮基的；最后，酯基的 C=O 键的信号占主导地位。因此，微区傅里叶变换红外光谱表明随着加热时间的延长，样品的氧化程度变高，羰基也随之增加。因此，样品中酮基和酯基的 C=O 键很有可能是红移的主要原因。

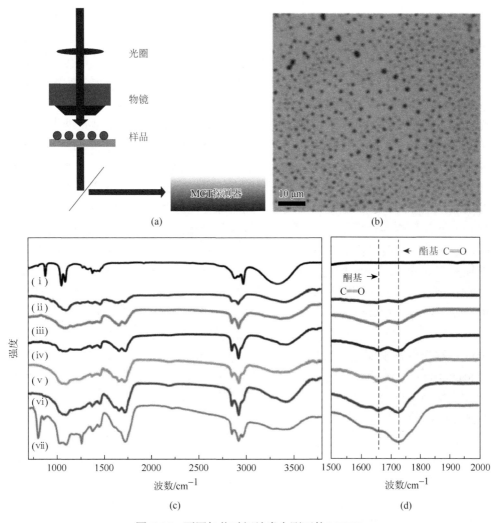

图 5-16 不同加热时间纳米金刚石的 MFTIR

(a) 实验装置示意图和 (b) 样品染色干燥后的光学显微照片; (c) 不同加热时间纳米金刚石的 FTIR 光谱: (ⅰ) 纯乙醇、(ⅱ) LAL 合成纳米金刚石、热处理 (ⅲ) 1 h、(ⅳ) 2 h、(ⅴ) 3 h、(ⅵ) 4 h 和 (ⅶ) 5 h 的样品, 和 (d) 1500~2000 cm^{-1} 的放大图

根据上面的讨论,我们从三方面解释纳米金刚石的荧光起源。第一,为什么纳米金刚石会有激发依赖的荧光发射? 一般来说, 对于半导体量子点和有机染料, 荧光峰的位置都是与激发波长无关的, 因为所有的激发电子都会弛豫到导带低端再跃迁[87]。然而, 纳米金刚石的荧光好像不符合这个规则, 明显与激发波长的位置有关。我们认为这是因为纳米金刚石胶体不能被简单地视为一个单组分的纯净物。从吸收光谱可以看出, 纳米金刚石胶体溶液是由许多芳香烃和多种生色团构成的。这些生色团和荧光基团有自己的最高占据分子轨道 (HOMO) 和最低未占

据分子轨道（LUMO），因此导致不同的带隙。根据氧化石墨烯荧光机制研究的报道，在酸处理后，氧化石墨烯局域的分子结构能被—OH 或—COOH 占据，—OH 主导的氧化石墨烯的发光中心在 500 nm 附近，而—COOH 主导的氧化石墨烯发光中心在 630 nm 处[88]。这些结果说明官能团对发射波长有很重要的影响。基于我们实验观察，C=O 基团的强度随着加热时间的增加而增加。同时，最强的荧光峰也表现出从蓝光区域红移到黄光区域。因此，我们可以简单地把纳米金刚石荧光的颜色归类成受三种基团影响。第一，—OH 控制蓝光的区域、酮基 C=O 控制绿光区域、脂基 C=O 控制黄光区域。在 LAL 合成纳米金刚石中，酮基 C=O 和脂基 C=O 的信号都比—OH 要弱，因此，发射蓝光。随着合成胶体加热时间的增加，酮基 C=O 的吸收峰增强，伴随着绿光的增强和蓝光的减弱。最后，脂基 C=O 信号变为主导，因此，样品发射黄色荧光。当用紫外光（380 nm）激发时，蓝光发射是最优化的激发光谱，样品发出蓝光。同时，该激发波长也能激发出绿光和黄光，尽管它们的荧光强度比蓝光要弱。因此，合成样品会显示出从 380 nm 到 700 nm 的宽谱，它的半高宽超过 150 nm，远大于一般半导体量子点[89]，也意味着同时激发多个带。

第二，为什么当用特定波长激发时，随着加热时间的延长，荧光发生红移？最近，发光红移现象同时在修饰单酰基甘油（monoacylglycerol）团簇的十聚体（decametric）纳米颗粒研究中被观察到[90]，并且 Lee 等[92]发现了一个向下凹的 LUMO 轨道并认为这是由 C=O 基团导致的。另外，酒精中两个相邻的氧原子连接到 C=O 的 π^* 轨道，通过非成键对（n）引起 H—O⋯C=O 相互作用，也就是 n（OH）→π^*（CO）相互作用。Lee 等的计算显示 H—O⋯C=O 在稳定 n（OH）→π^*（CO）跃迁上起到非常重要的作用，导致 LUMO 下沉及荧光红移。所以，这些结果也表明新的电子填充到了 LUMO 中。我们发现在不同激发波长激发下红移的间隔是不一样的，激发波长越大，红移就越小。Lee 等提到，随着 C=O 基团的增加，LUMO 的电子变多，因此带隙变小，发光红移。因此，该现象也表明 LUMO 的下降不是一致的，基团不同，下降的程度也不同。从我们的实验推断，下降的程度为—OH＞酮基 C=O＞脂基 C=O。从其他角度看，因为 C=O 基团控制着绿光区域和黄光区域，所以它会强烈地影响着蓝光区域，而对绿光区域和黄光区域的影响较小。

第三，为什么最佳激发波长和发射波长随着加热时间的延长，会有 100 nm 之大的红移量？事实上，这个现象很容易被误以为是发射光谱的整体位移。然而，如果发射光谱确实是整体位移的话，那么特定激发波长所对应的发射峰应该也有相似的位移。但是根据我们的实验观察，它们的位移量远小于 100 nm。这个差别可以归因于发射物质的相对强度的改变。我们知道，纳米金刚石上每一个官能团都有自己最佳的激发波长和发射波长。刚开始时，—OH 基团处于主导地位，如果用紫外光激发，样品会发出强烈的蓝光。然后，随着加热时间的延长，酮基的 C═O 增加，当用蓝光激发时，样品会发出绿光。同理，脂基的 C═O 占据主导地位，当用绿光激发时，样品发出黄光。因此，荧光的改变从蓝光到黄光并不只是简单的发射峰的整体红移，而是官能团相对含量的增加和减少引起的。换句话说，纳米金刚石荧光的本质是官能团的竞争和协作的关系。协作体现在激发依赖的荧光，当激发波长改变时，荧光可以从蓝光变到黄光；竞争体现在最佳激发波长和发射波长，也就是发射峰和发光峰最强的位置。不同基团如—OH、酮基 C═O、脂基 C═O 对荧光的影响，如图 5-17 所示。我们可以看到，荧光与金刚石晶核关系不大；相反，与表面的官能团关系非常大。尽管有研究认为荧光与 sp^3 键有关系，但是基于我们的研究，我们认为表面态起到了绝对主导的作用。

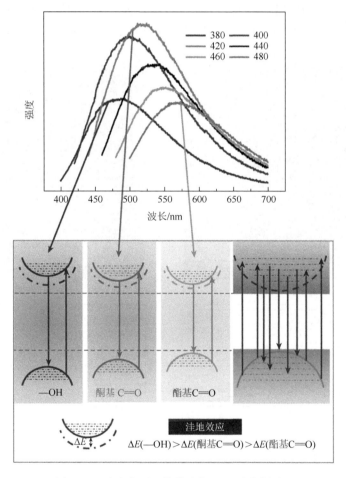

图 5-17 纳米金刚石的荧光起源（后附彩图）

—OH、酮基 C═O 和酯基 C═O 分别由蓝色、绿色和黄色发光表示，依赖于激发的荧光的本质在于这三组的相对强度，并且洼地效应是不同的：蓝色的 LUMO 比绿色和黄色的 LUMO 变化更大因此，我们把纳米金刚石的荧光起源归结于其表面不同官能团的竞争与协作。因为每一个基团都有自己最佳的激发波长和发射波长。当激发波长改变时，相对应的发射波长也会随之改变，从而显示出激发波长依赖性。

那么，如果表面基团减少到只有一种，荧光显示会与激发波长无关吗？我们应用 LAL 合成了表面主要为—OH 基团的纳米金刚石胶体，如图 5-18 所示。我们可以很清楚地看到紫外可见吸收光谱只有 247 nm 和 352 nm 两个峰，它们起源于多荧光团的 n-π*跃迁[91]，—OH 基团在 FTIR 光谱占据主导地位 [图 5-18（d）]。与在酒精中的吸收峰相比，样品吸收峰的个数种类已经大量减少。有趣的是，当激发波长变化时，发射波长基本没有移动。例如，激发波长从 360 nm 变化到 420 nm，发射波长仅仅红移了 6 nm，明显地小于在酒精中制备样品时移动的 53 nm。所以在水中制备的纳米金刚石可以被认为是不依赖激发波长的荧光，因为表面主要为一种基团，所以电子从导带底部跃迁到价带顶部，只发一种光。

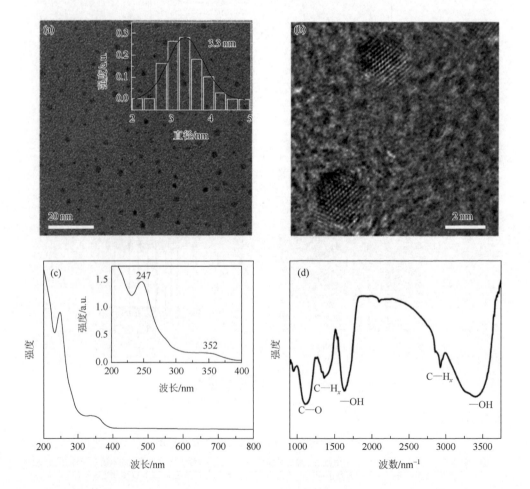

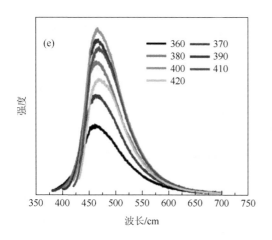 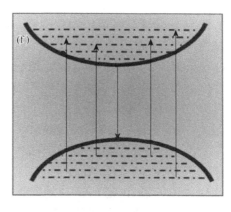

图 5-18 具有激发无关荧光的纳米金刚石（后附彩图）

（a）以水为环境液体 LAL 合成的纳米金刚石，尺寸约为 3.3 nm，具有良好分散性；（b）相应的 HRTEM 图像；（c）纳米金刚石的紫外可见吸收光谱，显示出比在酒精中合成的纳米金刚石少得多的吸收峰；（d）纳米金刚石的 MFTIR 光谱，表明—OH 基团在样品中占主导地位；（e）具有激发无关荧光的纳米金刚石发光光谱和（f）荧光机理示意图

综上所述，我们可以看到，纳米金刚石几乎所有的荧光起源都可以归结到表面官能团的增加、减少。需要注意的是，单纯的 C=O 和—OH 基团都不会发射荧光。有机溶剂，如乙醇和乙酸，包含—OH 和 C=O，都不是荧光材料。然而，当这些基团依附在纳米金刚石的表面时，它们可以与金刚石表面形成特殊的表面构型（表面态），大量的实验研究表明，这些特殊的表面态是荧光中心形成的关键。因此，我们的研究不仅对纳米金刚石荧光起源有了更深入的理解，而且提供了一种行之有效的方法来调控纳米金刚石的荧光发射。

5.2 LAL 中金刚石的相变

由于独特的物理和化学性质，金刚石在科学和技术上都有着巨大的应用价值，因此，人们对金刚石的合成始终怀有极大的兴趣。同时，研究碳的致密相，例如，石墨和金刚石，以及它们相互转化的机制，是凝聚态物理和材料科学中具有重要基础意义的经典课题。众所周知，六方石墨（hexagonal graphite，hex-g）是环境条件下最稳定的碳相，但是，大的活化势垒阻止了亚稳相的立方金刚石（cubic diamond，cub-d）自发转化为石墨。考虑到碳的热力学平衡相图的高温高压区是由金刚石相主导的[92, 93]，所以，人们使用高温高压来实现石墨向金刚石的转化，这种转变发生在碳相图中 5～12 GPa 的压强和 2000～3000 K 的温度区域，并且过渡金属催化剂有助于这种转化。冲击波法可以在没有催化剂的情况下诱导石墨转化为金刚石,这种转变发生在 15 GPa 的压强和 1000 K 的温度下[94],并且六方金刚石（hexagonal diamond，hex-d）即朗斯代尔石（Lonsdaleite），作为 cub-d 的亚稳相,也是通过 SW 合成的[95]。Bundy 和 Kasper[96]曾发表在高温下沿 c 轴通过静态高压诱导 hex-g 转化为 hex-d 的文章。然而,后来许多研究人员试图在静压下再现石墨转化为 hex-d 的尝试都失败了[97, 98]。因此,截至目前还没有完全证实石墨可以在静压下转化为 hex-d。研究人员报道了由 C_{60} 分子形成的富勒烯晶体可以在 20 GPa 的压力和环境温度下转化为金刚石[99-101]。

本节,我们讨论 LAL 中观察到的石墨与金刚石的相变。我们可以看到,LAL 不仅为研究石墨与金刚石相变提供了一个独特平台,而且为发展新的合成金刚石技术提供了有力工具。

5.2.1　金刚石-石墨相变的中间相

理论学家曾经提出,从菱形石墨（rhombohedral graphite，rh-g）到 cub-d 的高度对称的转化路径应该是最易发生的,即 rh-g 作为 hex-g 到 cub-d 转化的中间相[102, 103],该转化路径会使金刚石[111]取向平行于初始六方石墨的 c 轴。这些理论预测与 HTHP 中的催化剂辅助转化是一致的[96],可以形成 hex-g 和 cub-d 的相互取向关系[104]。而正交石墨（orthorhombic graphite，or-g）也被认为是 hex-g 向 cub-d 和 hex-d 的转化路径[105, 106],该路径会使金刚石[112]取向平行于初始 hex-g 的 c 轴,类似于 SW 实验结果[107]。然而,在石墨转化为金刚石的过程中,上述理论假设的中间相,即 rh-g 和 or-g 未直接被实验证实[102, 108, 109]。非常幸运的是,我们在 LAL 合成金刚石的过程中观察到了 rh-g,即石墨向金刚石转化的中间相[46]。

我们应用 LAL 方法在水中激光烧蚀固体石墨靶合成的金刚石纳米晶具有立方（cubic）和六方（hexagonal）两种结构即混合相结构,并且两种结构的取向关系为：$[110]_c // [1210]_h$、$[111]_c // [0001]_h$ 和 $[112]_c // [1010]_h$（图 5-2）。为了探测石墨向金刚石转变过程中可能的亚稳中间相,我们系统研究了这种相变的激光能量依赖性。我们分别使用能量值为 280 mJ、300 mJ 和 350 mJ 的脉冲激光来诱导石墨向金刚石的转化,并通过 XRD 谱对不同脉冲激光能量获得的样品进行了分析,图 5-19 显示了相应的 XRD 谱。我们可以看到,石墨靶的 XRD 峰都属于 hex-g,并且没有发现 rh-g 的衍射峰。然而,随着所施加脉冲激光能量的增加,我们在 330 mJ 脉冲激光能量制备样品的 XRD 谱中清楚地观察到 rh-g 的显著衍射峰。rh-g 最强的（003）衍射峰,$2\theta = 26.6°$；rh-g 的（101）衍射峰,$2\theta = 43.44°$,在 hex-g（101）和（102）峰之间。

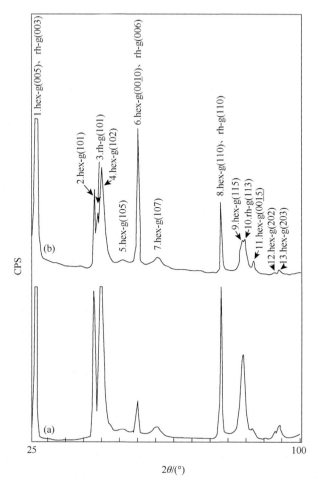

图 5-19 LAL 制备产物的 XRD 谱

(a) 固体石墨靶；(b) 330 mJ 脉冲激光制备产物

这些实验结果提供了在 LAL 作用下石墨转化为金刚石的过程中确实形成了亚稳态 rh-g 的证据。为了进一步研究中间相 rh-g，我们选择 rh-g 的（101）衍射峰作为标记线，观察该标记线的强度对所施加脉冲激光能量的依赖性，结果如图 5-20 所示。我们可以清楚地看到，rh-g 的（101）衍射峰的强度随着所施加脉冲激光能量的增加而迅速增加。这一结果意味着，随着所用脉冲激光能量的增加，hex-g 向 rh-g 转化的概率增加。因此，上述实验结果表明了两个重要的证据。首先，排除金属催化剂效应，石墨转化为金刚石在 LAL 过程中得到了证实；其次，在 LAL 过程中获得了作为 hex-g 到 cub-d 转化的亚稳态的中间相 rh-g。

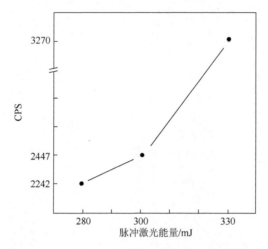

图 5-20　rh-g 的(101)衍射峰的脉冲激光能量依赖性

根据以上实验观察，在 LAL 过程中，石墨与金刚石的相变有两个途径，从而导致了两种结构金刚石的形成。一个途径是 hex-g 直接转化为 hex-d；另一个途径是 hex-g 首先转化为中间相 rh-g，然后转化为 cub-d[26]，图 5-21 给出了这个途径的结构演变示意图。换句话说，hex-g 的基面可以发生"船形"弯曲并直接形成 hex-d，而 hex-d 的 c 轴平行于 hex-g 初始的 c 轴。或者，石墨平面首先朝着 rh-g 堆叠滑动，然后堆叠平面可以变成"椅子"弯曲并产生 cub-d，而 cub-d 的[111]取向平行于初始 hex-g 的 c 轴（图 5-21），因此，hex-d 的新 c 轴平行于 cub-d 的[111]取向。所以，我们可以看到，亚稳态 rh-g 在石墨转化为金刚石过程中是一个中间相，也就是理论预测的石墨转化为金刚石的高度对称路径[102]。基于上述结构相变模型，两个金刚石相与 hex-g 的晶面取向关系的实验结果（图 5-2）与假设 rh-g 作为石墨—金刚石转化的中间相的理论预测非常一致[102,103]。此外，从上述理论模型和实验观察表明，在将 hex-g 转化为 cub-d 的过程中，hex-g 转变为 rh-g 的概率将随着 LAL 施加脉冲激光能量的增加而增加。因此，如果中间相 rh-g 保留在最终的产物中，那么 rh-g 在制备样品中的比例也应该随着施加脉冲激光能量的增加而增加。显然，这个推论也与实验结果非常一致（图 5-20）。

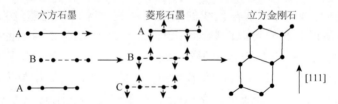

图 5-21　石墨向金刚石转化的结构路径

六方石墨→菱形石墨→立方金刚石（沿着侧面观察六边形平面的不同堆叠形式）

我们给出一个 LAL 过程中石墨与金刚石相变的热力学分析。如前所述，当脉冲激光烧蚀水中石墨靶时，在环境液体中会产生一个激光诱导的等离子体羽，液体的束缚作用会将等离子体羽推向一个高温高压高密度的极端热力学状态。在这个实验中，等离子体羽的温度在 3500～4000 K，压强在 10～15 GPa[110]。众所周知，Bundy 等[111]在 1994 年绘制了一个新的碳的压强-温度相图，如图 5-22 所示，实线代表平衡相边界，从低温低压到金刚石、石墨和液相的三相点（12 GPa，5000 K）的实线是相图中的 Berman-Simon 线（B-S 线）。从图 5-22 中可以看出，金刚石在 B-S 线上方的区域处于亚稳态，A 区（5～12 GPa，2000～3000 K）是用于石墨高温高压商业制备金刚石的压强-温度区域。值得注意的是，该区域似乎存在石墨通过中间相 rh-g 转化为金刚石的转化途径[96]。B 区（15 GPa 左右，1000 K 左右）表示石墨通过中间相 or-g 转化为金刚石，这是在 SW 法中发现的转化路径。此外，我们主要关注 P-T 相图中的 C 区（10～15 GPa，3500～4500 K），也就是 LAL 合成金刚石的相区，其中石墨通过中间相 rh-g 转化金刚石的途径可以在 LAL 过程中实现。我们的研究表明[110]，在上述区域中，碳原子越过势垒转化为金刚石的转化概率是不同的，C 区的转化概率最高，而 B 区的转化概率最低。因此，这些分析表明，LAL 过程中石墨向金刚石转变的两种途径，即 or-g 和 rh-g，分别可能发生在碳相图中的 B 和 C 区域。

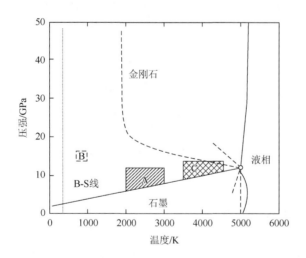

图 5-22 碳的压强-温度相图

A：HTHP；B：SW；C：LAL

在 HTHP 金属催化剂辅助石墨转化金刚石的过程中，人们发现最终 cub-d 的 [111]取向与初始 hex-g 的 c 轴平行[112]，因此，rh-g 作为中间相的转化路径似乎可以在 HTHP 中实现。然而，我们知道，金属催化剂的催化作用会显著降低相变的压强和温度，因此也会显著改变相变机制（例如，可能通过金属诱导的石墨基面屈曲）[104]。然而，在我们的结果中，石墨到金刚石的转化更纯粹，即不存在任何金属催化剂。因此，本节中的实验结果是上述结构相变模型的实验证据。考虑到 LAL 石墨向金刚石的转化发生在碳相图的 C 区，我们推测，如果不使用过渡金属催化剂，那么在 A 区可能无法实现以 rh-g 为中间相的从 hex-g 到 cub-d 的转变。

在 SW 过程中发生的石墨向金刚石的转化最终导致 cub-d 的[112]取向与初始 hex-g 的 c 轴平行，并且该相变发生在 B 区。因此，这一过程被认为是石墨通过 or-g 转化路径转化为金刚石的实验证明，但是，在相应的实验中从未获得 or-g[104]。有意义的是，最近使用恒压从头算分子动力学模拟证实了该相变模型[113]。这些结果再次表明，不同的石墨到金刚石的转化途径可能存在于碳相图中不同的压强-温度区[111]。然而，让人无法理解的是，为什么在 Scandolo 等[113]的计算与模拟中，从未出现过 rh-g 路径的相变呢？根据 HTHP、SW 和 LAL 的区别，我们认为在他们的计算与模拟中没有考虑石墨转化为金刚石的一个重要因素，即温度。换句话说，在分子动力学模拟中，由于压强的过度作用，温度没有发挥相关作用。事实上，SW 诱导转化的压强（B 区）并不比 HTHP（A 区）和 LAL（C 区）的压强大多少，但在 HTHP 和 LAL 上的转化温度远高于 SW。因此，对于石墨向金刚石的转化，压强有利于诱导 or-g 转化路径，温度更容易产生 rh-g 转化途径。

综上所述，通过 LAL 合成金刚石，我们观察到在碳相图的 C 区，通过菱形石墨相作为中间相，石墨向金刚石转变的新途径。首次实验证实了石墨通过菱形石墨转化为金刚石的理论预测，并且指出在碳相图的 C 区，石墨向金刚石的转化主要是由温度诱导的。这些研究为人们理解石墨与金刚石的相变提供了许多新的物理认知，同时也为发展新的金刚石合成技术提供了重要的理论基础。

5.2.2 金刚石-碳葱可逆相变

纳米金刚石，粒径大约为 3~5 nm，被发现于陨石、星云和星际间的尘埃中，同时它也能人工合成[54, 114-119]。与纳米金刚石非常类似，碳葱（carbon onion）也是一种常见的纳米尺度碳相，在太空中和通过人工合成的方法都能得到[120-126]。研究人员一直在考虑寻找一条能实现碳葱和纳米金刚石的相变路径，该路径的探索不仅有利于人们理解陨石中纳米金刚石的起源，还可能发现金刚石与其他碳相相变的可控路径。一些研究报道了碳葱与纳米金刚石的转变，例如，大剂量电子束在高温（800℃）和超高真空（10^{-6} Pa）环境中辐照碳葱，碳葱可以转变成纳米金刚石；而纳米金刚石能在高温（1000~1500℃）和高真空（10^{-5} Pa）环境下退火，纳米金刚石会转变为碳葱。但是，这些相变都是单向的。我们在 LAL 合成纳米金刚石过程中，首次发现了纳米金刚石和碳葱的可逆相变，这是一种金刚石相碳与石墨相碳的新相变，具有重要的科学意义[127]。

实验过程如 LAL 合成纳米金刚石所描述的那样，首先，将 3 mg 的爆炸法纳米金刚石（初始原料）放入 10 mL 酒精的小瓶并通过超声制备成悬浮液。然后，使用脉冲激光（波长 532 nm，脉宽 10 ns，频率 10 Hz，能量 150 mJ）聚焦在小瓶子中部。LAL 作用时间分别为 0 min、5 min、8 min、12 min、20 min、30 min、45 min、50 min 和 60 min。最后，我们在这 9 个时间节点分别提取样品进行 TEM 表征，结果如图 5-23 所示。非常奇妙的是小瓶子中溶液颜色的变化，从透明到深黑再到透明！这一现象恰好反映了初始原料爆炸法纳米金刚石在 LAL 过程中发生了系列相变。

我们注意到，初始原料爆炸法纳米金刚石团聚得非常厉害，包含了晶态的金刚石核及非晶碳壳层。当 LAL 开始后，悬浮液的颜色开始逐渐发生变化。在 LAL 作用 20 min 后，合成溶液变成了深黑色，这意味着产生了新的物质。从 TEM 图像来看，是因为产生了碳葱。我们可以看到不同形貌的碳葱，包括类球形的同轴石墨壳层、通过外部石墨壳层连接的细长的颗粒等。这些类球形碳葱的尺寸在 5~10 nm，HRTEM 图像显示它们的晶面间距为 0.34 nm，与石墨（002）面的间距一致。有趣的是，当 LAL 作用 60 min 时，深黑色的合成溶液逐渐变成黄色透明溶液；但 LAL 作用持续进行，合成溶液的颜色也不再会改变了［图 5-23（i）的 9 号瓶子］。TEM 表征结果表明，最后黄色透明溶液是纳米金刚石胶体溶液。显然，与初始原料纳米金刚石相比，新合成的纳米金刚石胶体的分散性有了很大的提高，而且外表面的非晶碳壳层也消失了。所以，通过合成胶体溶液的 TEM 表征，我们可以确信，在 LAL 过程中可以实现纳米金刚石与碳葱的可逆相变。

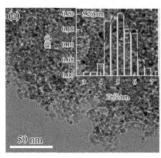 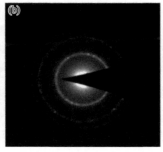 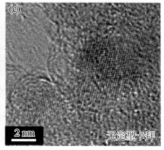

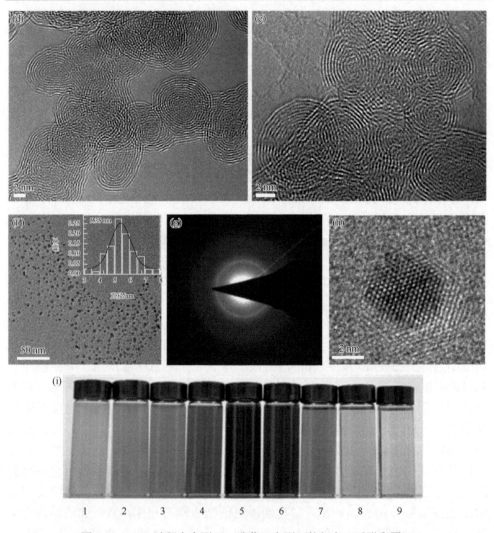

图 5-23　LAL 过程中金刚石—碳葱—金刚石的相变（后附彩图）

（a）初始原料爆炸法纳米金刚石 TEM 图像显示这些纳米颗粒是团聚的，插图为粒径分布，显示平均尺寸为 5.26 nm；（b）相应 SAED 图像；（c）HRTEM 图像；（d）、(e) 碳葱 TEM 图像；（f）LAL 合成纳米金刚石 TEM 图像，分散性得到改善，插图为粒径分布；（g）相应 SAED 图像；（h）HRTEM 图像；（i）LAL 作用过程中合成胶体溶液颜色的变化

如果我们仔细观察实验过程，那么可以看到更多奇异的相变现象。在 LAL 过程中，我们很明显地观察到悬浮液的颜色从不透明的灰白色到深黑色，最后变为黄色透明液体，表明有不同的产物生成。例如，我们在样品中发现了五重孪晶结构的纳米金刚石，这个结构经常在立方金刚石中观察到 [图 5-24（a）]。需要注意的是，图 5-24（c）为初始纳米金刚石向碳葱转变的中间产物 [图 5-23（i）的 4 号瓶子]，很明显，这些颗粒是核-壳纳米结构，与碳葱有很大的不同。我们

称这个中间产物称为巴基金刚石（bucky diamond）。同时，巴基金刚石在碳葱转变成纳米金刚石的过程中也被观察到［图 5-23（i）的 6 号瓶子］。因此，这些结果说明，相同的核-壳纳米结构巴基金刚石（核是金刚石，壳是石墨层）是两个相变过程的中间相。因此，这些实验证据表明，在 LAL 合成纳米金刚石过程中，初始原料纳米金刚石首先变为巴基金刚石，巴基金刚石再转变为碳葱；然后碳葱又转变为巴基金刚石；最后巴基金刚石转变为纳米新金刚石。因此，该可逆相变包含了一系列的碳的同素异形体，如碳葱和巴基金刚石，且巴基金刚石是碳葱与金刚石相变的中间相[29, 128, 129]。

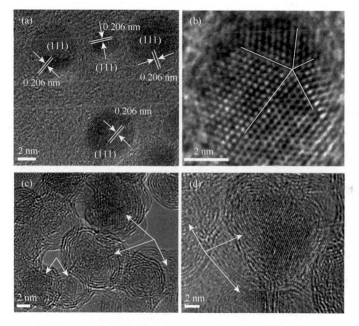

图 5-24　具有五重孪晶结构的纳米金刚石和巴基金刚石

（a）纳米金刚石的 HRTEM 图像，晶格间距为 0.206 nm，与立方金刚石的（111）晶面一致；（b）单个纳米金刚石的五重孪晶结构；（c）取自 4 号瓶子的合成产物的 HRTEM 图像，由碳葱壳和金刚石核组成的球形纳米颗粒（由白色箭头显示），晶核显示金刚石的(111)晶格条纹，晶面间距为 0.206 nm;（d）取自 6 号瓶子的合成产物的 HRTEM 图像，同样，由碳葱壳和金刚石核组成的球形纳米颗粒（由白色箭头显示）

为了更准确地阐明 TEM 表征揭示的金刚石-碳葱的可逆相变，我们研究了这一过程的拉曼光谱，如图 5-25 所示。我们可以看到，初始原料纳米金刚石的拉曼光谱显示了两个特征峰［图 5-25（a）的曲线 i］，一个是金刚石的峰，位于 1325 cm^{-1}；另一个较宽的峰位于 1600 cm^{-1}，被认为是石墨碳相的 G 峰。当 LAL 作用 10 min 后，金刚石的峰移向了 1360 cm^{-1}，这个峰的出现是因为巴基金刚石的产生［图 5-25（a）的曲线 ii］。为了确定这个推断，我们用 Gaussian-Lorentzian 曲线进行拟合，发现拟合后的巴基金刚石的曲线包含 1324 cm^{-1} 的峰［图 5-25（b）

的标记曲线],它来源于纳米金刚石,峰位比 1332 cm^{-1} 块状金刚石的信号要低,这是由于量子限域效应。当 LAL 作用 20 min 后,该峰(1324 cm^{-1})继续移动到 1400 cm^{-1},这是由于非晶碳的产生(D 带)。同时,G 带移动到 1584 cm^{-1},这是因为受到碳葱壳层结构曲率的影响[130]。然后,D 带波数下降,G 带的波数上升,这就意味着产生了新的巴基金刚石。最后,金刚石的峰重新生成,G 带移动到比初始原料纳米金刚石更高的位置[图 5-25(a)的曲线 v],这是由于—OH 官能团吸附在纳米金刚石的表面。因此,拉曼光谱表征提供了更详细的结构演化过程。

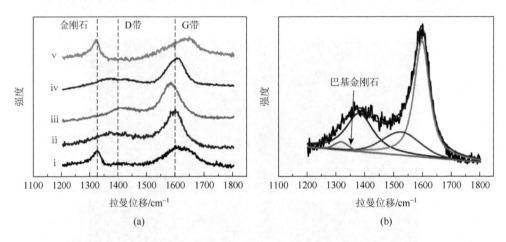

图 5-25 从初始原料纳米金刚石到 LAL 合成的纳米新金刚石相变的拉曼光谱分析(后附彩图)
(a)初始原料纳米金刚石(曲线 i)显示 1325 cm^{-1} 的金刚石峰和 1600 cm^{-1} 的加宽峰;巴基金刚石(曲线 ii)显示峰值移向 1360 cm^{-1};碳葱(曲线 iii)显示金刚石峰不断向 1400 cm^{-1} 移动(D 带),同时 G 带也明显下降到 1584 cm^{-1};巴基金刚石(曲线 iv)显示 D 带波数下降而 G 带向上移动到更高的波数;最终的纳米新金刚石(曲线 v)显示金刚石峰(1325 cm^{-1})再次生成,G 带移动到 1650 cm^{-1},高于初始原料纳米金刚石。(b)巴基金刚石曲线的详细拟合[(a)中的曲线 ii],绿色拟合曲线位于 1324 cm^{-1},可以归属于纳米金刚石

为了更深入地理解纳米金刚石-碳葱可逆相变机制,我们基于纳米热力学理论,提出了该相变过程的热力学模型。众所周知,碳葱在常温常压下是碳的稳定相,而金刚石是亚稳相。尽管两者的单位原子能量差别只有 0.02 eV/atom[131],但是它们之间的势垒比较高,大约为 0.5 eV/atom,需要高温高压的环境才能实现它们之间的转变。在 LAL 合成金刚石过程中,初始原料金刚石的表面包裹着非晶碳,它能吸收激光的能量而导致温度升高。所以,基于非晶碳和碳葱能吸收激光的能量[110],我们首先建立纳米颗粒在激光作用下的加热模型。与此不同的是,纳米金刚石带隙约为 5.5 eV,对于可见光来说是透明的。

不同碳材料与光相互作用如图 5-26(a)所示。纳米颗粒吸收激光能量加热的过程可以通过方程(5-1)描述[132]:

$$J\sigma_{\text{abs}}^{\lambda} = m\int_{T_0}^{T} C_p(T)\mathrm{d}T \tag{5-1}$$

式中，$m = \rho \dfrac{\pi d^3}{6}$ 表示纳米颗粒的质量，其中，d 是粒径，ρ 是密度；J 是激光能量密度；T 设定为 300 K；$\sigma_{\mathrm{abs}}^{\lambda}$ 是颗粒的吸收截面，该界面的大小强烈依赖于激光波长；$C_p(T)$ 是比热容，为温度函数[133]，关系为 $C_p = -30.7\exp(-T/486.7) + 25.4$。

对于一个球形纳米颗粒，吸收截面可以表示为[134]

$$\sigma_{\mathrm{abs}}^{\lambda} = \frac{\pi d^2 Q_{\mathrm{abs}}^{\lambda}}{4} \tag{5-2}$$

式中，$Q_{\mathrm{abs}}^{\lambda}$ 表示光吸收效率，指数 λ 表示吸光效率和吸光截面都依赖于激光的波长。事实上，$Q_{\mathrm{abs}}^{\lambda}$ 可以表示为相对折射率 m 的函数：

$$Q_{\mathrm{abs}}^{\lambda} = 4x\,\mathrm{Im}\left(\frac{m^2-1}{m^2+2}\right)\left[1 + \frac{4x^3}{3}\,\mathrm{Im}\left(\frac{m^2-1}{m^2+2}\right)\right] \tag{5-3}$$

$$x = ka = \frac{2\pi \tilde{n}_1 a}{\lambda} \tag{5-4}$$

$$m = \frac{\tilde{n}}{\tilde{n}_1} \tag{5-5}$$

式中，x 是纳米颗粒的直径；a 是半径；$\tilde{n}$ 和 $\tilde{n}_1$ 分别是颗粒与介质的折射率（在我们的研究中，取酒精的折射率为 1.36）。为了计算一个球形纳米颗粒的吸收效率，我们需要知道颗粒的两个特征，折射率 $n(\lambda)$ 和消光系数 $k(\lambda)$，或者复折射率的实部和虚部 $\tilde{n}(\lambda) = n(\lambda) + ik(\lambda)$，$n(\lambda)$ 和 $k(\lambda)$ 这些依赖激光波长的值可以从相关资料中找到[135]。我们取非晶碳和 532 nm 激光波长分别作为参考材料和波长：$\tilde{n}(532\ \mathrm{nm}) = 2.32626 + 0.85359i$，如果 $(4x^3/3)\,\mathrm{Im}[(m^2-1)/(m^2+2)] \ll 1$，该情况对于我们的材料体系是适用的，因为颗粒的粒径非常小（5 nm）。因此，吸收效率可以简化为

$$Q_{\mathrm{abs}} = 4x\,\mathrm{Im}\left(\frac{m^2-1}{m^2+2}\right) \tag{5-6}$$

除此之外，密度的差异也应该考虑在内。这里，我们取非晶碳的密度为 2.1 g/cm³[136]。

下面，我们考虑碳葱结构的受热膨胀。考虑一个包含 N 个不同半径碳笼（carbon cages，CC）的碳葱，每个碳笼在高温下都会发生膨胀，体积膨胀系数（α_V）与德拜温度（Debye temperature）有关，并且基于 Lindemann 方程和 Gründisen 理论[137, 138]，体积膨胀系数可以表示为 $\alpha_V = c/\left(\theta_{\mathrm{D}}^2 V^{2/3} A_r\right)$，其中，$c$、$V$ 和 A_r 分别表示常数、摩尔体积和与 C^{12} 相比的原子重量。根据以上的关系，我们可以得到

$$\frac{\alpha_V(\mathrm{CC})}{\alpha_V(\mathrm{G})} = \frac{\theta_{\mathrm{D}}^2(\mathrm{G})}{\theta_{\mathrm{D}}^2(\mathrm{CC})} \tag{5-7}$$

式中，$\alpha_V(G)$ 是石墨的体积膨胀系数。而 $\theta_D^2 \propto E_c \propto \gamma$ [139,140]，E_c 和 γ 表示的是结合能和表面能。因此，我们得到

$$\frac{\alpha_V(CC)}{\alpha_V(G)} = \frac{\gamma(G)}{\gamma(CC)} \tag{5-8}$$

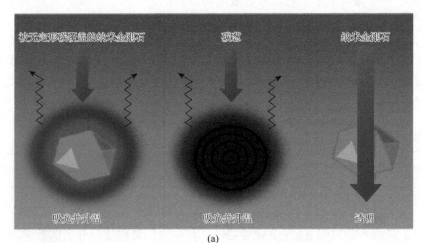

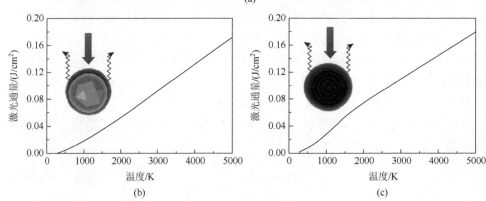

图 5-26 各种碳材料和激光相互作用的示意图

（a）由于没有带隙，可见光照射下的被无定形碳覆盖的纳米金刚石（左）和碳葱（中）覆盖的纳米金刚石可以被加热，然而，由于 5.5 eV 的大带隙，没有任何覆盖物的纳米金刚石（右）对可见光是透明的；（b）被无定形碳覆盖的纳米金刚石的激光能量和温度的关系；（c）碳葱的激光能量与温度的关系

近似地，尺度依赖的表面能可以表达成 $\gamma(G)/\gamma(CC) = 1/(1+2h/R)$，其中，$h$ 和 R 分别代表了碳原子和碳笼的半径，结合方程（5-8），可以得到

$$\alpha_V(CC) = \alpha_V(G)(1 - 2h/R) \tag{5-9}$$

根据体积膨胀系数的公式 $\alpha_V = (1/V)(dV/dT)$ [141]，我们得到依赖尺度与温度的碳葱的体积公式

$$V(CC) = V_0[1 + \alpha_V(G)(1 - 2h/R)T] \quad (5\text{-}10)$$

从以上的公式可以看出，在碳葱中每一个碳笼膨胀的体积是不一样的，这将会导致层间距的变化，最终会影响到层与层的范德瓦耳斯力的变化。考虑一个只有两层碳笼的碳葱，内层碳笼的半径为 R_1，外层碳笼的半径为 R_2，压强（P）作用于内层碳笼和外层碳笼，压强分别为[142]

$$P_{\text{inner}} = -\frac{\rho_\infty^2}{4\pi R_1^2} \int_{S_{\text{inner}}^{(0)}} \left(\int_{S_{\text{outer}}^{(0)}} F_1 dS_{\text{outer}}^{(0)} \right) dS_{\text{inner}}^{(0)} \quad (5\text{-}11)$$

$$P_{\text{outer}} = \frac{\rho_\infty^2}{4\pi R_2^2} \int_{S_{\text{outer}}^{(0)}} \left(\int_{S_{\text{inner}}^{(0)}} F_2 dS_{\text{inner}}^{(0)} \right) dS_{\text{outer}}^{(0)} \quad (5\text{-}12)$$

式中，ρ_∞ 是石墨片的原子密度；$S_{\text{inner}}^{(0)}$ 和 $S_{\text{outer}}^{(0)}$ 分别是内层和外层碳笼未变形的状态时的面积；F_1 和 F_2 是作用在每个原子的范德瓦耳斯力，可以从伦纳德-琼斯势（L-J 势）的一阶导数中得到，$V(d) = 4\varepsilon\left[\left(\sigma/d\right)^{12} - \left(\sigma/d\right)^6\right]$，其中，$\varepsilon$ 和 σ 是 L-J 势的参数。

根据方程（5-10）得到依赖于尺寸和温度的碳笼的热膨胀体积，如图 5-27（a）所示。我们注意到，石墨的热膨胀系数可以表达为 $\alpha_V(G) = 2\alpha_a + \alpha_c$，其中，$\alpha_a$ 和 α_c 分别表示为 a 轴和 c 轴的热膨胀系数，与比热容和德拜函数有关。图 5-27（a）的插图表示的是石墨在 a 轴及 c 轴的比热容（C_V）。因此，碳笼的体积热膨胀系数随着温度的增加和尺寸的减少而增加，也就是说，在碳葱中最小的碳笼在固定的温度下有着最大的膨胀速度。

我们考虑一个双层的碳葱包含两个球状层，比如，C_{60} 在 C_{180} 里面，我们能够计算温度依赖的层间距的变化，发现随着温度的升高，C_{60} 和 C_{180} 的间距越来越小

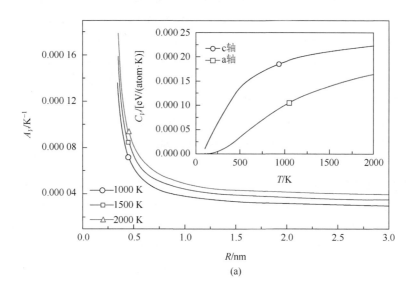

(a)

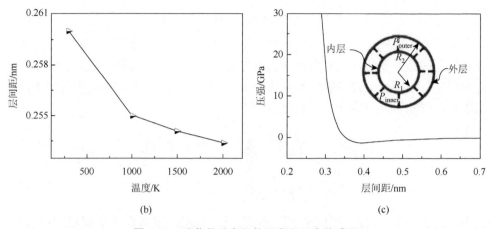

图 5-27 碳葱是纳米级的温度和压力传感器

(a) 碳笼中体积热膨胀系数的尺寸和温度相关性,插图是石墨在 c 轴和 a 轴方向上的比热容; (b) C_{180} 与 C_{60} 的温度相关的层间距; (c) 双层碳葱中压强 (内部) 对层间距的依赖性,插图是考虑的双层碳葱 (C_{180} 中的 C_{60})

[图 5-27 (b)]。因此,由于碳笼半径的变化,层间的压强也会改变。图 5-27 (c) 显示压强与层间距的关系。很明显,较小的层间距导致较大的压强,当温度达到 2000 K 时,碳葱内部会产生 10 GPa 的压强。因此,LAL 过程中的高温高压环境导致了碳葱向金刚石的转变。

基于我们的实验观察及上述纳米热力学分析,LAL 过程中的纳米金刚石-碳葱的可逆相变可以总结为在纳米金刚石转变成碳葱的过程中,金刚石外表面的非晶碳吸收激光的能量,温度迅速升高,激光诱导的高温驱动纳米金刚石转变成碳葱,同时高温会破坏纳米金刚石外表面的非晶碳,生成巴基金刚石,作为相变的中间相;随后,激光辐照碳葱使之温度上升,使得碳葱间的石墨的层间距变小,由此导致高压的产生,这种高温高压的热力学环境诱发碳葱转化为金刚石。由于金刚石的带隙为 5.5 eV,因此即使继续受到激光的辐照,它也会稳定存在。图 5-28 描绘了整个可逆相变过程。需要说明的是这一系列科学发现的重要性在于:首先为人们认识宇宙尘埃中纳米金刚石的起源提供了重要线索,即它们可能是碳葱通过相变而生成的,因为宇宙空间中弥散着大量的碳葱;其次是揭示了金刚石与碳的同素异形体如碳葱等之间丰富的相变通道,丰富了人们关于金刚石与石墨相碳之间相变的知识。

天文学家观察发现纳米金刚石存在于原始的陨石中,但是它的起源是个谜。研究人员提出了许多关于陨石中纳米金刚石起源的说法,但是几乎每个说法都有缺陷。化学气相沉积是现在大家广为接受的说法,但是实验室的化学气相沉积的条件与太空的条件有很大的不同。太空中存在着大量的碳葱,它们来自高温下自动重排的独立碳碎片,这个高温环境常见于温暖的星际中,以及恒星残骸碎片的

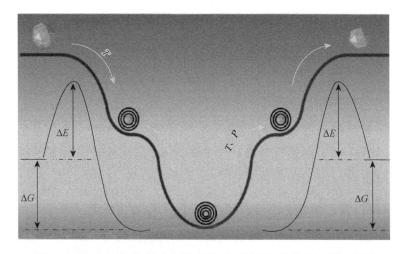

图 5-28 LAL 过程中纳米金刚石-碳葱可逆相变的物理机制示意图

ΔG 表示金刚石和碳葱的能量差，ΔE 表示从金刚石向碳葱相变所需的活化能

区域。因此，通过高能辐射，比如，电子、光子可能诱导纳米金刚石-碳葱的可逆相变，这为陨石中纳米金刚石的起源提出了新的解释。换句话说，星际间的纳米金刚石来源于太空中的碳葱，碳葱存在的地方，就有可能产生纳米金刚石。

5.2.3 金刚石-新金刚石相变

新金刚石（new diamond）具有面心立方结构碳相，由于它的电子衍射花样与传统的金刚石非常相似而得名[143-145]。21 世纪，新金刚石吸引了很多人的关注，因为其在机械工程、微电子、光电子等领域有很大的潜在应用[143-147]。但是，新金刚石的制备存在着巨大的挑战，因为它的形成能比传统的金刚石还要高，所以很难合成[148]。我们在 LAL 过程中金刚石与碳葱可逆相变基础上，通过设计环境液体，实现了金刚石向新金刚石的转变，发展了一种利用纳米金刚石作为原料合成新金刚石的方法[149]。

LAL 合成纳米金刚石充满了神奇。当我们将 LAL 过程中金刚石与碳葱可逆相变实验的环境液体换成水而其他条件不变时，得到了完全不同的结果，如图 5-29 所示，图 5-29（a）～图 5-29（c）是初始原料纳米金刚石相关图像，图 5-29（d）～图 5-29（i）是最终合成的纳米新金刚石相关图像。HRTEM 图像显示初始原料纳米金刚石是由晶态金刚石核及非晶碳壳层构成的。而 LAL 合成纳米颗粒的粒径为 30～40 nm，互相连接在一起。新金刚石的结构参数与金刚石非常类似，但是它的电子衍射与金刚石还是有所不同的。图 5-29（f）显示出面心立方新金刚石的 4 个较强衍射环，分别代表（111）、（200）、（220）和（311）晶面。值得一提的是，金刚石是不存在（200）这样的指数晶面的[150]。在新金刚石的 HRTEM 图像中［图 5-29（i）］，我们可以清楚地看到，新金刚石是嵌入到非晶碳中的，这与金刚石的稳定性是相关的，我们会在下面详细讨论。单个新金刚石 HRTEM 图像显示出 0.313 nm 和 0.272 nm 的晶面间距，与面心立方结构的（111）和（200）两个晶面符合很好，而且，两个晶面的夹角为 125°，与晶胞参数为 0.545 nm 的面心立方结构完全吻合。因此，两套不同的 TEM 表征数据证明了金刚石向新金刚石发生转变。为了详细地观察相变的过程，我们分析了不同激光辐照时间（0 min、5 min、10 min、15 min、20 min、45 min、60 min、70 min 和 90 min）的样品。有趣的是，我们在中间相的 HRTEM 图像中看到了不同形貌的碳葱，如图 5-30 所示，包括准球形颗粒、具有连接外部类石墨层和闭合准球形内部壳的细长颗粒等，这些纳米颗粒的晶面间距是 0.34 nm，对应于石墨的（002）晶面。所以，在 LAL 过程中，初始原料纳米金刚石首先转化为碳葱，然后碳葱再转化成新金刚石。类似地，我们看到小瓶子中液体颜色的变化，从不透明到近黑色再到近似透明［图 5-30（e）］。需要注意的是，无论环境液体是酒精还是水，LAL 过程的中间产物都是深黑色的，说明从初始原料纳米金刚石到最终产物，碳葱都作为中间相。

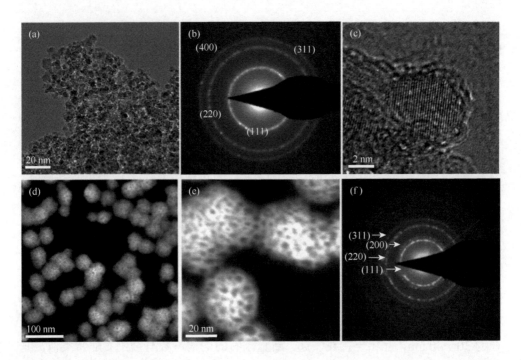

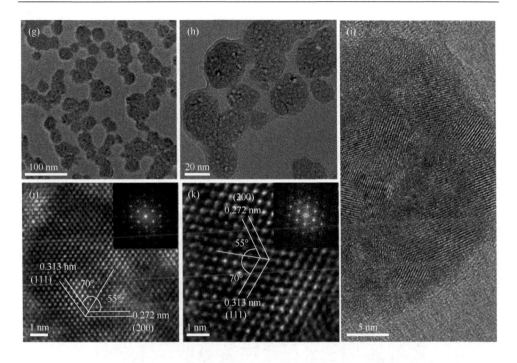

图 5-29 纳米金刚石和纳米新金刚石的 TEM 表征

(a) 尺寸约为 5 nm 的聚集态初始原料纳米金刚石的典型形态；(b) 4 个强衍射环代表金刚石的(111)、(220)、(311)和(400)晶面；(c) 覆盖无定形碳壳层的初始原料纳米金刚石的 HRTEM 图像；(d) 和 (e) 纳米新金刚石的低倍和高倍 STEM 图像；(f) 相应的 SAED 图像，4 个较强衍射环表示纳米新金刚石的(111)、(200)、(220)和(311)晶面；(g) 和 (h) 纳米新金刚石的低分辨和高分辨 TEM 图像；(i) 纳米新金刚石的 HRTEM 图像，这些纳米新金刚石被发现于非晶碳中；(j) 和 (k) 单个纳米新金刚石的 HRTEM 图像，晶面间距测量值分别为 0.313 nm 和 0.272 nm，与新金刚石的(111)和(200)晶面间距 D 一致

为了仔细区分金刚石与新金刚石，我们采用电子能量损失谱（electron energy loss spectrum，EELS）技术分别对实验中出现的初始原料纳米金刚石、中间相碳葱和 LAL 合成纳米新金刚石进行了表征，如图 5-31 所示。我们可以看到，图 5-31(a) 的曲线 i 给出了 291.9 eV 的纳米金刚石的特征峰，它起源于纳米金刚石 1 s 核能级的跃迁（1 s→σ^*）[151]，并且在 284.8 eV 的位置还有一个小峰，应该来自于 1 s→π^* 跃迁。这个结果说明了初始原料纳米金刚石含有 sp^2 碳原子。同时，我们注意到，尽管碳葱是有序的石墨层结构，但是它的 EELS 中的 K-edge 与非晶碳的很相似[152]，而出现在 284.8 eV 的峰则对应于 1 s 壳层到 π^* 的跃迁，在约 295 eV 以上没有明显的峰，这就说明了 sp^3 键已经转变成 sp^2 键。这一现象与 Tomita 等[153]的实验结果非常类似。而在纳米新金刚石的 EELS 中［图 5-30(a) 的曲线 iii］，我们看到有一个出现在 284.8 eV 的峰，这应该归因于它的非晶碳成分，与 TEM 表征的结果是一致的。然而，在 1 s→σ^* 跃迁的图谱分析上，无论是峰的位置还是强度的对比，纳米新金刚石与传统金刚石都有很大的不一样，纳米金刚石的三个 1 s→σ^* 跃迁的特

征峰 291.9 eV、297.7 eV 和 305.3 eV 分别移动到纳米新金刚石的 291.7 eV、302.8 eV 和 314.2 eV。所以，通过 EELS 分析，我们可以区分实验中的金刚石相和新金刚石相。我们还进行了基于密度泛函理论的计算与模拟技术来模拟纳米金刚石和纳米新金刚石的 EELS，如图 5-31（b）和图 5-31（c）。首先，我们计算了纳米金刚石的 EELS，它与实验数据几乎完全一致，这说明我们的计算与模拟方法是可靠的。然后，我们用同样的方法计算了纳米新金刚石的 EELS，发现它的相对强度与形状与实验数据十分吻合。因此，实验表征和理论计算与模拟结果证实了 LAL 可以合成面心立方结构的纳米新金刚石。

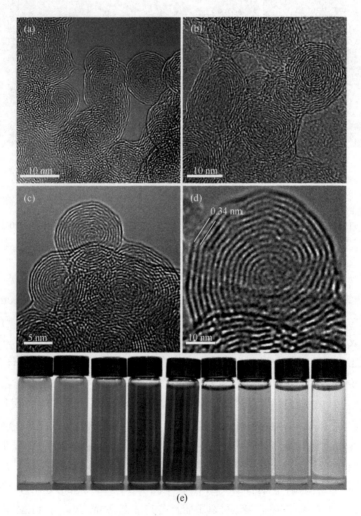

图 5-30　作为相变中间相的碳葱（后附彩图）

（a）～（d）聚集的球形碳葱；（e）合成液体颜色随 LAL 作用时间的变化

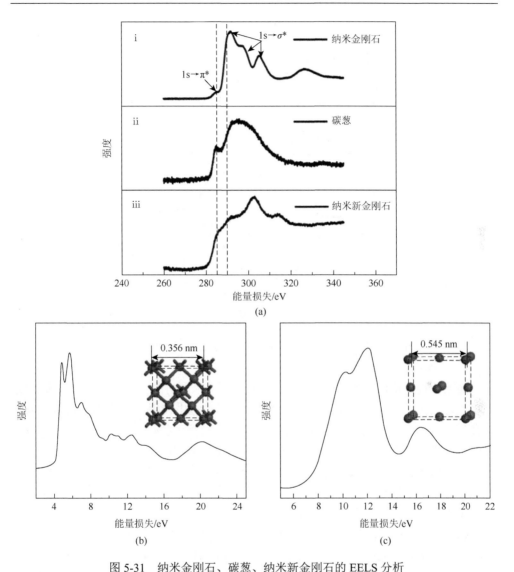

图 5-31 纳米金刚石、碳葱、纳米新金刚石的 EELS 分析

（a）纳米金刚石、碳葱和纳米新金刚石的实验比较；（b）和（c）分别对纳米金刚石和纳米新金刚石的 EELS 进行理论模拟，插图为相应的晶体结构

更进一步地，为了理解 LAL 过程中纳米金刚石向纳米新金刚石的转变，我们给出了初始纳米金刚石、碳葱和纳米新金刚石的拉曼光谱分析，如图 5-32 所示。我们很清楚地看到，初始纳米金刚石在 1325 cm^{-1} 有明显峰，这是金刚石的典型的拉曼信号，并且在 1600 cm^{-1} 也有信号（G 峰）。LAL 作用 20 min 后，金刚石的峰右移到 1410 cm^{-1}，原因是无序的 sp^2 碳相出现（D 峰）[130]，而且，受碳葱结构的曲率的影响，G 峰左移到 1585 cm^{-1}[131]。最后，D 峰移动到 1363 cm^{-1}。这些结果

说明，尽管纳米新金刚石表面的非晶碳对其特征峰有一定的影响，但是 D 峰的左移还是与纳米新金刚石产生的 sp^3 碳信号有关。

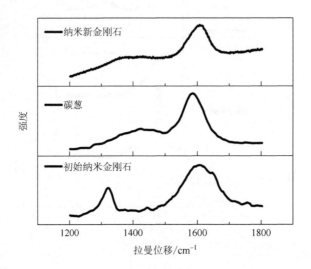

图 5-32 初始纳米金刚石到纳米新金刚石相变的拉曼光谱分析

如上所述，我们不难发现，在 LAL 合成碳纳米材料时，环境液体对合成产物有着重要影响。当我们在 LAL 合成纳米金刚石过程中使用乙醇作为环境液体时，观察到了纳米金刚石与碳葱的可逆相变。而当我们用水替代乙醇作为环境液体时，实现了纳米金刚石向纳米新金刚石的转变。所以，我们有必要对环境液体对 LAL 合成的作用进行分析。我们采用 FTIR 技术分析了在不同环境液体合成产物的表面，如图 5-33 所示。

我们看到，在酒精中合成纳米金刚石的表面有 C=O、—COOR、C—O—C、—OH 和 C—H 基团，而在水中—OH 信号更强，C—H 和 C—O 键的信号减弱。根据 Gogotsi 的说法[154, 155]，为了保持 sp^3 键的稳定，在酒精中合成的纳米金刚石表面吸附了不同的官能团，尤其是羧基和羰基，这些官能团很容易吸附在碳骨架上面[154]。然而，水环境中缺乏相应的基团，上述基团在水中的产物中都没有发现。因此，裸露的（非官能化的）纳米金刚石表面不稳定，这样表面的 sp^3 碳团簇会发生向 sp^2 的转变[155]。我们在 EELS 和 HRTEM 分析中都探测到了非晶碳信号，所以，表面非晶碳的存在是保持纳米新金刚石稳定的关键因素。另外，值得一提的是，我们以水为环境液体使用 LAL 方法合成了一系列新碳纳米材料[156, 157]，并且应用 ArF 激光在水中烧蚀石墨靶，合成了 C_8 及普通的纳米新金刚石[158]。这些研究都说明了水环境在 LAL 合成亚稳碳纳米相中起到很关键的作用。尽管水的作用还没被完全认识清楚，但是水中羟基的存在可能是合成亚稳碳纳米相一个重要的因素。我们知道，碳葱是常态下的稳定相，而纳米金刚石和纳米新金刚石都是亚稳相，因此，通过简单的热退火，纳米金刚石可以转化成碳葱。根据我们的计算，纳米新金刚石的形成能比传统金刚石还要高。因此，需要高温高压的热力学环境才能使碳葱变成纳米新金刚石[126]。

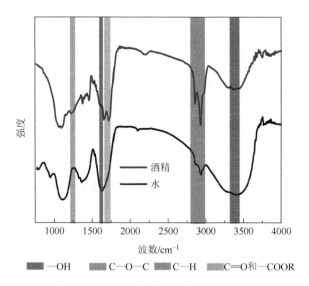

图 5-33　LAL 分别在乙醇和水中合成产物的 FTIR 光谱（后附彩图）

在 LAL 过程中纳米金刚石与纳米新金刚石相变的机制与纳米金刚石与碳葱可逆相变过程非常相似。例如，非晶碳吸收激光的能量导致其温度升高，高温驱动金刚石相变成碳葱；随后，碳葱吸收激光的能量，温度也上升，并且导致碳葱内部碳笼的膨胀，由于内部碳笼膨胀系数大，晶面间距变小，由此导致范德瓦耳斯力变大，继而导致层间压力变大，高温和高压又驱动碳葱向金刚石相变。根据我们的估算，金刚石如果在碳葱内成核的话，那么温度和压强分别需要 400℃和 10 GPa。而在 LAL 过程中，碳葱内部的温度和压强分别达到 2000℃和 20 GPa。所以，考虑到如此高的温度和压强及水的影响，在 LAL 过程中非常有可能合成出具有比金刚石相更高形成能的新碳相，也就是新金刚石相。图 5-34 展现了在 LAL 过程中纳米金刚石到纳米新金刚石的整个相变过程。在纳米金刚石向碳葱转变过程中，首先，纳米金刚石表面的非晶碳吸收了激光能量，起到了加热的作用，使得纳米金刚石变成碳葱。然后，碳葱吸收激光能量产生温升，高温环境下的碳葱晶面间距减小，诱导高压产生。最后，碳葱在高温高压驱动下转化成纳米新金刚石。这些发现为从不同的碳原料中制备新金刚石提供了一种有效的方法。

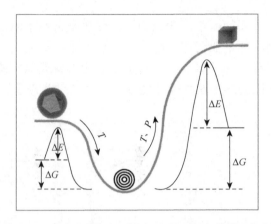

图 5-34 纳米金刚石与纳米新金刚石相变的物理机制

ΔG 表示两相的能量差；ΔE 表示从一个相到另一个相所必需的活化能

5.3 新碳相纳米材料

碳是自然界一种很常见却很重要的元素之一，它以多种形式存在于地球和宇宙空间，是已知的所有生命系统都不可或缺的，而碳材料已经广泛地应用于工业的许多领域。在科学上，如果以碳原子存在的形式来对碳材料进行分类的话，那么自然界应该存在三种碳材料。第一种是以三维 sp^3 杂化结构碳原子构成的金刚石及基于 sp^3 杂化的六方金刚石和 C_8 等金刚石的同素异形体[156, 159]；第二种是以二维 sp^2 杂化结构碳原子构成的石墨及基于 sp^2 杂化的富勒烯、碳纳米管和石墨烯等石墨的同素异形体。众所周知，对这两种碳及相应的同素异形体如金刚石、石墨、富勒烯、碳纳米管、石墨烯等[160-162]的研究不仅对人类的科学认识产生了重要影响，而且极大地推动了技术的进步。然而，自然界还应该存在第三种碳即由一维 sp 杂化结构碳原子构成的卡拜（carbyne）。所谓一维 sp 杂化结构就是指由碳原子构成的线型链状结构，其中碳原子分别以碳-碳单键和碳-碳叁键相连（图 5-35）。20 世纪 60 年代，科学家在来自宇宙尘埃的光谱中捕捉到碳-碳单键和碳-碳叁键存在的信息，认为除了存在三维 sp^3 杂化结构和二维 sp^2 杂化结构以外，还应该存在一维 sp 杂化结构。在坠入地球的一块陨石中发现一种呈白色粉末状的不同于金刚石和石墨的新六方碳相，并由此而推测这应该就是第三种碳。这一发现激起了人们对新碳相材料研究的极大热情，自此，来自物理、化学和材料等领域的科学家开始探索在实验室合成卡拜。同时，理论物理学家和化学家预测了卡拜在物理和化学上有着许多奇异的性质并使得它在诸多领域会像富勒烯和石墨烯等同素异形体一样有着重要应用。但是，尽管科学家们尝试了无数次的物理制备和化学合成，还没有令人信服的实验证据表明人们能够在实验室合成卡拜。难道卡拜只是一个碳的传说吗？这种猜测成了碳材料研究者脑海里挥之不去的疑云。实际上，卡拜在实验室很难合成的主要原因是它是碳的一种亚稳纳米相，它的热力学稳定区处于高温和高压状态[163]，并且其独特的线型碳-碳单键和碳-碳叁键分子链的形成对所采用的碳前驱物有特殊要求。所谓亚稳相纳米材料是指在热力学常态下处于亚稳相的一类材料，由于它们通常表现出稳定材料所不具备的优异性能而备受人们关注。亚稳相纳米材料是指纳米尺度下具有亚稳结构或亚稳形貌的材料。亚稳纳米相生长有着异于宏观尺度下的材料生长行为，导致了许多新型亚稳相纳米材料的出现，近年来受到了国际材料科学界的广泛关注。所以，在实验和理论上对亚稳相纳米材料生长的基本科学认识，是发展新型亚稳相纳米材料的核心科学问题。

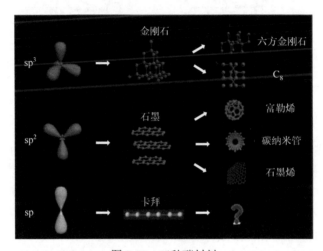

图 5-35　三种碳材料

在上述 LAL 的热力学和动力学描述中，我们知道，LAL 创造的局域高温高压高密度的极端热力学环境为探索新型亚稳相纳米材料提供了得天独厚的条件。我们充分利用了 LAL 在合成亚稳相纳米材料方面的优势，探索了它在新碳相纳米材料制备中的应用，合成了一系列新碳相纳米材料。本节将介绍 LAL 方法在三种典型新碳相纳米材料合成中的应用。

5.3.1　C_8 纳米晶

在纷繁多样的碳的同素异形体研究中，有一种体心立方结构的亚稳碳材料引起许多科学家的关注。1978 年，Strel'Nitskii 等[164]首先发表在高压等离子体羽沉积的非晶碳膜中发现了一种新碳结构即体心立方结构，其晶格常数是 4.28 Å。这种新碳结构被命名为 C_8，因为在它的晶胞中有 8 个碳原子和 16 个等效碳原子，该结构具有扭曲的四面体形状[165]，如图 5-36 所示。

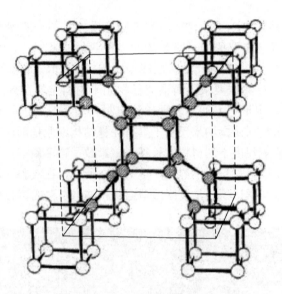

图 5-36　C_8 晶胞结构示意图

阴影是 16 个等效碳原子

这种结构表明，C_8 中应该存在比正常的 C—C 键短得多的扭曲的 C—C 键，这意味着扭曲的 C—C 键可能在 C_8 中引起扭曲的 sp^3 键[165-168]。Johnston[171]从理论上研究了 C_8 的结构和性质，发现 C_8 的密度为 4.1 g/cm^3，比立方或六方金刚石的密度 (3.51 g/cm^3) 还要大。众所周知，在所有已知材料中，金刚石具有最高的原子序数密度 [0.295 atom/(cm^3·mol)]，而 C_8 的原子序数密度为 0.338 atom/(cm^3·mol)。所以，如果 C_8 真的存在的话，那么它应该被称为超密度碳。虽然这种结构在原子数较少时会很不稳定，但是理论研究认为在一个富电子体系当中，这样的超密度碳结构是能够稳定存在的[165,169]。然而，截至目前，C_8 的存在一直受到强烈怀疑，因为所报道的晶体结构分析存在一些差异[165,170]。此外，也没有实质性的实验数据来证实 Strel'Nitskii 等的发现（现有文献中没有精确的电子衍射图案和单个 C_8 的形态[170]）。

Liu 等应用 LAL 方法在水中使用激光烧蚀非晶碳靶成功地合成了 C_8 纳米晶，如图 5-37 所示。通过对图 5-37（a）的分析，我们发现，合成的产物中有大量类球状微纳颗粒，并且混杂了类立方体状的纳米颗粒（图中箭头所指及插图）。通过对样品组分表征证明这些类立方体状或带有棱角状的纳米颗粒的成分是纯碳。考虑到合成的是新碳相，所以我们采用拉曼光谱技术对碳的成键进行了深入分析，如图 5-37（b）和图 5-37（c）所示。我们可以看到，非晶碳靶只有在 1600 cm^{-1} 左右出现一个特征峰（G 模），表明非晶碳靶中有大量 sp^3 杂化键存在；而在 C_8 纳米晶的拉曼光谱中有一个新峰出现在 1348 cm^{-1}，它可以被标定为碳的 D 模；在 1580 cm^{-1} 附近出现的峰则可以被标定为碳的 G 模。通过对两个拉曼光谱的比较我们可以清楚地看到，合成产物和非晶碳靶的拉曼光谱有明显的差异，即合成产物的 G 模，相对于非晶碳靶的 G 模出现了向低波数的移动，这就意味着合成产物中存在与非晶碳靶不同的 C—C 键结构。

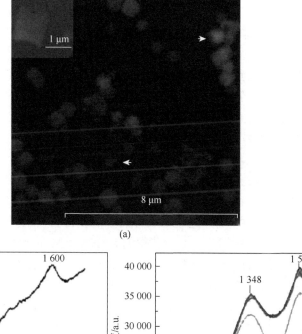

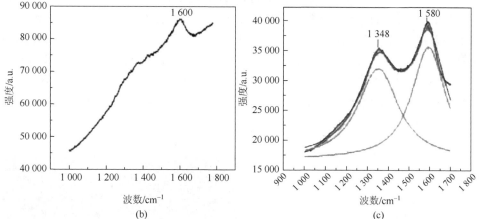

图 5-37 （a）LAL 合成 C_8 纳米晶的 SEM 图像，插图为类立方体状纳米颗粒的高倍 SEM 图像；（b）非晶碳靶的拉曼光谱；（c）C_8 纳米晶的拉曼光谱，其中的拟合曲线是分峰光谱

众所周知，碳的拉曼光谱主要就是由 G 模和 D 模决定的，为了能准确地确定合成产物的 C—C 键结构与非晶碳靶的差异，我们通过对比这两种拉曼光谱的峰移来分析其中的 C—C 键结构[171, 172]。首先，非晶碳靶本身就含有大量的 sp^3 键，因此它的拉曼光谱中 G 模能够从 1600 cm^{-1} 移动至 1580 cm^{-1}，意味着合成产物中必然含有一些 sp^3 键成分和其他 C—C 键结构的混合态[171-174]。其次，与非晶碳靶的 G 模特征峰不同，合成产物的 G 模特征峰显示出很好的对称性，这表明在制备

样品中可能存在粒径非常小的 sp^2 杂化 C—C 键,或者说存在一些不同于非晶碳 sp^3 杂化的其他类型的 C—C 键杂化形式[173,174]。已有研究指出,在 1348 cm^{-1} 出现的 D 模特征峰是由碳的非 sp^2 成分所导致的,同时该特征峰也可以被归为由非 sp^2 成分的碳结构所导致的 C—C 键振动峰[171,172,174]。另外,在合成产物的拉曼光谱分析中,我们发现,它的拉曼光谱可以被分解为两个特征峰:第一个特征峰在 1583 cm^{-1} 附近,这一个特征峰与碳结构的 E_{2g} 声学模所导致的特征峰十分吻合,因此,该峰的出现意味着合成产物中含有少量的 sp^2 成分[175,176];第二个特征峰出现在 1335 cm^{-1} 附近,该特征峰不是由 E_{2g} 声学模所导致的,因此,我们可以将其归结为由碳的 sp^3 杂化键及扭曲的 sp^3 杂化键所导致的混合模[176]。也就是说,合成产物应该包含有 C—C 键的 sp^2、sp^3 及扭曲的 sp^3 杂化键。

为了标定合成产物的晶体结构,我们采用 TEM 技术对一个类立方体状的纳米颗粒进行了系统表征,如图 5-38 所示,粒径大约 40 nm,相应的 EDS 分析清楚显示该纳米颗粒成分是纯碳(其中铜的峰来自 TEM 样品铜网)。为了对合成纳米颗粒的晶体结构进行分析,我们对其进行了一系列的 SAED 表征,具体做法是,首先,保持一个晶轴不动,通过倾斜样品台而获得的一系列 SAED 图像,如图 5-38（b）和图 5-38（d）;然后,对所获得的 SAED 图像进行仔细地标定。非常有趣的是,我们发现所有的电子衍射斑点都可以归结为一种体心立方结构,并且通过对电子衍射斑点对应的晶面间距进行反算后得到所测量的这种碳纳米晶的晶格常数为 4.19Å。考虑到测量及计算上的误差,这一结果与理论预测的体心立方结构超密度碳即 C_8 的晶格常数 4.28 Å 比较吻合[165]。所以,SAED 表征证明了 LAL 合成的新碳相很有可能就是理论预测的体心立方结构碳。在我们的 SAED 标定中,实验误差要小于 4%。

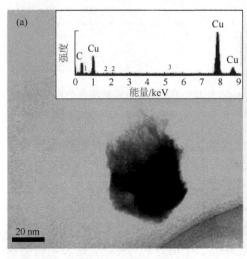
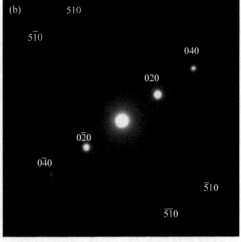

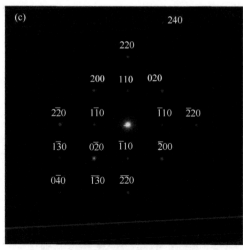
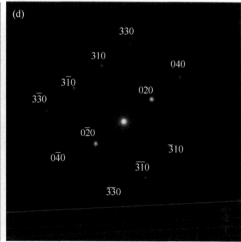

图 5-38 （a）类立方体状碳纳米晶的 TEM 图像，插图是相应的 EDS 分析；（b）纳米晶旋转到[020]晶轴上获得的 SAED 斑点；（c）保持[020]晶轴不动，旋转 1.87°后获得的 SAED 斑点；（d）保持[020]晶轴不动，旋转 5.15°后获得的 SAED 斑点

因此我们将获得的晶面间距实验值与体心立方相高密度碳的理论预测值进行比较，结果如表 5-1 所示。通过比较，我们可以清楚地看到，实验值与理论预测值吻合得非常好，这一结果进一步证明了 LAL 合成的新碳相纳米晶就是理论预测的亚稳相体心立方结构碳，即 C_8。此外，我们还对样品进行了高分辨 TEM 表征，如图 5-39 所示，结果表明合成纳米晶中的晶格条纹相面间距约为 0.215 nm，这一实验值也与理论预测的超密度碳的（020）晶面的面间距值（0.214 nm）吻合得很好。

表 5-1 实验获得的实验值与理论预测值对比表

(hkl)	D_{exp}[①]/Å	D_{calc}[②]/Å
020	2.11	2.14
040	1.06	1.07
110	2.97	3.03
130	1.33	1.35
200	2.11	2.14
220	1.48	1.51
310	1.32	1.35
330	0.99	1.01
510	0.83	0.84

①D_{exp} 表示实验值；②D_{calc} 表示理论预测值。

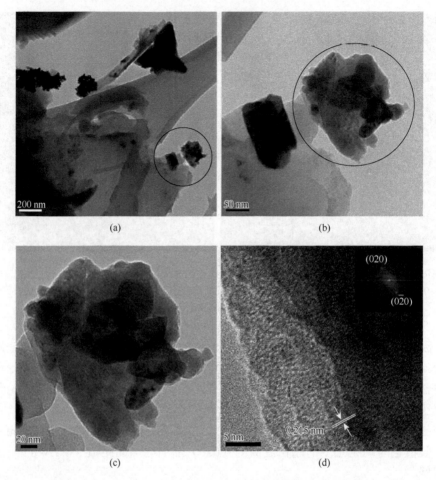

图 5-39 （a）～（c）LAL 合成新碳相纳米晶的 TEM 图像和（d）相应的 HRTEM 图像

为了能够在晶体结构理论上对实验结果有深刻的理解，我们运用约化晶胞理论，对实验获得的三组 SAED 斑点阵列进行了晶格重构，并得到了一个约化简立方晶胞模型，如图 5-40 所示。具体来说，图 5-40（a）中 ABCD 围成的方格表示一个垂直于实际[0k0]晶向的约化晶胞晶面，通过对其进行测量，我们得到它所对应的晶格常数是 2.1112 Å，这一结果与我们实验结果中计算得到的相应的晶格常数 2.1042 Å 吻合得很好。由于在晶格重构过程中我们使用的是约化晶胞理论，因此重构出来的晶格所具有的结构就是对应的实际晶体所属的布拉维晶格对应的晶系结构。而在我们的理论分析中，重构出来的约化晶胞具有立方结构。因此，这就在晶体结构理论上确认了实验所获得的碳纳米晶具有立方结构［图 5-40（b）］，这也与 SAED 和 HRTEM 分析所获得的结果完全一致。

新碳相纳米晶即 C_8 在 LAL 过程中的形成机制与 LAL 合成纳米金刚石类似，

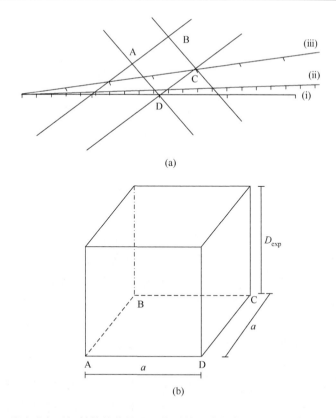

图 5-40 （a）通过对电子衍射的约化晶胞晶格重构后获得的 ABCD 方格表明约化晶胞的一个具有正方形结构的晶面；（b）基于（a）的晶胞晶格重构得到立方结构的约化晶胞

可以分为三个过程。①激光诱导的等离子体羽首先会从非晶碳靶表面喷射出来并被环境液体（水）束缚在非晶碳和水之间的界面处，等离子体羽中包含有各种碳结构基团，例如，各种 sp、sp^2、sp^3 杂化的 C—C 键团簇、具有很高活性的碳离子或团簇等。②由于激光诱导的等离子体羽被水环境强烈束缚，因此等离子体羽内部会因为激光导入能量的聚集而形成一个局域高温高压高密度的极端热力学状态微区，在这一状态下的各种碳成分会互相碰撞和反应[177, 178]。特别需要指出的是，高温高压高密度状态下碳的 sp^3 杂化键具有更高的稳定性[179]。因此，一些原本处于 sp、sp^2 杂化的碳会变得不稳定而向 sp^3 杂化转变，在这一转化过程中扭曲的 sp^3 C—C 键就可能以亚稳结构状态出现在等离子体羽中。另外，在 LAL 过程的化学反应中，等离子体羽与水界面处的水分子本身也会被离化而处于一种高活性状态而进入等离子体羽内部。因此，这些来自于水的活性基团会对等离子体羽中的 C—C 键再结晶过程产生影响，诱导其趋向一种亚稳相结晶态[180]。这样，亚稳相体心立方碳就可能在这一过程中成核。众所周知，高温高压高密度热力学状态是亚稳结构喜欢的状态之一，因此，在等离子体羽中生成的亚稳相体心立方碳反而会处于一种稳定状态。③随着环境液体束缚中的等离子体羽的迅速淬灭，原本在高温高压高密度状态中生成的亚稳纳米相不能很快地释放掉自身的内能，因此会被"冻结"而保留在最终产物中。

以上令人信服的实验证据表明，我们采用 LAL 方法首次合成了具有体心立方结构的超密度碳纳米晶（C_8）。我们的研究表明，LAL 技术为合成具有亚稳相的新型碳纳米材料开辟了一条新途径。

5.3.2　C_8-like 碳纳米方块

随着纳米技术的发展，各种碳纳米材料如碳纳米管、球状石墨纳米晶、纳米金刚石、石墨烯等，因其所具有的多样性结构及各种优异性能而被物理学、化学及材料科学等高度关注[181-183]，并且它们在信息、生物、能源、环境等领域的各个方面都展现出了广阔的应用前景[184,185]。在长期的研究中人们发现，碳纳米材料的结构、维度、形貌、尺寸等因素对它们的性能有着重要的影响，因此对碳纳米材料的可控制备成为了碳纳米科技发展的重要方向，也成为了探索碳纳米材料性能及应用研究的基础。

我们在 LAL 石墨与金刚石相变研究和 C_8 纳米晶合成中发现，环境液体对于 LAL 中的基本物理和化学过程有很大的影响。例如，在脉冲激光烧蚀石墨靶中如果使用乙醇作为环境液体，那么可以实现金刚石与碳葱的可逆相变；如果使用水作为环境液体则观察到纳米金刚石向纳米新金刚石的转变。所以，我们考虑通过对环境液体的设计来探索 LAL 在新碳相纳米材料合成中的应用。因此，在 LAL 合成 C_8 纳米晶的基础上，我们尝试在环境液体中加入少许无机盐来调控合成纳米颗粒的形貌。在新的 LAL 合成中，除了环境液体不同，其他合成条件都与合成 C_8 纳米晶时一样。这里的环境液体是由去离子水、乙醇和丙酮，与低浓度的无机盐溶液［KCl 和 NaCl 的盐溶液（纯度大于 99.5%）］按一定配比配制而成的混合液。这样，我们应用 LAL 方法成功合成出了一种新碳相，即亚稳相 C_8-like 碳纳米方块，如图 5-41 所示，并且发现这种碳纳米方块具有蓝光发射能力，有望成为一种新型的宽带半导体发光材料[157]。

在 SEM 图像中，我们可以看到，合成的碳纳米方块形貌相当规整，可以清楚地看到一个立方体的三维轮廓，对单个碳纳米方块进行 SEM 表征后可以看出它的 8 个角都是圆角。XRD 表征［图 5-41（c）］显示合成产物具有晶体结构，出现了 4 个尖锐的衍射峰，但是，样品的 XRD 衍射峰无法与现有粉末衍射标准联合委员会（JCPDS）卡片库中任何一种碳单质晶体的 XRD 衍射峰匹配。因此，合成的碳纳米方块是一种新碳相。

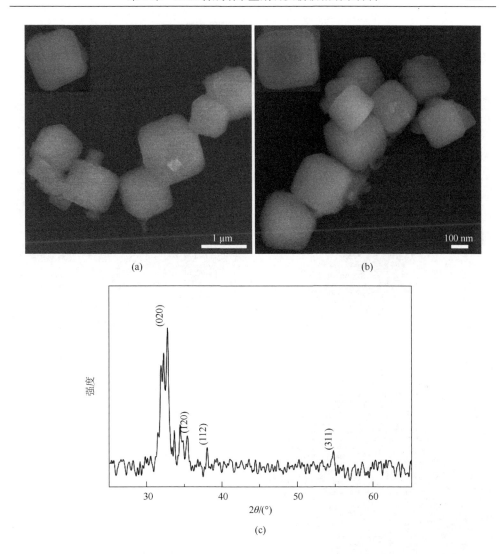

图 5-41 （a）和（b）LAL 合成的碳纳米方块不同放大倍数的 SEM 图像；（c）相应的 XRD 谱，分析表明合成产物为纯碳

我们采用 TEM 技术对合成的碳纳米方块的形貌和晶体结构进行了系统表征。图 5-42（a）和图 5-42（b）给出了低倍 TEM 形貌，我们可以清楚看到合成的碳纳米方块的大小较为均匀，尺寸大约在 200～500 nm。图 5-42（b）的插图给出了一个具有三维轮廓的碳纳米方块 TEM 形貌，可以更清楚地看到立方体形状。图 5-42（c）是单个碳纳米方块的高倍 TEM 形貌，可以看出它有着很好的正方对称性，表现出立方体特征。相应的 EDS 分析 [图 5-42（d），分析误差小于 2%] 表明纳米方块的主要成分是碳单质（92%），而 Cu 峰和少量的 O 峰应该是来

自 TEM 样品铜网，以及吸附在样品上的氧分子，痕量的 Si、Na、K 和 Cl 峰则是来自没有被完全去除而吸附在样品表面的无机盐杂质。

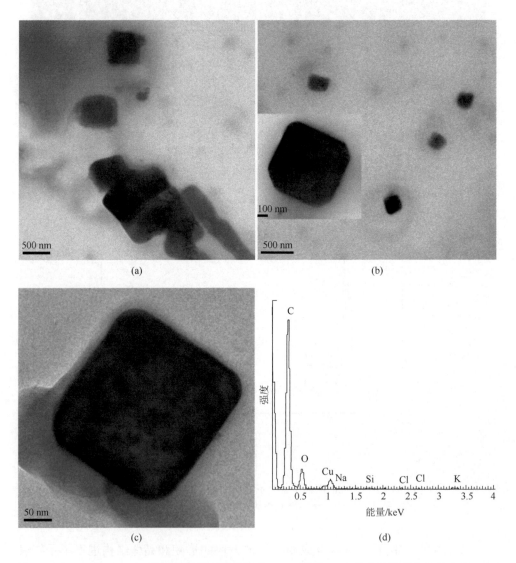

图 5-42 （a）和（b）LAL 合成碳纳米方块的低倍 TEM 图像；（c）单个碳纳米方块的高倍 TEM 图像；（d）单个碳纳米方块的 EDS 分析

我们对合成的碳纳米方块进行系统 SAED 表征来确定它的晶体结构。图 5-43（a）～图 5-43（d）是 4 张具有代表性的 SAED 图像，它们是通过将样品固定一个晶轴不变，然后转动 TEM 中的双倾台，使电子束从一个晶面转向另一个晶面，在这个过程中记录下来的 4 组 SAED 图像。这样,我们可以通过对其中的电子衍射斑点的变化推断出这种碳纳米方块的晶体结构。在对所获得的电子衍射图像进行仔细分析后,我们发现,所有 SAED 图像中计算得到的晶面间距 D 与现有数据库中所有碳单质晶体的晶面 D 数据都不相符。为了能够对这种未知晶体结构做出标定,我们利用所获得的 SAED 数据进行了晶格重构。采用约化晶胞理论计算方法[186, 187],我们计算出这种碳纳米方块具有立方结构,其晶胞中晶格常数为 5.462 Å。进一步,通过电子衍射图像分析,我们判断这种碳纳米方块的包裹面分别为两个(200)晶面、两个(020)晶面和两个(002)晶面。正因为有这样一种晶体结构,整个晶体看起来才会呈现出立方体形状,图 5-43(e)给出了一个相应的碳纳米方块的晶体示意模型。同时,我们对所获得的所有 SAED 图像中的斑点进行了标定[图 5-43(a)～图 5-43(d)],并且发现所有的 SAED 斑点标定均能自洽。所以,这一结果验证了我们通过约化晶胞理论重构出来的正交结构是正确的。最后,结合上面的电子衍射分析,我们通过布拉格衍射公式计算出了合成产物 XRD 谱中的 4 个特征峰分别属于立方结构碳纳米方块的(020)、(120)、(112)和(311)晶面的衍射峰[图 5-41(c)]。

为了进一步验证对这种新碳相纳米晶标定的正确性,我们对碳纳米方块进行了高分辨 TEM 分析,如图 5-43(f)给出了碳纳米方块的一个面的 HRTEM 晶格条纹相。我们测量得到的晶面间距分别为 0.275 nm 和 0.274 nm,这两个结果和我们通过对所构造的正交结构进行计算后得到的理论晶面间距即(020)晶面的晶面间距 2.731 Å 接近,而之前的 SAED 分析也表明了这种碳纳米方块的暴露面为{200}晶面族。此外,我们对图中晶格条纹相进行傅里叶变换后衍射点,结果表明这套衍射点确实分别可被标定为(200)和(020)晶面的衍射点,这也与我们的理论判断吻合。因此,通过样品的高分辨 TEM 分析我们直接观察到了与约化晶胞理论模型推断出来的晶格数据一致的晶格条纹相,这就进一步验证了我们给出的碳纳米方块的立方结构是正确的。

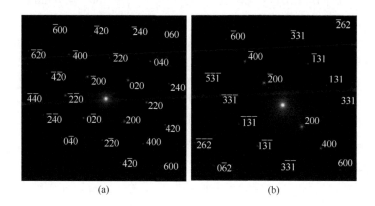

(a)　　　　　　　　　　　(b)

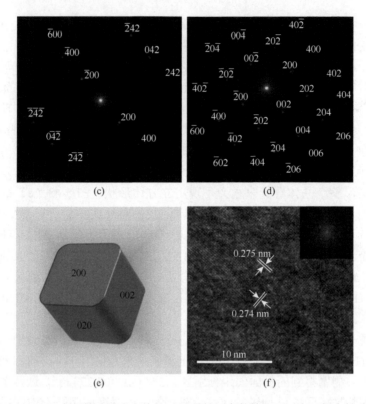

图 5-43 （a）～（d）在保持碳纳米方块的[200]晶轴不变的情况下，在双倾台转动过程中，记录下来的系列（4 组）SAED 图像；（e）碳纳米方块的晶体示意模型；（f）碳纳米方块中（002）晶面的 HRTEM 晶格条纹相

在确定了 LAL 合成的新碳相是当时还未被报道过的亚稳相立方结构之后，我们饶有兴趣地对这种碳纳米方块进行了初步的物理性能表征，包括阴极射线发光（CL）光谱、光致发光（photoluminescence，PL）谱和紫外可见吸收光谱等，如图 5-44 所示。我们可以清楚地从 CL 光谱中看到，在 360～450 nm 波长的范围内出现了一个清晰的蓝光发射宽峰，并且在这一宽峰之中可以分辨出 3 个小发光峰分别出现在 366 nm、393 nm 和 412 nm。因此，这一结果显示合成的碳纳米方块具有蓝光发射能力。在碳纳米方块的 PL 谱中，360～450 nm 范围内也出现了一个宽大的发光峰，并且在这一宽峰中可以分辨出有 3 个小发光峰分别出现在 367 nm、393 nm 和 432 nm 的位置。显然，PL 谱的结果与 CL 光谱的结果基本一致。因此，通过 CL 光谱和 PL 谱的比较，我们可以得出这样一个结论，LAL 合成的碳纳米方块是一种宽带隙蓝光发射的半导体。需要注意的是，在 PL 测试中我们使用的激发光是 230 nm 的紫外光，因此，在大约 462 nm 位置出现的小发光峰属于激发光的二次谐振发光峰，它与碳纳米方块的发光无关。此外，在碳纳米方块的紫外

可见吸收光谱中[图 5-44（d）]，我们可以看到 3 个宽吸收峰出现在 250～340 nm（A、B 和 C）。通过与前面的 PL 谱和 CL 光谱分析进行比较，我们认为 A 的吸收峰是碳纳米方块的本征吸收峰，因为从物理本质上来说，材料的光吸收特性是由材料本身的能带结构决定的。但是，在纳米尺度下，由于纳米尺寸效应及材料本身可能具有的偏振的影响，纳米材料的形貌可能对其光吸收能带产生调制作用[188]。因此，我们认为 B 和 C 的吸收峰是碳纳米方块的立方体形状效应诱导的分布极化电场作用的结果。同样，正是这样一种极化吸收导致了在 PL 谱中出现了在 393 nm 和 432 nm 这两个位置的小发光峰。

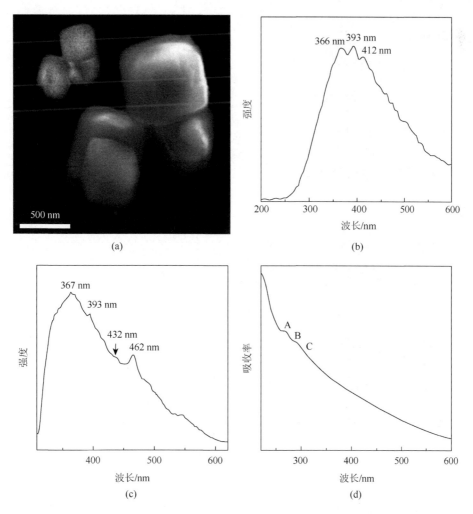

图 5-44　（a）单分散碳纳米方块的 SEM 形貌，左上角为其相应的 CL 发光相；（b）相应的 CL 光谱；（c）相应的 PL 谱；（d）相应的紫外可见吸收光谱

为揭示所制备的碳纳米方块独特的立方结构及蓝光发射背后的物理机制，我们运用第一性原理方法对碳纳米方块的晶体结构和能带结构进行计算与模拟。在对所构筑的立方晶格及其晶胞总能的收敛计算中，我们采用了全势线性缀加平面波方法（FP-LAPW），并对交换关联能采用了广义梯度近似的处理方法[189]，所有的计算都是在 WIEN2k 下完成的。晶胞结构的计算与模拟结果如图 5-45（a）和图 5-45（b）所示。我们看到，立方结构晶胞具有一种新型的体心立方结构，

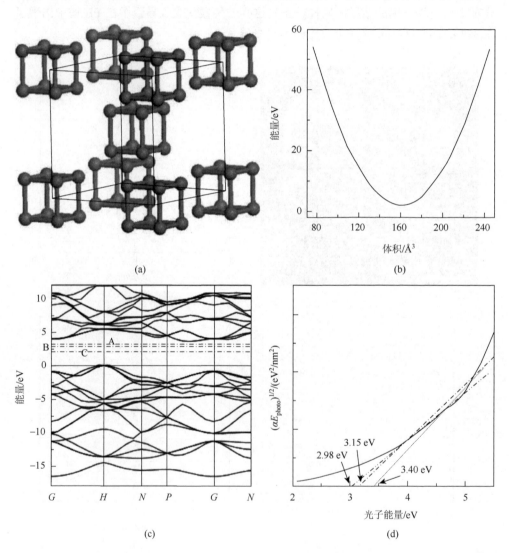

图 5-45 （a）C_8-like 结构的晶胞示意图；（b）C_8-like 结构中单个晶胞能量；（c）C_8-like 碳纳米方块的能带结构；（d）基于紫外可见吸收光谱计算得到的陶氏曲线，通过对其中三个不同的斜率截取分析可推导得到三个光学带隙

即 C_8-like 结构,同时,相应的晶胞总能收敛曲线清楚地表明,这种 C_8-like 结构在晶格常数大约为 5.432 Å 时达到总能最低,也就是结构最稳定状态或最可能存在状态。这一结果与我们通过 TEM 分析获得的碳纳米方块的晶格常数 5.462 Å 非常一致。所以,计算与模拟结果进一步确认了实验发现,LAL 合成的新碳相是一种 C_8-like 结构的亚稳相。

C_8-like 碳纳米方块的能带结构如图 5-45(c)所示。显然,我们能够发现,能带结构中的发光带隙与实验测量获得的发光行为是相符的。但是,理论上 C_8-like 碳纳米方块具有间接带隙特征。由此我们推断,实验中观测到的蓝光发射应该来自在间接带隙能带基础上产生的表面态,相应的三个表面态发光带隙(A、B 和 C)用虚线在图 5-45(c)中表示。此外,我们通过紫外可见吸收光谱计算了 C_8-like 碳纳米方块的吸收光子能量,计算中我们运用陶氏公式[190]

$$(\alpha h v)^n = B(h v - E_g) \tag{5-13}$$

计算结果如图 5-45(d)所示,三个光学带隙分别为 3.40 eV、3.15 eV 和 2.98 eV,与我们通过第一性原理计算得到的三个表面态发光带隙是一致的。

如上所述,LAL 合成 C_8-like 碳纳米方块的基本过程与 LAL 合成 C_8 纳米晶相似,只是在 C_8-like 碳纳米方块合成中的环境液体里加入了少许无机盐溶液。所以,在这里,我们主要讨论在 LAL 过程中激光诱导的等离子体羽中无机盐离子对方块形状的碳颗粒的成核及生长的影响。具体来说,我们实验中使用的无机盐属于立方晶系结构,它们在结晶的时候往往趋向于形成立方体形状。因此,这些在等离子体羽内部和外部的无机盐离子,对等离子体羽中的非晶碳成分的成核及生长过程产生了模板作用。在亚稳纳米相的成核过程中,由于晶面生长存在各向异性,因此等离子体羽内部的无机盐离子会选择晶体能量最低的晶面进行附着,也就是形成导向媒介的作用。这种优势吸附的导向作用会降低所附着晶面的表面能,从而阻碍了垂直于该晶面方向的晶体生长,并最终使所形成的碳纳米颗粒呈方块形貌[191]。这样,由无机盐立方模板效应导致的晶核生长使得原本处于无定形状态的碳团簇趋向于形成一个立方结构,最终形成的产物被引导生成立方结构的碳纳米晶。

综上所述,我们应用 LAL 方法首次合成了一种新碳相纳米材料,即 C_8-like 碳纳米方块,它是一种宽带隙发蓝光的半导体,在纳米发光器件中有着巨大的潜在应用价值。

5.3.3 白碳纳米晶

如前所述,第三种碳即由一维 sp 杂化结构碳原子构成的卡拜(carbyne)的研究由来已久。20 世纪 60 年代,El Goresy 和 Donnay[192]率先在 Ries 陨石坑中发现了一种新的碳的同素异形体。后来，在高温低压的环境下，Whittaker[163]在石墨升华过程中合成了"白色"的碳，并发现合成样品的晶面间距与 Ries 陨石坑中的碳结构相同。之后，Kasatochkin 等[193]的研究和 Nakamizo 等[194]的拉曼光谱测量表明这些相似的碳含有碳-碳叁键，证明其很有可能是以链状的构型存在。Carbyne 结构单元类似于聚炔烃（polyyne），因此得名[195]。由于 carbyne 具有非常独特的结构，很多理论学家对 carbyne 的性质和可能的应用进行了大量的预测。Kudryavtsev 等[196, 197]认为 carbyne 是具有 1.5~2.2 eV 带隙的半导体，完全不同于其他两种常见的碳的同素异形体，如金刚石（绝缘体）和石墨（导体）。另外，由于 carbyne 的带隙为 1.5~2.2 eV，功函数为 4.9 eV，这些性质使其作为一种半导体材料非常适合应用于 LED、太阳电池和光采集天线等光电子学领域[197]。不仅如此，根据理论计算，carbyne 的硬度是已知最硬材料的两倍[198]。所以，这些对 carbyne 优异性能的预测激起了人们对其研究的极大热情。

虽然近几十年来不断有科学家声称成功合成了 carbyne 样品，但是，直到现在，carbyne 的存在与否仍然存在很大争议。例如，有机化学合成学家认为 carbyne 很难稳定存在，因为相邻碳链之间的碳原子很容易发生反应，如交联反应，这些反应会破坏 carbyne 内部的一维线性结构[199]。存在争议的主要原因还有是否能够合成 carbyne 晶体[200]，那些宣称已经合成的 carbyne 样品，实际上都是"类卡拜或聚炔烃"材料，通常定义为富碳链状聚合物或聚炔烃，而不是真正的 carbyne[201-203]。真正的 carbyne 是一种像石墨或金刚石一样的碳的同素异形体，它应该是一种无机碳材料。所以，科学家认为还没有任何令人信服的实验证据表明人们能够在实验室合成 carbyne[204, 203]。实际上，carbyne 在实验室很难合成的主要原因是它是碳的一种亚稳相，它的热力学稳定区处于高温和高压状态[163]，并且其独特的线型碳-碳单键和碳-碳叁键分子链的形成对所采用的碳前驱物有特殊要求。

2015 年，我们应用 LAL 方法，通过选择合适的环境液体和固体靶，利用贵金属高温脱氢反应和 LAL 创造的高温高压高密度极端热力学环境，在液相中实现了 carbyne 分子的成核与生长，合成出了 carbyne 晶体，一种白色粉末晶体（图 5-46）[204]，这是国际上第一次合成 carbyne 及其凝聚相 carbyne 晶体。重要的是，我们首次给出了 carbyne 晶体的包括完整的拉曼光谱在内的各种特征谱和完美的晶体结构数据。这些令人信服的实验事实充分表明除了金刚石和石墨以外，第三种碳，这个曾经的"碳传说"，可以在实验室合成！显然，这一成果堪称碳材料研究的重大突破。根据它的研究历史和物质形态，我们称第三种碳为"白碳"。自此，自然界的三种碳分别是三维 sp^3 杂化结构碳原子构成的金刚石、二维 sp^2 杂化结构碳原子构成的石墨和一维 sp 杂化结构碳原子构成的白碳。

图 5-46 carbyne 纳米晶胶体（左）和粉末（右）照片（后附彩图）

我们在环境液体乙醇中使用纳秒激光烧蚀固体金靶合成出了 carbyne 纳米晶，如图 5-46 所示。并且，我们应用拉曼光谱、红外吸收光谱、紫外可见吸收光谱、荧光光谱等对合成的 carbyne 纳米晶胶体进行了系统表征。

众所周知，拉曼光谱是用于研究碳纳米材料结构的重要手段。理论研究发现，在 2100 cm^{-1} 附近，存在由碳-碳叁键产生的拉曼带[195, 206]，而在 1060 cm^{-1} 附近则存在碳-碳单键的拉曼信号[207]。非常让人兴奋的是，在合成的胶体溶液中，我们观察到在 1050 cm^{-1} 和 2175 cm^{-1} 处的两个尖锐的特征峰，这两个峰分别被标定为碳-碳单键和碳-碳叁键的特征峰［图 5-47（a）］，而且这两个特征峰的半高宽较小，表明了合成产物具有高纯度和高结晶度。重要的是，这是国际上首次给出 carbyne 完整的拉曼光谱。同时，carbyne 的碳-碳叁键在傅里叶变换红外光谱仪（FTIR）中也能观察到，即 FTIR 中的在 2157 cm^{-1} 处的吸收峰［图 5-47（b）］来自碳-碳叁键[195]。因此，这些光谱表征证明了合成产物是由交替的单键和叁键组成的一维 sp 杂化的 carbyne。同时，合成胶体溶液紫外可见吸收光谱，如图 5-48 所示。

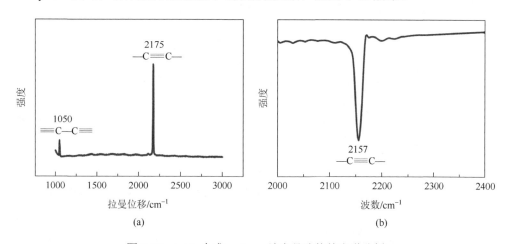

图 5-47 LAL 合成 carbyne 纳米晶胶体的光谱分析

（a）拉曼光谱，在 1050 cm^{-1} 和 2175 cm^{-1} 处的峰分别来自碳-碳单键和碳-碳叁键；（b）FTIR，在 2157 cm^{-1} 处的峰归因于碳-碳叁键的拉伸振动

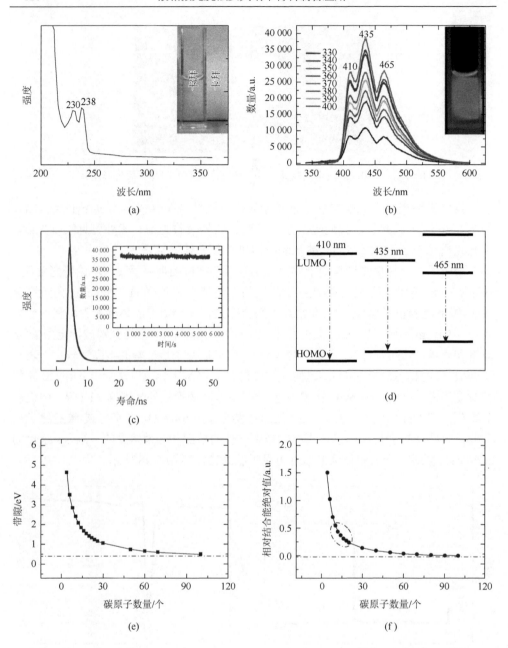

图 5-48 LAL 合成 carbyne 纳米晶胶体的紫外可见吸收光谱和荧光光谱（后附彩图）

（a）紫外可见吸收光谱，插图为无色透明溶液的光学照片；（b）荧光光谱，插图是用 370 nm 的光激发的紫蓝色荧光；（c）发光寿命测量为 1.3 ns，插图表示使用 450 W 氙灯辐照 1.5 h 未观察到光漂白现象；（d）不同长度碳链的三种荧光行为；（e）带隙对碳原子数量的依赖；（f）作为比较，垂直坐标表示相对结合能的绝对值，负结合能对应于稳定的构型，绝对值越大，碳链的稳定性就越高，结合能的绝对值随着碳原子数量的增加而降低，但降低速度逐渐变慢，最终达到特定值

我们看到，位于 230 nm 和 238 nm 处的吸收峰来自于 $\pi \to \pi^*$ 的跃迁吸收[208]，溶液颜色是无色透明的［图 5-48（a）的插图］。需要注意的是，这两个吸收峰类似于氢饱和的 8 个原子碳链的结果[209]。当我们使用 330~400 nm 波长的激发光照射合成溶液时，我们观察到强烈的紫蓝色荧光［图 5-48（b）］。显然，410 nm、435 nm 和 465 nm 处的荧光峰并没有随着激发光波长的变化而变化，这表明这些发光峰应该是 carbyne 的本征发射。有趣的是，紫蓝色荧光可以用肉眼观察到［图 5-48（b）插图］，这是国际上首次观察到 carbyne 的荧光发射。

我们知道，LAL 合成的纳米结构常常有一定的尺度范围。因此，这里合成的 carbyne 中的碳链应该有一定的链长。根据我们的理论计算［图 5-48（e）］，carbyne 中碳链的最高占据分子轨道（HOMO）和最低未占据分子轨道（LUMO）的带隙是随着链长变化而变化的，并且存在明显的有限间隙趋势，估计约为 0.48 eV。所以，carbyne 晶体中碳链的带隙由相应的链长决定。因此，我们看到的 410 nm、435 nm 和 465 nm 处的三个荧光峰应该是来自具有 8~12 个碳原子的碳链［图 5-48（d）］。为了找出为什么这些特殊长度的碳链在样品中占主导地位，我们计算了长度依赖的碳链稳定性。图 5-48（f）说明了碳链的相对结合能绝对值随着碳原子数量的增加而降低，然而，递减的速度却逐渐变慢，并且最终达到一个稳定值。换句话说，随着获得更多的能量，碳链倾向于增长得更长。因此，LAL 创造的局域高温高压高密度极端热力学环境会驱动碳链生长得更长。如果我们仔细观察的话，可以发现，在碳链为 10 个原子附近出现了一个特殊区域［图 5-48（f）虚线圈］，这似乎是一个分界区域，它表明碳链的相对结合能绝对值随着少于 10 个原子而显著变化，但是，随着多于 10 个原子则缓慢变化。所以，这种单调下降的趋势表明，碳链越长，它的相对结合能绝对值就越小，它就越不稳定。也就是说，在 LAL 过程中碳链趋向于生长得更长，但是，从相对结合能绝对值来看，长碳链似乎是不稳定的。因此，它们之间必须保持一个平衡。在我们的实验中，LAL 可以提供约 10 个原子的碳链所需要的有限热力学驱动力，但由于能垒的原因，它无法提供相对结合能绝对值更小的长碳链生长的热力学驱动力。

随后，我们通过反复将合成的 carbyne 溶液滴到硅和载玻片衬底上来生产 carbyne 晶体，并在衬底表面获得白色粉末（图 5-46）。XRD 谱［图 5-49（a）］显示出干净、尖锐、强烈的衍射峰，表明合成晶体具有良好的结晶度。这些衍射峰被标记为 carbyne 晶体六角结构。显然，这些白色粉末即 carbyne 晶体是多晶的，并且沿着 c 轴有一个明显的优选方向。图 5-49（b）的 SEM 图像显示 carbyne 晶体由堆叠的薄片组成。一般来说，片状晶体易于沿 c 轴生长，这一结果与 XRD 谱一致。相应的 EDS［图 5-49（b）的插图］显示样品含有碳、氧和金。显然，氧源于样品表面吸附。图 5-49（c）为相应的 TEM 图像，显示了 10~30 nm 宽和 50~100 nm 长的 carbyne 纳米棒。因此，我们可以在 TEM 和 SEM 图像中分别观察到的不同结晶形态，这主要归因于来自 carbyne 胶体溶液的不同类型的结晶。仔细观察可以看到，carbyne 纳米棒表面的微小球形颗粒是金，这也就是为什么在 SEM 和 TEM 的 EDS 中检测到 Au 信号［图 5-49（b）的插图和图 5-49（j）］。此外，TEM 相应的 EDS 显示，样品几乎完全由碳组成（95.62%），氧（2.43%）和铜（0.76%）的峰则分别来自吸附的氧分子和样品铜格栅。我们采用 TEM 技术研究了 30 多个 carbyne 晶体，并根据入射电子束的方向将相应 carbyne 晶体的 SAED 图像和 HRTEM 图像分为三类（[1-20]、[012]和[001]）。如图 5-49（d）~图 5-49（i），其中两类是矩形点阵，一类是六角点阵。结合 XRD 的结果，我们搭建出 carbyne 晶体六方结构，晶胞参数为 $a = b = 5.78$ Å、$c = 9.92$ Å、$\gamma = 120°$，如图 5-50（a）所示。

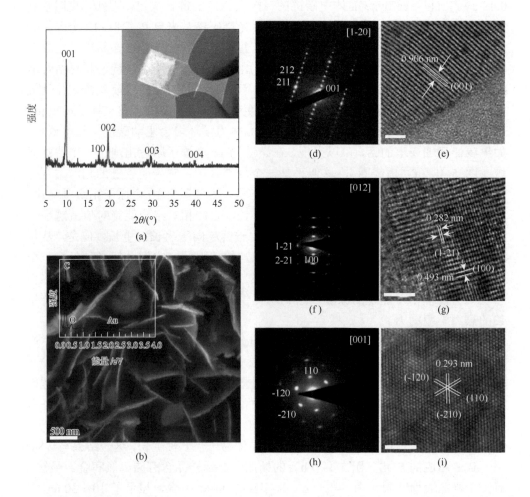

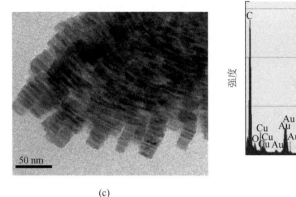
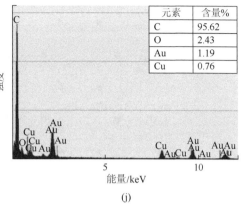

图 5-49　carbyne 晶体的形态和结构

(a) XRD 谱，峰的形状和强度表明结晶度好并沿 c 轴有明显的优先方向，插图显示了玻璃基材上的白色粉末；(b) SEM 图像显示由薄片堆叠在一起的晶体，相应 EDS 显示样品中含有 C、O 和 Au；(c) TEM 图像；(d) ～ (i) carbyne 晶体的 SAED 图像和对应的 HRTEM 图像；(j) 纳米晶的元素构成

为了确定 carbyne 的晶体结构，我们考虑：首先，拉曼光谱和红外吸收光谱表明，合成晶体只含有 sp 杂化，因此，它是由一维碳链组成的；其次，XRD 和 SAED 图谱证实了合成晶体的六方结构，因此，它们可以用于直接确定 x-y 平面中的晶格参数 a 和 b；最后，从 XRD 谱中我们可以得到晶格参数 c 的长度（即 9.10 Å）。如果碳链是直的（没有扭结），则 c 的长度应该是碳-碳单键和碳-碳叁键的长度之和（约 2.6 Å）。然而，这与我们的实验结果不符。为了解决这个难题，我们使用了扭结链构象[210]。这样，我们就构建了图 5-50（a）所示的晶体结构。c 轴的长度是碳链中碳原子数 n 和扭结角（垂直方向和扭结键的角度）的函数。如果我们在碳链中使用 8 个碳原子，扭转角为 30°，则晶格参数 a 和 b 将与实验值相同（5.78 Å），晶格参数 c 的长度为 9.92 Å，这是 4 个碳-碳叁键和 2 个碳-碳单键的长度加上两个扭结键的总和。该长度略长于实验值（9.10 Å），因为碳-碳单键（1.33 Å）和碳-碳叁键（1.24 Å）的键长是在结构优化后从碳链中获得的，所以，碳链的扭折会导致链长的变化[211]。同时，我们发现相邻碳原子的距离为 3.34 Å，接近石墨的层间距。因此，连接碳链的力可以被认为是范德瓦耳斯力。这些特性完全符合 Heimann 等[210]对线性碳多型体的描述，其中碳链平行于 c 轴排列，并且通过分子间相互作用连接。综上所述，我们构建的晶体结构是合理的。为了确定所提出的晶体结构是否稳定，我们使用 SIESTA 代码[212]计算了晶体结构的稳定性，该代码基于标准 Kohn-Sham 自洽密度泛函理论。我们可以看到，结构优化后获得的平衡构型表明所提出的六方结构是稳定的［图 5-50（b）］，扭曲的碳链在晶体结构中意外地变形为螺旋状。因此，carbyne 晶体类似于平行的螺旋阵列，而这种螺旋状形状类似于理论结果中确定的弯曲碳链[211]。

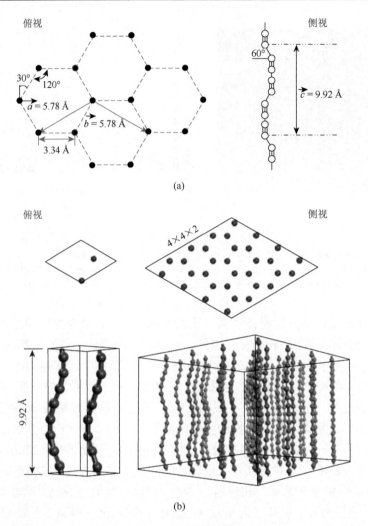

图 5-50 carbyne 晶体结构

（a）构造的六方结构（a = 5.78 Å，b = 5.78 Å，c = 9.92 Å，γ = 120°），相邻碳原子间距为 3.34 Å，扭结由两个碳原子共轭，形成 C—C，扭结角（垂直方向和扭结键的角度）为 30°；(b) 基于第一性原理计算构建的 carbyne 晶体的平衡构型

实际上，碳链可以很容易地用一个小弧线弯曲得到，最细的碳链（sp 碳链，其横截面中只包含一个原子）应该比碳纳米管更容易弯曲。因此，这些结果表明，具有弯曲结构的 sp 碳链是 carbyne 晶体的一个重要特征。优化结构中的碳-碳单键和碳-碳叁键的长度分别变为 1.27 Å 和 1.30 Å，这也验证了我们的理论推测，即碳链的弯曲会导致键长的变化。我们比较了 carbyne 晶体与金刚石和石墨的晶体参数和稳定性（表 5-2）。显然，在这些碳的同素异形体中，carbyne 晶体具有最高的结合能，这意味着它是一种亚稳态碳相。

表 5-2　carbyne、石墨、金刚石晶体参数和稳定性对比

结构	晶系	杂化	晶格参数	d_{CC}/Å	结合能/(eV/atom)
carbyne	六方晶系	sp	$a=b=5.78$Å，$c=9.92$Å $\alpha=\beta=90°$，$\gamma=120°$	1.30，1.27	−6.347
石墨	六方晶系	sp^2	$a=b=2.46$Å，$c=6.80$Å $\alpha=\beta=90°$，$\gamma=120°$	1.42	−7.844
金刚石	立方晶系	sp^3	$a=b=c=3.56$Å $\alpha=\beta=\gamma=90°$	1.54	−7.730

我们提出了 LAL 合成白碳纳米晶的动力学和热力学过程。在关于 LAL 中化学反应的论述中我们指出，脉冲激光诱导的等离子体羽包含大量的分别来自环境液体和固体靶的原子、分子、离子和活性基团等。当环境液体的束缚作用将等离子体羽驱动到一个高温高压高密度的极端热力学状态时，一系列的高温化学反应会在等离子体羽中发生。而在 LAL 合成白碳纳米晶中，我们认为乙醇分子的脱氢是一个很关键的动力学过程，如图 5-51 所示。当等离子体羽处于高温高压高密度极端热力学状态时，乙醇分子羟基基团上的氢首先跟来自金靶的活泼的 Au 离子反应，生成 Au—H 中间体。然后，在 Au 催化剂的作用下，乙醇分子上的一对氢原子（α—H 和 β—H）会被剥去，这样，乙醇分子发生第一次脱氢。在这个过程中，初始乙醇分子中的 C—C 会转变成 C═C。紧接着，在 Au 催化剂的作用下，

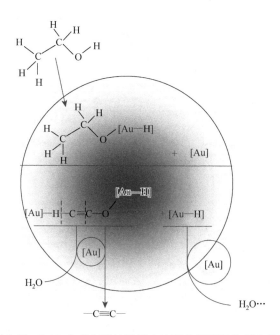

图 5-51　LAL 合成白碳纳米晶中乙醇分子脱氢的反应机理

乙醇分子的另外一对氢（α—H 和 β—H）也会脱除，产生 C≡C。而在这个时候，生成的 C≡C 还连接着 Au—H 中间体和氧原子，但是，这些 Au—H 中间体可以与亲核物质反应，比如—OH 和 O 等，把氢从 Au 表面除去并产生水。最后，单独的 C≡C 就会产生，它们是形成一维碳链的基本结构单元。

事实上，LAL 最终产物中的这种"C≡C"构型类似于双原子碳（C_2分子）。研究表明，C_2 分子的化学键已被证明是一种叁键结构[213]，这表明 C_2 分子很可能就是 LAL 合成白碳纳米晶中"C≡C"构型的物理实体。因此，我们推断 C_2 分子在碳链形成过程中起到了重要作用。为此，我们应用原位激光发射光谱系统来探测 LAL 合成白碳纳米晶过程中是否确实产生了 C_2 分子，结果如图 5-52 所示。我们看到，发射光谱中乙醇在 558.1 nm 和 563.3 nm 处有两个峰，它们起源于 C_2 谱带电子在 $d^3\Pi_g$ 态和 $d\,a^3\Pi_u$ 态之间的跃迁[66]。显然，这个重要结果为 C_2 分子的形成提供了直接的实验证据。此外，我们也测量了环境液体为甲醇、丙酮、正丙醇和异丙醇的激光发射光谱，但是并不能探测到 C_2 分子的信号。

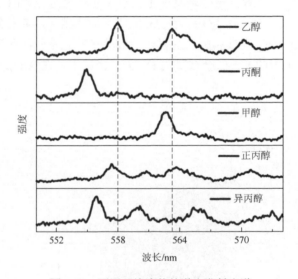

图 5-52　不同环境液体的激光发射光谱

上述实验结果表明，乙醇分子在 LAL 合成 carbyne 中的成核和生长过程起着重要作用。就分子结构而言，乙醇有两个碳原子和一个羟基，这种具有两个碳原子的构型恰好与双原子碳（C_2 分子）的构型相匹配。这样，我们就很容易理解为什么甲醇、丙酮和异丙醇作为 LAL 的环境液体不能形成 carbyne。此外，乙醇分子中的羟基也很重要，因为整个 LAL 过程的第一步是 Au 离子与乙醇分子中的羟基反应生成 Au—H 中间体。因此，我们认为，LAL 合成白碳纳米晶的环境液体至少由两个成对的碳原子和一个羟基组成。所以，在这种考虑下，乙醇似乎是 LAL 合成 carbyne 环境液体的最佳选择。实际上，不仅环境液体很重要，靶材也是整个 LAL 合成过程中的关键因素，因为它在上述一系列化学反应中起着催化剂的重要作用。例如，我们选择铂作为靶材来代替金，尽管铂具有与金相似的催化性能，但是在合成产物中没有发现 carbyne。所以，在我们的合成中，金优于铂是因为金表面被吸附的氢覆盖[214]，这导致碳-氢键的断裂更容易进行。

如上所述，环境液体的束缚效应会将脉冲激光诱导的等离子体羽推向一个高温高压高密度的极端热力学状态。在 LAL 合成白碳纳米晶中，等离子体羽的温度为 2500～5000 K[215]，压强为 2～5 GPa[205]。

Whittaker[163]发表的 carbyne 热力学平衡相图表明，carbyne 的热力学稳定区为 2600～3800 K 和 4～6 GPa（图 5-53）。我们可以清楚地看到，LAL 提供的热力学环境非常适合 carbyne 的成核和生长，是它的稳定相区。因此，上述 LAL 系列化学反应的产物中 C≡C 结构单元可以在 LAL 创造的热力学环境中容易地成核和生长。大量的碳链就在等离子体羽淬灭过程中形成。这里需要说明的一点是，在 LAL 合成白碳纳米晶中存在一个"激光能量阈值"。这意味着在其他条件都相同时，只有激光能量超过阈值才能合成出 carbyne。在我们的合成中，这个能量阈值大约是每脉冲 200 mJ。我们知道，LAL 产生的等离子体羽的温度和压强强烈依赖激光能量[29]，如果激光能量降低到阈值以下，那么产生的等离子体羽的温度和压强将相应下降，达不到合成 carbyne 的热力学条件。

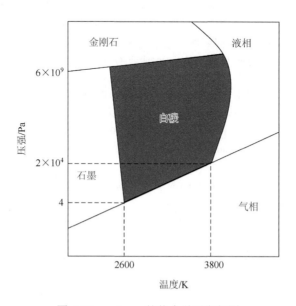

图 5-53　carbyne 的热力学平衡相图

我们已经获得了充分的实验证据证实,在实验室中,使用 LAL 方法在环境条件下合成了有限碳链长度的第三种碳及凝聚相呈白色粉末状的六方白碳晶体,并且展示了来自白碳本征发射的强烈紫蓝色荧光。这些研究再一次证明了 LAL 方法在探索新碳纳米材料中的巨大潜力。

参 考 文 献

[1] Li H D, Yang H B, Yu S, et al. Synthesis of ultrafine gallium nitride powder by the direct current arc plasma method[J]. Applied Physics Letters, 1996, 69 (9): 1285-1287.

[2] Xie Y, Qian Y T, Zhang S Y, et al. Coexistence of wurtzite GaN with zinc blende and rocksalt studied by X-ray power diffraction and high-resolution transmission electron microscopy[J]. Applied Physics Letters, 1996, 69(3): 334-336.

[3] Lee S T, Peng H Y, Zhou X T, et al. A nucleation site and mechanism leading to epitaxial growth of diamond films[J]. Science, 2000, 287 (5450): 104-106.

[4] Hao X P, Cui D L, Shi G X, et al. Synthesis of cubic boron nitride at low-temperature and low-pressure conditions[J]. Chemistry of Materials, 2001, 13 (8): 2457-2459.

[5] Jiao S, Sumant A, Kirk M A, et al. Microstructure of ultrananocrystalline diamond films grown by microwave Ar-CH_4 plasma chemical vapor deposition with or without added H_2[J]. Journal of Applied Physics, 2001, 90 (1): 118-122.

[6] Krauss A R, Auciello O, Ding M Q, et al. Electron field emission for ultrananocrystalline diamond films[J]. Journal of Applied Physics, 2001, 89 (5): 2958-2967.

[7] Komatsu S, Kurashima K, Shimizu Y, et al. Condensation of sp^3-bonded boron nitride through a highly nonequilibrium fluid state[J]. The Journal of Physical Chemistry B, 2004, 108 (1): 205-211.

[8] Ogale S B, Malshe A P, Kanetkar S M, et al. Formation of diamond particulates by pulsed ruby laser irradiation of graphite immersed in benzene[J]. Solid State Communications, 1992, 84 (4): 371-373.

[9] Graebner J E, Jin S, Kammlott G W, et al. Large anisotropic thermal conductivity in synthetic diamond films[J]. Nature, 1992, 359: 401-403.

[10] Chrisey D B, Hubler G K. Pulsed laser deposition of thin solid films[M]. New York: Wiley-Interscience, 1994.

[11] Holland L, Ojha S M. The growth of carbon films with random atomic structure from ion impact damage in a hydrocarbon plasma[J]. Thin Solid Films, 1979, 58 (1): 107-116.

[12] Vora H, Moravec T J. Structural investigation of thin films of diamondlike carbon[J]. Journal of Applied Physics, 1981, 52 (10): 6151-6157.

[13] Sawabe A, Inuzuka T. Growth of diamond thin films by electron assisted chemical vapor deposition[J]. Applied Physics Letters, 1985, 46 (2): 146-147.

[14] Kurihara K, Sasaki K, Kawarada M, et al. High rate synthesis of diamond by dc plasma jet chemical vapor deposition[J]. Applied Physics Letters, 1988, 52 (6): 437-438.

[15] Weissmantel C, Bewilogua K, Dietrich D, et al. Structure and properties of quasi-amorphous films prepared by ion beam techniques[J]. Thin Solid Films, 1980, 72 (1): 19-32.

[16] Miyasato T, Kawakami Y, Kawano T, et al. Preparation of sp^3-rich amorphous carbon film by hydrogen gas reactive RF-sputtering of graphite, and its properties[J]. Japanese Journal of Applied Physics, 1984, 23 (4A): L234.

[17] Kitabatake M, Wasa K. Growth of diamond at room temperature by an ion-beam sputter deposition under hydrogen-ion bombardment[J]. Journal of Applied Physics, 1985, 58 (4): 1693-1695.

[18] Dijkkamp D, Venkatesan T, Wu X D, et al. Preparation of Y-Ba-Cu oxide superconductor thin films using pulsed laser evaporation from high T_c bulk material[J]. Applied Physics Letters, 1987, 51 (8): 619-621.

[19] Horwitz J S, Grabowski K S, Chrisey D B, et al. In situ deposition of epitaxial $PbZr_xTi_{(1-x)}O_3$ thin films by pulsed laser deposition[J]. Applied Physics Letters, 1991, 59 (13): 1565-1567.

[20] Delgado J C, Andreu J, Sardin G, et al. Hydrogenated amorphous silicon films obtained by a low pressure dc glow discharge[J]. Applied Physics A, 1988, 46: 207-213.

[21] Sato T, Furuno S, Iguchi S, et al. Diamond-like carbon films prepared by pulsed-laser evaporation[J]. Applied Physics A, 1988, 45: 355-360.

[22] Collins C B, Davanloo F, Lee T J, et al. Noncrystalline films with the chemistry, bonding, and properties of diamond[J]. Journal of Vacuum Science & Technology B, 1993, 11 (5): 1936-1941.

[23] Bourdon E B D, Kovarik P, Prince R H. Microcrystalline diamond phase by laser ablation of graphite[J]. Diamond and Related Materials, 1993, 2 (2-4): 425-431.

[24] Seth J, Padiyath R, Rasmussen D H, et al. Laser-plasma deposition of diamond phase at low temperatures[J]. Applied Physics Letters, 1993, 63 (4): 473-475.

[25] Polo M C, Cifre J, Sanchez G, et al. Pulsed laser deposition of diamond from graphite targets[J]. Applied Physics Letters, 1995, 67 (4): 485-487.

[26] Yang G W, Wang J B, Liu Q X. Preparation of nano-crystalline diamonds using pulsed laser induced reactive quenching[J]. Journal of Physics: Condensed Matter, 1998, 10 (35): 7923.

[27] Wang J B, Zhang C Y, Zhong X L, et al. Cubic and hexagonal structures of diamond nanocrystals formed upon pulsed laser induced liquid-solid interfacial reaction[J]. Chemical Physics Letters, 2002, 361 (1-2): 86-90.

[28] Pearce S R J, Henley S J, Claeyssens F, et al. Production of nanocrystalline diamond by laser ablation at the solid/liquid interface[J]. Diamond and Related Materials, 2004, 13 (4-8): 661-665.

[29] Yang G W. Laser ablation in liquids: Applications in the synthesis of nanocrystals[J]. Progress in Materials Science, 2007, 52 (4): 648-698.

[30] Yoshikawa M, Nagai N, Matsuki M, et al. Raman scattering from sp^2 carbon clusters[J]. Physical Review B, 1992, 46 (11): 7169.

[31] Tamor M A, Vassell W C. Raman "fingerprinting" of amorphous carbon films[J]. Journal of Applied Physics, 1994, 76 (6): 3823-3830.

[32] Grimsditch M H, Ramdas A K. Brillouin scattering in diamond[J]. Physical Review B, 1975, 11 (8): 3139.

[33] Dillon R O, Woollam J A, Katkanant V. Use of Raman scattering to investigate disorder and crystallite formation in as-deposited and annealed carbon films[J]. Physical Review B, 1984, 29 (6): 3482.

[34] Nemanich R J, Lucovsky G, Solin S A. Infrared active optical vibrations of graphite[J]. Solid State Communications, 1977, 23 (2): 117-120.

[35] Meenakshi V, Sayeed A, Subramanyam S V. Conductivity and structural studies on disordered amorphous conducting carbon films[J]. Materials Science Forum, 1996, 223-224: 307-310.

[36] Yoshikawa M, Mori Y, Maegawa M, et al. Raman scattering from diamond particles[J]. Applied Physics Letters, 1993, 62 (24): 3114-3116.

[37] Scandolo S, Bernasconi M, Chiarotti G L, et al. Pressure-induced transformation path of graphite to diamond[J]. Physical Review Letters, 1995, 74 (20): 4015.

[38] Nemanich R J, Glass J T, Lucovsky G, et al. Raman scattering characterization of carbon bonding in diamond and diamondlike thin films[J]. Journal of Vacuum Science & Technology A, 1988, 6 (3): 1783-1787.

[39] Maruyama K, Makino M, Kikukawa N, et al. Synthesis of hexagonal diamond in a hydrogen plasma jet[J]. Journal of Materials Science Letters, 1992, 11: 116-118.

[40] Prawer S, Nugent K W, Jamieson D N. The Raman spectrum of amorphous diamond[J]. Diamond and Related Materials, 1998, 7 (1): 106-110.

[41] Beeman D, Silverman J, Lynds R, et al. Modeling studies of amorphous carbon[J]. Physical Review B, 1984, 30 (2): 870.

[42] Mao H K, Hemley R J. Optical transitions in diamond at ultrahigh pressures[J]. Nature, 1991, 351: 721-724.

[43] Yang G W, Wang J B. Carbon nitride nanocrystals having cubic structure using pulsed laser induced liquid-solid interfacial reaction[J]. Applied Physics A, 2000, 71: 343-344.

[44] Zhang C Y, Wang C X, Yang Y H, et al. A nanoscaled thermodynamic approach in nucleation of CVD diamond on nondiamond surfaces[J]. The Journal of Physical Chemistry B, 2004, 108 (8): 2589-2593.

[45] Simon A, Kántor Z. Micro-RBS characterisation of the chemical composition and particulate deposition on pulsed laser deposited $Si_{1-x}Ge_x$ thin films[J]. Nuclear Instruments and Methods in Physics Research Section B: Beam Interactions with Materials and Atoms, 2002, 190 (1-4): 351-356.

[46] Yang G W, Wang J B. Pulsed-laser-induced transformation path of graphite to diamond via an intermediate rhombohedral graphite[J]. Applied Physics A, 2001, 72: 475-479.

[47] Greiner N R, Phillips D S, Johnson J D, et al. Diamonds in detonation soot[J]. Nature, 1988, 333: 440-442.

[48] Banhart F, Ajayan P M. Carbon onions as nanoscopic pressure cells for diamond formation[J]. Nature, 1996, 382: 433-435.

[49] Mochalin V N, Shenderova O, Ho D, et al. The properties and applications of nanodiamonds[J]. Nature Nanotechnology, 2012, 7: 11-23.

[50] Yang G W. Laser ablation in liquids: Principles and applications in the preparation of nanomaterials[M]. New York: CRC Press, 2012.

[51] Yang W S, Auciello O, Butler J E, et al. DNA-modified nanocrystalline diamond thin-films as stable, biologically active substrates[J]. Nature Materials, 2002, 1: 253-257.

[52] Härtl A, Schmich E, Garrido J A, et al. Protein-modified nanocrystalline diamond thin films for biosensor applications[J]. Nature Materials, 2004, 3: 736-742.

[53] Baker M. Nanotechnology imaging probes: Smaller and more stable[J]. Nature Methods, 2010, 7: 957-962.

[54] Chang Y R, Lee H Y, Chen K, et al. Mass production and dynamic imaging of fluorescent nanodiamonds[J]. Nature Nanotechnology, 2008, 3: 284-288.

[55] Becher C. Fluorescent nanoparticles: Diamonds from outer space[J]. Nature Nanotechnology, 2014, 9: 16-17.

[56] Chen T M, Yang F, Liu P, et al. General top-down strategy for generating single-digit nanodiamonds for bioimaging[J]. Nanotechnology, 2020, 31 (48): 485601.

[57] Shafeev G A, Obraztsova E D, Pimenov S M. Laser-assisted etching of diamonds in air and in liquid media[J]. Applied Physics A, 1997, 65: 29-32.

[58] Sun Y P, Zhou B, Lin Y, et al. Quantum-sized carbon dots for bright and colorful photoluminescence[J]. Journal of the American Chemical Society, 2006, 128 (24): 7756-7757.

[59] Webster T J. Safety of nanoparticles[M]. New York: Springer, 2008.

[60] Schrand A M, Huang H J, Carlson C, et al. Are diamond nanoparticles cytotoxic? [J]. The Journal of Physical

Chemistry B, 2007, 111 (1): 2-7.

[61] Höök M, Zittel W, Schindler J, et al. Global coal production outlooks based on a logistic model[J]. Fuel, 2010, 89 (11): 3546-3558.

[62] Given P H. The distribution of hydrogen in coals and its relation to coal structure[J]. Fuel, 1960, 39 (2): 147-153.

[63] Heredy L A, Kostyo A E, Neuworth M B. Studies on the structure of coals of different rank: Hydrogen distribution of depolymerization products[J]. Advances in Chemistry, 1966, 55 (31): 493-502.

[64] Levine D G, Schlosberg R H, Silbernagel B G. Understanding the chemistry and physics of coal structure (A Review) [J]. Proceedings of the National Academy of Sciences, 1982, 79 (10): 3365-3370.

[65] Derbyshire F, Marzec A, Schulten H R, et al. Molecular structure of coals: A debate[J]. Fuel, 1989, 68 (9): 1091-1106.

[66] Xiao J, Liu P, Yang G W. Nanodiamonds from coal under ambient conditions[J]. Nanoscale, 2015, 7 (14): 6114-6125.

[67] Ferrari A C, Robertson J. Raman spectroscopy of amorphous, nanostructured, diamond-like carbon, and nanodiamond[J]. Philosophical Transactions of the Royal Society A: Mathematical, Physical and Engineering Sciences, 2004, 362 (1824): 2477-2512.

[68] Merel P, Tabbal M, Chaker M, et al. Direct evaluation of the sp^3 content in diamond-like-carbon films by XPS[J]. Applied Surface Science, 1998, 136 (1-2): 105-110.

[69] Tang L B, Ji R B, Cao X K, et al. Deep ultraviolet photoluminescence of water-soluble self-passivated graphene quantum dots[J]. ACS Nano, 2012, 6 (6): 5102-5110.

[70] Hu S L, Niu K Y, Sun J, et al. One-step synthesis of fluorescent carbon nanoparticles by laser irradiation[J]. Journal of Materials Chemistry, 2009, 19 (4): 484-488.

[71] Tan D Z, Zhou S F, Xu B B, et al. Simple synthesis of ultra-small nanodiamonds with tunable size and photoluminescence[J]. Carbon, 2013, 62: 374-381.

[72] Hu S L, Guo Y, Dong Y G, et al. Understanding the effects of the structures on the energy gaps in carbon nanoparticles from laser synthesis[J]. Journal of Materials Chemistry, 2012, 22 (24): 12053-12057.

[73] Li H T, He X D, Kang Z H, et al. Water-soluble fluorescent carbon quantum dots and photocatalyst design[J]. Angewandte Chemie International Edition, 2010, 49 (26): 4430-4434.

[74] Baker S N, Baker G A. Luminescent carbon nanodots: Emergent nanolights[J]. Angewandte Chemie International Edition, 2010, 49 (38): 6726-6744.

[75] Eda G, Lin Y Y, Mattevi C, et al. Blue Photoluminescence from chemically derived graphene oxide[J]. Advanced Materials, 2010, 22 (4): 505-509.

[76] Exarhos A L, Turk M E, Kikkawa J M. Ultrafast spectral migration of photoluminescence in graphene oxide[J]. Nano Letters, 2013, 13 (2): 344-349.

[77] Fang Y X, Guo S J, Li D, et al. Easy synthesis and imaging applications of cross-linked green fluorescent hollow carbon nanoparticles[J]. ACS Nano, 2012, 6 (1): 400-409.

[78] Galande C, Mohite A D, Naumov A V, et al. Quasi-molecular fluorescence from graphene oxide[J]. Scientific Reports, 2011, 1: 85.

[79] Mueller M L, Yan X, McGuire J A, et al. Triplet states and electronic relaxation in photoexcited graphene quantum dots[J]. Nano Letters, 2010, 10 (7): 2679-2682.

[80] Qu S N, Wang X Y, Lu Q P, et al. A biocompatible fluorescent ink based on water-soluble luminescent carbon nanodots[J]. Angewandte Chemie International Edition, 2012, 51 (49): 12215-12218.

[81] Shen J H, Zhu Y H, Yang X L, et al. Graphene quantum dots: Emergent nanolights for bioimaging, sensors, catalysis and photovoltaic devices[J]. Chemical Communications, 2012, 48 (31): 3686-3699.

[82] Wu X L, Fan J Y, Qiu T, et al. Experimental evidence for the quantum confinement effect in 3C-SiC nanocrystallites[J]. Physical Review Letters, 2005, 94 (2): 026102.

[83] Tan D Z, Zhou S F, Shimotsuma Y, et al. Effect of UV irradiation on photoluminescence of carbon dots[J]. Optical Materials Express, 2014, 4 (2): 213-219.

[84] Reffner J A, Martoglio P A, Williams G P. Fourier transform infrared microscopical analysis with synchrotron radiation: The microscope optics and system performance[J]. Review of Scientific Instruments, 1995, 66 (2): 1298-1302.

[85] Nakanishi K, Solomon P H. Infrared absorption spectroscopy[M]. Sanfrancisco: Holden-Day, 1977.

[86] Silverstein R M, Bassler G C. Spectrometric identification of organic compounds[J]. Journal of Chemical Education, 1962, 39 (11): 546.

[87] Kasha M. Characterization of electronic transitions in complex molecules[J]. Discussions of the Faraday Society, 1950, 9: 14-19.

[88] Cushing S K, Li M, Huang F Q, et al. Origin of strong excitation wavelength dependent fluorescence of graphene oxide[J]. ACS Nano, 2014, 8 (1): 1002-1013.

[89] Bruchez Jr M, Moronne M, Gin P, et al. Semiconductor nanocrystals as fluorescent biological labels[J]. Science, 1998, 281 (5385): 2013-2016.

[90] Lee K M, Cheng W Y, Chen C Y, et al. Excitation-dependent visible fluorescence in decameric nanoparticles with monoacylglycerol cluster chromophores[J]. Nature Communications, 2013, 4: 1544.

[91] Zhang M, Bai L L, Shang W H, et al. Facile synthesis of water-soluble, highly fluorescent graphene quantum dots as a robust biological label for stem cells[J]. Journal of Materials Chemistry, 2012, 22 (15): 7461-7467.

[92] Bundy F P. Pressure-temperature phase diagram of elemental carbon[J]. Physica A: Statistical Mechanics and its Applications, 1989, 156 (1): 169-178.

[93] Bundy F P, Hall H T, Strong H M, et al. Man-made diamonds[J]. Nature, 1955, 176: 51-55.

[94] DeCarli P S, Jamieson J C. Formation of diamond by explosive shock[J]. Science, 1961, 133 (3467): 1821-1822.

[95] Aust R B, Drickamer H G. Carbon: A new crystalline phase[J]. Science, 1963, 140 (3568): 817-819.

[96] Bundy F P, Kasper J S. Hexagonal diamond: A new form of carbon[J]. The Journal of Chemical Physics, 1967, 46 (9): 3437-3446.

[97] Endo S, Idani N, Oshima R, et al. X-ray diffraction and transmission-electron microscopy of natural polycrystalline graphite recovered from high pressure[J]. Physical Review B, 1994, 49 (1): 22.

[98] Yagi T, Utsumi W, Yamakata M, et al. High-pressure in situ X-ray-diffraction study of the phase transformation from graphite to hexagonal diamond at room temperature[J]. Physical Review B, 1992, 46 (10): 6031.

[99] Regueiro M N, Monceau P, Hodeau J L. Crushing C_{60} to diamond at room temperature[J]. Nature, 1992, 355: 237-239.

[100] Banhart F, Ajayan P M. Carbon onions as nanoscopic pressure cells for diamond formation[J]. Nature, 1996, 382: 433-435.

[101] Wesolowski P, Lyutovich Y, Banhart F, et al. Formation of diamond in carbon onions under MeV ion irradiation[J]. Applied Physics Letters, 1997, 71 (14): 1948-1950.

[102] Fahy S, Louie S G, Cohen M L. Theoretical total-energy study of the transformation of graphite into hexagonal diamond[J]. Physical Review B, 1987, 35 (14): 7623.

[103] Kertesz M, Hoffmann R. The graphite-to-diamond transformation[J]. Journal of Solid State Chemistry, 1984, 54 (3): 313-319.

[104] Clarke R, Uher C. High pressure properties of graphite and its intercalation compounds[J]. Advances in Physics, 1984, 33 (5): 469-566.

[105] Kurdyumov A V. Mechanism of phase transformations of carbon and boron nitride at high pressures[C]//Dokl. Akad. Nauk SSSR; (USSR), 1975, 221 (2).

[106] Kurdyumov A V, Ostrovskaya N F, Pilyankevich A N, et al. Mechanism of lonsdalite-graphite phase transformation[C]//Soviet Physics Doklady, 1978, 23: 278.

[107] Wheeler E J, Lewis D. The structure of a shock-quenched diamond[J]. Materials Research Bulletin, 1975, 10 (7): 687-693.

[108] Utsumi W, Yagi T. Light-transparent phase formed by room-temperature compression of graphite[J]. Science, 1991, 252 (5012): 1542-1544.

[109] Takano K J, Harashima H H H, Wakatsuki M W M. New high-pressure phases of carbon[J]. Japanese Journal of Applied Physics, 1991, 30 (5A): L860.

[110] Wang J B, Yang G W. Phase transformation between diamond and graphite in preparation of diamonds by pulsed-laser induced liquid-solid interface reaction[J]. Journal of Physics: Condensed Matter, 1999, 11 (37): 7089.

[111] Bundy F P, Bassett W A, Weathers M S, et al. The pressure-temperature phase and transformation diagram for carbon; Updated through 1994[J]. Carbon, 1996, 34 (2): 141-153.

[112] Vereshchagin L F, Kalashnikov Y A, Feklichev E M, et al. Mechanism of the polymorphic conversion of graphite to diamond[C]//Soviet Physics Doklady, 1965, 10: 534-537.

[113] Scandolo S, Bernasconi M, Chiarotti G L, et al. Pressure-induced transformation path of graphite to diamond[J]. Physical Review Letters, 1995, 74 (20): 4015.

[114] Lewis R S, Ming T, Wacker J F, et al. Interstellar diamonds in meteorites[J]. Nature, 1987, 326: 160-162.

[115] Guillois O, Ledoux G, Reynaud C. Diamond infrared emission bands in circumstellar media[J]. The Astrophysical Journal, 1999, 521 (2): L133.

[116] Goto M, Henning T, Kouchi A, et al. Spatially resolved 3 μm spectroscopy of Elias 1: Origin of diamonds in protoplanetary disks[J]. The Astrophysical Journal, 2009, 693 (1): 610.

[117] Marks N A, Lattemann M, McKenzie D R. Nonequilibrium route to nanodiamond with astrophysical implications[J]. Physical Review Letters, 2012, 108 (7): 075503.

[118] Krishna V, Stevens N, Koopman B, et al. Optical heating and rapid transformation of functionalized fullerenes[J]. Nature Nanotechnology, 2010, 5: 330-334.

[119] Maze J R, Stanwix P L, Hodges J S, et al. Nanoscale magnetic sensing with an individual electronic spin in diamond[J]. Nature, 2008, 455: 644-647.

[120] Ugarte D. Curling and closure of graphitic networks under electron-beam irradiation[J]. Nature, 1992, 359: 707-709.

[121] Iijima S, Wakabayashi T, Achiba Y. Structures of carbon soot prepared by laser ablation[J]. The Journal of Physical Chemistry, 1996, 100 (14): 5839-5843.

[122] Sano N, Wang H, Alexandrou I, et al. Properties of carbon onions produced by an arc discharge in water[J]. Journal of Applied Physics, 2002, 92 (5): 2783-2788.

[123] Lau D W M, McCulloch D G, Marks N A, et al. High-temperature formation of concentric fullerene-like structures

within foam-like carbon: Experiment and molecular dynamics simulation[J]. Physical Review B, 2007, 75（23）: 233408.

[124] Chhowalla M, Wang H, Sano N, et al. Carbon onions: Carriers of the 217.5 nm interstellar absorption feature[J]. Physical Review Letters, 2003, 90（15）: 155504.

[125] Li A, Chen J H, Li M P, et al. On buckyonions as an interstellar grain component[J]. Monthly Notices of the Royal Astronomical Society: Letters, 2008, 390（1）: L39-L42.

[126] Kuznetsov V L, Chuvilin A L, Butenko Y V, et al. Onion-like carbon from ultra-disperse diamond[J]. Chemical Physics Letters, 1994, 222（4）: 343-348.

[127] Xiao J, Ouyang G, Liu P, et al. Reversible nanodiamond-carbon onion phase transformations[J]. Nano Letters, 2014, 14（6）: 3645-3652.

[128] Kuznetsov V L, Chuvilin A L, Moroz E M, et al. Effect of explosion conditions on the structure of detonation soots: Ultradisperse diamond and onion carbon[J]. Carbon, 1994, 32（5）: 873-882.

[129] Tomita S, Burian A, Dore J C, et al. Diamond nanoparticles to carbon onions transformation: X-ray diffraction studies[J]. Carbon, 2002, 40（9）: 1469-1474.

[130] Cebik J, McDonough J K, Peerally F, et al. Raman spectroscopy study of the nanodiamond-to-carbon onion transformation[J]. Nanotechnology, 2013, 24（20）: 205703.

[131] Obraztsova E D, Fujii M, Hayashi S, et al. Raman identification of onion-like carbon[J]. Carbon, 1998, 36（5-6）: 821-826.

[132] Niu K Y, Zheng H M, Li Z Q, et al. Laser dispersion of detonation nanodiamonds[J]. Angewandte Chemie International Edition, 2011, 50（18）: 4099-4102.

[133] Wang H Q, Pyatenko A, Kawaguchi K, et al. Selective pulsed heating for the synthesis of semiconductor and metal submicrometer spheres[J]. Angewandte Chemie International Edition, 2010, 49（36）: 6361-6364.

[134] Barin I, Sauert F, Schultze-Rhonhof E, et al. Thermochemical data of pure substances. 2nd ed[J]. Weinheim, Federal Republic of Germany, 1993.

[135] Bohren C F, Huffman D R. Absorption and scattering of light by small particles[M]. New York: John Wiley & Sons, 2008.

[136] Palik E D. Handbook of optical constants of solids II[M]. New York: Academic Press, 1991.

[137] Lide D R. CRC handbook of chemistry and physics[M]. New York: CRC Press, 2004.

[138] Swalin R A. Thermodynamics of Solids[M]. New York: Wiley, 1972.

[139] Zhao Y H, Lu K. Grain-size dependence of thermal properties of nanocrystalline elemental selenium studied by X-ray diffraction[J]. Physical Review B, 1997, 56（22）: 14330.

[140] Ouyang G, Zhu Z M, Zhu W G, et al. Size dependent Debye temperature and total mean square relative atomic displacement in nanosolids under high pressure and high temperature[J]. The Journal of Physical Chemistry C, 2010, 114（4）: 1805-1808.

[141] Ouyang G, Tan X, Yang G W. Thermodynamic model of the surface energy of nanocrystals[J]. Physical Review B, 2006, 74（19）: 195408.

[142] Kwon Y K, Berber S, Tománek D. Thermal contraction of carbon fullerenes and nanotubes[J]. Physical Review Letters, 2004, 92（1）: 015901.

[143] Hirai H, Kondo K. Modified phases of diamond formed under shock compression and rapid quenching[J]. Science, 1991, 253（5021）: 772-774.

[144] Jarkov S M, Titarenko Y N, Churilov G N. Electron microscopy studies of FCC carbon particles[J]. Carbon, 1998,

36（5-6）：595-597.

[145] Konyashin I, Zern A, Mayer J, et al. A new carbon modification：'n-diamond' or face-centred cubic carbon? [J]. Diamond and Related Materials, 2001, 10（1）：99-102.

[146] Murrieta G, Tapia A, De Coss R. Structural stability of carbon in the face-centered-cubic (Fm3m) phase[J]. Carbon, 2004, 42（4）：771-774.

[147] Terranova M L, Manno D, Rossi M, et al. Self-assembly of N-diamond nanocrystals into supercrystals[J]. Crystal Growth & Design, 2009, 9（3）：1245-1249.

[148] Baldissin G, Bull D J. N-diamond: Dynamical stability of proposed structures[J]. Diamond and Related Materials, 2013, 34：60-64.

[149] Xiao J, Li J L, Liu P, et al. A new phase transformation path from nanodiamond to new-diamond via an intermediate carbon onion[J]. Nanoscale, 2014, 6（24）：15098-15106.

[150] Fatow M, Konyashin I, Babaev V, et al. Carbon modification with the fcc crystal structure[J]. Vacuum, 2002, 68（1）：75-78.

[151] Gogotsi Y, Welz S, Ersoy D A, et al. Conversion of silicon carbide to crystalline diamond-structured carbon at ambient pressure[J]. Nature, 2001, 411：283-287.

[152] Mykhaylyk O O, Solonin Y M, Batchelder D N, et al. Transformation of nanodiamond into carbon onions: A comparative study by high-resolution transmission electron microscopy, electron energy-loss spectroscopy, X-ray diffraction, small-angle X-ray scattering, and ultraviolet Raman spectroscopy[J]. Journal of Applied Physics, 2005, 97（7）：074302.

[153] Tomita S, Fujii M, Hayashi S, et al. Electron energy-loss spectroscopy of carbon onions[J]. Chemical Physics Letters, 1999, 305（3-4）：225-229.

[154] Wang L, Zhu S J, Wang H Y, et al. Common origin of green luminescence in carbon nanodots and graphene quantum dots[J]. ACS Nano, 2014, 8（3）：2541-2547.

[155] Barnard A S, Russo S P, Snook I K. Structural relaxation and relative stability of nanodiamond morphologies[J]. Diamond and Related Materials, 2003, 12（10-11）：1867-1872.

[156] Liu P, Cui H, Yang G W. Synthesis of body-centered cubic carbon nanocrystals[J]. Crystal Growth & Design, 2008, 8（2）：581-586.

[157] Liu P, Cao Y L, Wang C X, et al. Micro-and nanocubes of carbon with C_8-like and blue luminescence[J]. Nano Letters, 2008, 8（8）：2570-2575.

[158] Mortazavi S Z, Parvin P, Reyhani A, et al. Generation of various carbon nanostructures in water using IR/UV laser ablation[J]. Journal of Physics D: Applied Physics, 2013, 46（16）：165303.

[159] Frondel C, Marvin U B. Lonsdaleite, a hexagonal polymorph of diamond[J]. Nature, 1967, 214：587-589.

[160] Kroto H W, Heath J R, O'Brien S C, et al. C_{60}: Buckminsterfullerene[J]. Nature, 1985, 318：162-163.

[161] Iijima S. Helical microtubules of graphitic carbon[J]. Nature, 1991, 354：56-58.

[162] Novoselov K S, Geim A K, Morozov S V, et al. Electric field effect in atomically thin carbon films[J]. Science, 2004, 306（5696）：666-669.

[163] Whittaker A G. Carbon: A new view of its high-temperature behavior[J]. Science, 1978, 200（4343）：763-764.

[164] Strel'Nitskii V E, Padalka V G, Vakula S I. Properties of the diamond-like carbon film produced by the condensation of a plasma stream with an Rf potential[J]. SovPhysTechPhys, 1978, 23（2）：222.

[165] Johnston R L, Hoffmann R. Superdense carbon, C_8: Supercubane or analog of γ-Si? [J]. Journal of the American Chemical Society, 1989, 111（3）：810-819.

[166] Stankevich I V, Nikerov M V, Bochvar D A. The structural chemistry of crystalline carbon: Geometry, stability, and electronic spectrum[J]. Russian Chemical Reviews, 1984, 53 (7): 640.

[167] Biswas R, Martin R M, Needs R J, et al. Complex tetrahedral structures of silicon and carbon under pressure[J]. Physical Review B, 1984, 30 (6): 3210.

[168] Biswas R, Martin R M, Needs R J, et al. Stability and electronic properties of complex structures of silicon and carbon under pressure: Density-functional calculations[J]. Physical Review B, 1987, 35 (18): 9559.

[169] Burdett J K, Lee S. Moments and the energies of solids[J]. Journal of the American Chemical Society, 1985, 107 (11): 3063-3082.

[170] Winkler B, Milman V. Structure and properties of supercubane from density functional calculations[J]. Chemical Physics Letters, 1998, 293 (3-4): 284-288.

[171] Robertson J. Diamond-like amorphous carbon[J]. Materials Science and Engineering: R: Reports, 2002, 37 (4-6): 129-281.

[172] Ferrari A C, Robertson J. Interpretation of Raman spectra of disordered and amorphous carbon[J]. Physical Review B, 2000, 61 (20): 14095.

[173] Prawer S, Nugent K W, Lifshitz Y, et al. Systematic variation of the Raman spectra of DLC films as a function of sp^2: sp^3 composition[J]. Diamond and Related Materials, 1996, 5 (3-5): 433-438.

[174] Ferrari A C. Determination of bonding in diamond-like carbon by Raman spectroscopy[J]. Diamond and Related Materials, 2002, 11 (3-6): 1053-1061.

[175] Meguro T, Hida A, Suzuki M, et al. Creation of nanodiamonds by single impacts of highly charged ions upon graphite[J]. Applied Physics Letters, 2001, 79 (23): 3866-3868.

[176] Grimsditch M H, Ramdas A K. Brillouin scattering in diamond[J]. Physical Review B, 1975, 11 (8): 3139.

[177] Berthe L, Fabbro R, Peyre P, et al. Shock waves from a water-confined laser-generated plasma[J]. Journal of Applied Physics, 1997, 82 (6): 2826-2832.

[178] Berthe L, Fabbro R, Peyre P, et al. Wavelength dependent of laser shock-wave generation in the water-confinement regime[J]. Journal of Applied Physics, 1999, 85 (11): 7552-7555.

[179] Schwan J, Ulrich S, Roth H, et al. Tetrahedral amorphous carbon films prepared by magnetron sputtering and dc ion plating[J]. Journal of Applied Physics, 1996, 79 (3): 1416-1422.

[180] Sommer A P, Pavláth A E. The subaquatic water layer[J]. Crystal Growth & Design, 2007, 7 (1): 18-24.

[181] Pichler T. Molecular nanostructures: Carbon ahead[J]. Nature Materials, 2007, 6: 332-333.

[182] Geim A K, Novoselov K S. The rise of graphene[J]. Nature Materials, 2007, 6: 183-191.

[183] Fagan J A, Becker M L, Chun J, et al. Length fractionation of carbon nanotubes using centrifugation[J]. Advanced Materials, 2008, 20 (9): 1609-1613.

[184] Meng S, Maragakis P, Papaloukas C, et al. DNA Nucleoside Interaction and Identification with carbon nanotubes[J]. Nano Letters, 2007, 7 (1): 45-50.

[185] Wang C F, Choi Y S, Lee J C, et al. Observation of whispering gallery modes in nanocrystalline diamond microdisks[J]. Applied Physics Letters, 2007, 90 (8): 081110.

[186] Guo K X. Reduced cells and the indexing of electron diffraction patterns[J]. Acta Phisica Sinnica, 1978, 27: 160-168.

[187] Guo K X. On the extinction problem in the automatic indexing of electron diffraction patterns[J]. Acta Phisica Sinnica, 1978, 27: 473-475.

[188] Chen X Y, Cui H, Liu P, et al. Shape-induced ultraviolet absorption of CuO shuttlelike nanoparticles[J]. Applied

Physics Letters, 2007, 90 (18): 183118.

[189] Perdew J P, Burke K, Ernzerhof M. Generalized gradient approximation made simple[J]. Physical Review Letters, 1996, 77 (18): 3865.

[190] Pankove J I. Optical processes in semiconductors[M]. Chelmsford: Courier Corporation, 1975.

[191] Yang S W, Gao L. Controlled synthesis and self-assembly of CeO_2 nanocubes[J]. Journal of the American Chemical Society, 2006, 128 (29): 9330-9331.

[192] El Goresy A, Donnay G. A new allotropic form of carbon from the Ries crater[J]. Science, 1968, 161 (3839): 363-364.

[193] Kasatochkin V I, Korshak V V, Kudryavtsev Y P, et al. On crystalline structure of carbyne[J]. Carbon, 1973, 11 (1): 70-72.

[194] Nakamizo M, Kammereck R, Walker Jr P L. Laser raman studies on carbons[J]. Carbon, 1974, 12 (3): 259-267.

[195] Heimann R B, Evsyukov S E, Kavan L. Carbyne and carbynoid structures[M]. Dordreche: Kluwer Academic Publishers, 1999.

[196] Kudryavtsev Y P, Heimann R B, Evsyukov S E. Carbynes: Advances in the field of linear carbon chain compounds[J]. Journal of Materials Science, 1996, 31: 5557-5571.

[197] Kudryavtsev Y P, Evsyukov S, Guseva M, et al. Carbyne-a linear chainlike carbon allotrope[J]. Chemistry and Physics of Carbon, 1997, 25: 1-69.

[198] Liu M J, Artyukhov V I, Lee H, et al. Carbyne from first principles: Chain of C atoms, a nanorod or a nanorope[J]. ACS Nano, 2013, 7 (11): 10075-10082.

[199] Lagow R J, Kampa J J, Wei H C, et al. Synthesis of linear acetylenic carbon: The "sp" carbon allotrope[J]. Science, 1995, 267 (5196): 362-367.

[200] Sladkov A M. Carbyne-a new allotropic form of carbon[J]. Soviet Scientific Rviews Supplement Series B, 1981, 3: 75-110.

[201] Chalifoux W A, Tykwinski R R. Synthesis of extended polyynes: Toward carbyne[J]. Comptes Rendus Chimie, 2009, 12 (3-4): 341-358.

[202] Chalifoux W A, Tykwinski R R. Synthesis of polyynes to model the sp-carbon allotrope carbyne[J]. Nature Chemistry, 2010, 2: 967-971.

[203] Haley M M. Carbon allotropes: On the road to carbyne[J]. Nature Chemistry, 2010, 2: 912-913.

[204] Pan B T, Xiao J, Li J L, et al. Carbyne with finite length: The one-dimensional sp carbon[J]. Science Advances, 2015, 1 (9): e1500857.

[205] Berthe L, Sollier A, Peyre P, et al. The generation of laser shock waves in a water-confinement regime with 50 ns and 150 ns XeCl excimer laser pulses[J]. Journal of Physics D: Applied Physics, 2000, 33 (17): 2142.

[206] Ravagnan L, Siviero F, Lenardi C, et al. Cluster-beam deposition and in situ characterization of carbyne-rich carbon films[J]. Physical Review Letters, 2002, 89 (28): 285506.

[207] Mullazzi E, Brivio G P, Faulques E, et al. Experimental and theoretical Raman results in trans polyacetylene[J]. Solid State Communications, 1983, 46 (12): 851-855.

[208] Gibtner T, Hampel F, Gisselbrecht J P, et al. End-cap stabilized oligoynes: Model compounds for the linear sp carbon allotrope carbyne[J]. Chemistry: A European Journal, 2002, 8 (2): 408-432.

[209] Cataldo F. Polyynes: A new class of carbon allotropes. About the formation of dicyanopolyynes from an electric arc between graphite electrodes in liquid nitrogen[J]. Polyhedron, 2004, 23 (11): 1889-1896.

[210] Heimann R B, Kleiman J, Salansky N M. A unified structural approach to linear carbon polytypes[J]. Nature,

1983, 306: 164-167.
- [211] Hu Y H. Bending effect of sp-hybridized carbon (carbyne) chains on their structures and properties[J]. The Journal of Physical Chemistry C, 2011, 115 (5): 1843-1850.
- [212] Ordejón P, Artacho E, Soler J M. Self-consistent order-N density-functional calculations for very large systems[J]. Physical Review B, 1996, 53 (16): R10441.
- [213] Su P F, Wu J F, Gu J J, et al. Bonding conundrums in the C_2 molecule: A valence bond study[J]. Journal of Chemical Theory and Computation, 2011, 7 (1): 121-130.
- [214] Schwank J. Catalytic gold: Applications of elemental gold in heterogeneous catalysis[J]. Gold Bulletin, 1983, 16 (4): 103-110.
- [215] Sakka T, Oguchi H, Ogata Y H. Emission spectroscopy of ablation plumes in liquid for analytical purposes[J]. Journal of Physics: Conference Series, 2007, 59 (1): 559.

第 6 章　新颖亚稳相纳米材料的 LAL 探索

在环境条件（常温常压）下合成亚稳相纳米材料（具有热力学上的亚稳结构或晶体学上的亚稳形貌）是一项在材料科学中具有挑战性的工作。所谓亚稳相材料是指在热力学常态下处于亚稳相的一类材料，由于它们通常会表现出相应的稳定相材料所不具备的优异性能而备受人们关注。例如，传统的高温高压相材料如金刚石及相关材料就属于亚稳相材料，它们在高温高压下是稳定相，而在常温常压下则是亚稳相。然而，随着材料科学的发展，尤其是诸如非晶、准晶及纳米晶等亚稳相材料的发现和发展，亚稳相材料体系得到了极大的丰富。与传统的高温高压亚稳相材料相比，目前，非晶、准晶及纳米晶等材料被认为是新颖亚稳相材料。一般来说，绝大部分亚稳相材料有一个共同的特点就是它们在热力学常态下处于亚稳相或不稳定相，具有向稳定相转化的趋势。所以，纳米材料也是一种典型的亚稳相材料。实际上，与亚稳结构相比，在纳米材料物理研究中人们更关注的是亚稳形貌。例如，无论是在材料稳态生长的热力学与动力学上，还是在材料形成的晶体学上，纳米晶的稳定形貌一般为球状，而其他的非球状规则与不规则形貌（如线状、带状、管状等）均可以归为亚稳形貌。众所周知，纳米材料的形貌对其物性有着本质影响，许多奇异的纳米尺寸效应都是由其独特的形貌所诱导的。所以，发展新制备技术去探索新颖亚稳相纳米材料就成为当前纳米材料科学家关注的热点之一。

我们在前面已经介绍了 LAL 的基本物理和化学过程，并且指出其热力学过程最鲜明的特色就是环境液体的束缚效应会将脉冲激光诱导的等离子体羽推向一个高温高压高密度的极端热力学状态，这是亚稳相喜欢的热力学环境，为亚稳相的成核与生长提供了所需要的热力学条件。而其动力学过程的最大特点就是脉冲激光诱导的等离子体羽在环境液体的束缚下快速淬灭，这对于亚稳相的合成有两个重要影响，其一就是在高温高压高密度极端热力学状态下形成的亚稳相会被"冻结"在最终的产物中，其二就是超短的淬灭时间使得在等离子体羽中生成的亚稳相最终保持纳米尺度范围。因为等离子体羽的淬灭时间就是亚稳相的生长时间，所以 LAL 方法是探索新颖亚稳相纳米材料的有力工具。

通过应用 LAL 方法合成纳米金刚石及新碳相纳米材料的系统研究，使我们深深体会到 LAL 方法在合成亚稳相纳米材料方面的普适性，也展现了 LAL 方法在探索新颖亚稳相纳米材料方面的应用潜力。本章我们将介绍应用 LAL 方法去合成不同类型的新颖亚稳相纳米材料。

6.1 闪锌矿硅纳米晶

硅作为最重要的半导体材料在现代微电子工业领域处于核心材料的地位。但是，由于硅是间接带隙半导体，无法直接发光，不具备好的光电性，所以长期以来在光电子学领域难觅硅器件的踪影。然而，随着纳米科技的发展，人们发现当空间尺寸小至纳米量级时，在量子限制条件下硅的晶格对称性会被打破，从而使不同的动量态相互混杂，进而诱发有效的发光与光学增益，这就使得纳米硅材料可能具备发光性能。另外，微纳器件的发展也进一步拓展了硅的应用。例如，由于纳米尺寸的硅具有独特的物理性能（发光、场电子发射、量子效应等），因此可以用于多种量子器件，如单电子存储器[1, 2]、量子点晶体管[3, 4]及光电器件[5, 6]等。因此，纳米硅有望成为下一代纳米电子器件的基石[7]。近年来已经有许多研究人员开展纳米硅的制备研究，例如，通过化学气相沉积[8, 9]和物理气相沉积[10]等技术进行硅纳米晶薄膜的制备。我们应用 LAL 方法在国际上首次合成了一种新颖的亚稳相纳米硅，也就是亚稳相闪锌矿结构硅纳米方块（闪锌矿硅纳米晶）[11]。

在亚稳相闪锌矿结构硅纳米方块的 LAL 合成中，我们选择沉积 100 nm 非晶碳层的金刚石结构的硅片为靶材，二次去离子水、无水乙醇和丙酮与低浓度的无机盐溶液按一定配比配制成的混合液为环境液体。这里，我们选择的无机盐溶液为 KCl 和 NaCl 溶液，其目的是利用简单无机盐离子代替常用的化学溶剂法制备方法中使用的表面活性剂或有机高分子溶剂。一方面是因为无机盐溶液成分简单，容易进行调控，另一方面是因为在晶体形核过程中，无机盐离子可能比有机表面活性剂或高聚物更能影响晶核的结晶相及最终形貌。这样，我们合成出了新颖的硅纳米方块，如图 6-1 所示。

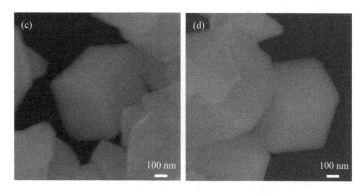

图 6-1　LAL 合成硅纳米方块的 SEM 图像

(a) 和 (b) 低倍；(c) 和 (d) 高倍

我们在低倍放大 SEM 图像中可以观察到大量有规则几何形状的纳米颗粒,并且从中可以看到很多呈规则立方体状的颗粒。而在高倍放大 SEM 图像中,我们可以清楚地看到,这种颗粒具有规则的六面体形貌,其边长大约为 200～500 nm。而且,我们注意到这些颗粒大都呈现出长方体形状,这一点与常见的金属纳米方块是有所差异的。我们知道,只要是立方结构的金属纳米方块,它们在形貌上大都会呈现为规则的立方体形状。

由于 LAL 合成的样品较为稀疏和分散,很难利用 XRD 技术进行晶体结构表征。因此,我们采用微区激光拉曼光谱分析手段对合成产物进行结构表征,从而判断 LAL 合成样品的晶体结构是否与硅靶的金刚石结构一样。样品的微区激光拉曼光谱如图 6-2 所示。通过对比样品与靶材的拉曼光谱,我们可以发现,样品的特征峰位于 504.0 cm^{-1},而单晶硅靶的特征峰位于 520.7 cm^{-1},显然二者有较大的差异。关于硅纳米颗粒与硅晶体的拉曼光谱有所不同的研究,有研究人员认为硅纳米颗粒的特征峰位移可以归结为小尺度下的量子效应,或者是由于表征时激光加热所产生的加热效应[12-14]。但是,通过我们实验后的仔细分析发现,LAL 合成的硅纳米方块尺寸分布大约在 200～500 nm,这与文献中所提出的颗粒尺寸在小于 10 nm 的情况下会出现量子效应,以及激光加热效应发生在硅纳米颗粒尺寸小于 50 nm 的情况下不符。而且我们在测试中使用的激光功率仅有 1 mW,与文献报道中的激光功率相比,这么小的激光功率所产生的加热效应是不足以对样品特征峰位移产生显著影响的。因此,硅纳米方块的特征峰位移无法归结为量子效应或加热效应。所以,我们观测到的特征峰位移表明 LAL 合成硅纳米方块的晶体结构应该与普通硅晶体的金刚石结构不同。我们认为,这种硅纳米方块的特征峰位移很可能是由硅纳米方块晶体结构内部扭曲的硅-硅键或位错造成的,另外,这种硅纳米方块的立方体形貌也很可能是造成特征峰位移的一个很重要的原因[15]。从拉曼光谱测试所得到的结果中我们判断，所制备得到的硅纳米方块可能存在与单晶硅靶的金刚石结构不同的晶体结构。

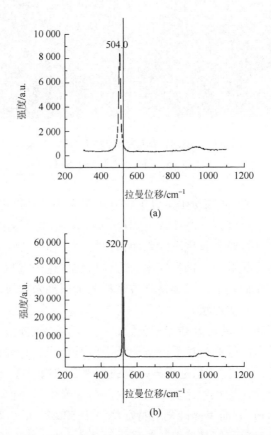

图 6-2　(a) LAL 合成硅纳米方块的拉曼光谱；(b) 单晶硅靶的拉曼光谱

接下来我们采用 TEM 技术对 LAL 合成产物的晶体结构进行了系统表征，如图 6-3 所示。我们从样品的 EDS 中看到，合成的纳米方块的成分主要为硅（误差在 2%）。因此，这些结果表明，合成的纳米方块可看作是纯硅。合成的硅纳米方块的三个相应的 SAED 图像如图 6-3 (c)～图 6-3 (e) 所示，它们是通过将电子束垂直于立方体的不同晶面来获得的代表性的 SAED 数据。

显然，这些 SAED 图像揭示了合成的硅纳米方块是单晶，并且主要由两个 (111) 面、两个 ($2\bar{2}0$) 面和两个 ($11\bar{2}$) 面构成，如图 6-3 (f) 所示。此外，我们根据 SAED 数据计算了硅纳米方块的面间距 D_{exp}。表 6-1 列出了硅纳米方块面间距的实验值、硅的闪锌矿结构的理论值（JCPDS 卡片文件号 800018）和具有金刚石结构的单晶硅的理论值（JCPDS 卡片文件号 895012）的比较。显然，硅纳米

方块的 D_{exp} 与硅闪锌矿结构的 $D_{calc,zb}$ 完全一致。因此，这些结果表明，LAL 合成的硅纳米方块具有闪锌矿结构。

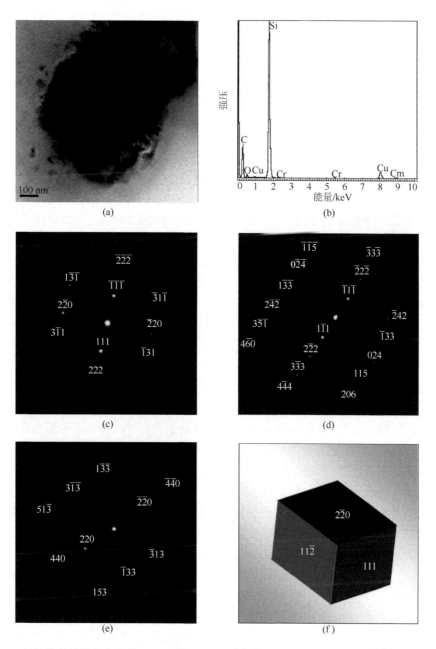

图 6-3 （a）单个硅纳米方块的 TEM 图像；（b）对应的 EDS；（c）～（e）对应的 SAED 图像；（f）在 SAED 分析的基础上重建的硅立方体模型

表 6-1　硅纳米方块实验值及硅闪锌矿结构和金刚石结构单晶硅理论值

(hkl)	D_{exp}[①]/Å	$D_{calc, zb}$[②]/Å	$D_{calc, d}$[③]/Å
111	3.1063	3.1130	3.1355
131	1.6291	1.6257	1.6374
133	1.2366	1.2370	1.2458
220	1.9002	1.9063	1.9201
222	1.5537	1.5565	1.5677
242	1.0998	1.1006	1.1085
311	1.6304	1.6257	1.6374
422	1.1002	1.1006	1.1085

① D_{exp} 是我们实验值；② $D_{calc, zb}$ 是硅闪锌矿结构理论值；③ $D_{calc, d}$ 是金刚石结构单晶硅理论值。

为了进一步确认硅纳米方块的晶体结构是闪锌矿结构，我们对样品进行了 HRTEM 表征，如图 6-4 所示。通过硅纳米方块一个晶面的高分辨晶格条纹相，我们对面间距进行仔细测量，面间距分别是 0.310 nm 和 0.189 nm，这分别与闪锌矿结构的硅的（111）面和（2$\bar{2}$0）面的面间距吻合得很好[16]，而与金刚石结构的硅晶体相应晶面的面间距则相差较大。随后，我们对所获得到的高分辨晶格条纹相进行了快速傅里叶变换分析，并对得到的转换衍射点进行了分析，结果表明这些衍射点分别对应于闪锌矿结构的硅的（111）面和（2$\bar{2}$0）面。这表明在 HRTEM 中观测到的晶格条纹相分别属于硅纳米方块的（111）面和（2$\bar{2}$0）面，这与 SAED 分析的结果一致，确认了多晶格条纹相所属晶面的标定。此外，我们从 HRTEM 分析中还可以看到，硅纳米方块的表面有一个无定形氧化层存在（图 6-4 中白色虚线框标注部分），以及在晶体内部存在的晶格扭曲结构（图 6-4 中用白色的圆圈标注）。无定形氧化层的观测结果验证了我们对 EDS 中氧元素来源的判断，即氧元素确实来自硅纳米方块的表面氧化层。晶格扭曲结构也意味着在硅纳米方块内部很可能存在着内部应力或轻微的层错或位错结构[17]。

电子能量损失谱被认为是表征纳米材料结构特征的"指纹"技术，所以，我们应用 EELS 系统表征了硅纳米方块的结构特征，如图 6-5 所示。首先，我们比较硅纳米方块的低能 EELS 和金刚石结构单晶硅的低能 EELS ［图 6-5（a）和图 6-5（c）］。我们可以清楚看到，硅纳米方块低能 EELS 中，除了在 17 eV 出现一个等离基元峰外，还在 22 eV 的位置上出现一个等离基元峰的肩，而金刚石结构单晶硅只有在 17 eV 出现一个对称的等离基元峰。因此，两个 EELS 对比结果显示，硅纳米方块的结构确实与金刚石结构单晶硅有明显不同。同时我们判断硅纳米方块的 22 eV 的等离基元峰的肩应该就是闪锌矿结构硅的特征峰。然后，我们进一步比较二者的高能 EELS，如图 6-5（b）和图 6-5（d）所示。我们

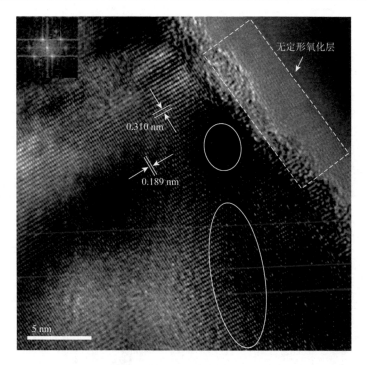

图 6-4 单个硅纳米方块边缘的 HRTEM 图像

插图是相应的快速傅里叶变换（FFT）分析

观察到两个 EELS 都在 101.5 eV 左右出现了属于硅元素的 L 边吸收峰，而没有出现属于其他元素的特征峰。这就在元素上确定了合成的纳米方块是纯硅。同时我们还可以观测到，金刚石结构单晶硅的高能 EELS 中在约 108 eV 位置出现一个属于结构效应造成的小等离基元峰，而在硅纳米方块中，该小等离基元峰位移到了 104 eV。这一结果表明，在高能 EELS 层面上明确显示了硅纳米方块所具有的结构确实与金刚石结构单晶硅存在细微差异。

我们利用第一性原理计算方法在理论上对硅的两种结构的低能 EELS 进行了计算模拟，结果如图 6-5（e）和图 6-5（f）所示。显然，理论计算与模拟结果显示出了硅两种结构的低能 EELS 存在显著的差别。对于金刚石结构单晶硅，只存在一个等离基元峰，且其对应的能量值在 17 eV 附近，与标准谱符合得很好。同时，对于闪锌矿结构硅，计算与模拟结果显示其除了在 17 eV 附近也有一个比较强的峰以外，在 19 eV 附近还存在有一个峰肩，这个模拟结果和实验中所得到的数据吻合得较好。从 EELS 的实验数据和理论模拟两方面说明，LAL 合成的硅纳米方块具有闪锌矿结构，从而在理论上确定了硅纳米方块所具有的亚稳结构。

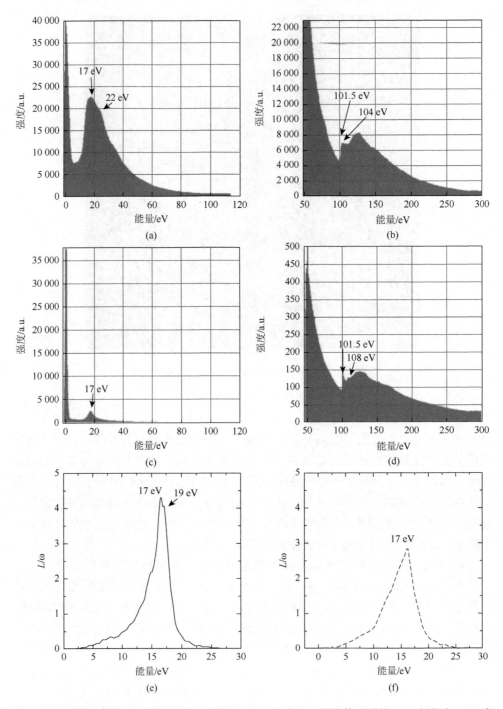

图 6-5 硅纳米方块的（a）低能和（b）高能 EELS；金刚石结构单晶硅的（c）低能和（d）高能 EELS；（e）硅纳米方块和（f）金刚石结构单晶硅低能 EELS 的计算模拟

众所周知，LAL 过程是一种远离热力学平衡的非平衡反应过程，不仅能够实现反应中生成的亚稳相的捕获，而且还可以通过设计反应参数有目的地去合成新亚稳相。基于此，我们提出了 LAL 合成亚稳相闪锌矿结构硅纳米方块的机制。

首先，脉冲激光与覆有非晶碳层的单晶硅靶作用，在靶材表面与环境液体之间产生一个激光诱导的等离子体羽。由于单晶硅靶表面有一层很薄的非晶碳，因此激光烧蚀会首先使非晶碳形成等离子体羽，然后使单晶硅靶气化，从而使得等离子体羽内部会同时包含少量的非晶碳团簇、液体环境中等离子体羽气化而混入的无机盐离子、大量硅团簇及微晶碎片。

其次，因为环境液体的束缚作用，在后续激光能量作用下，激光诱导的等离子体羽会被推进到被液体环境包围的局域空间内，同时，脉冲激光在内部产生的冲击波会将其驱动到高温高压高密度的极端热力学状态。在这样一个微反应区内，各种硅团簇和微晶碎片就有可能激烈碰撞并生成亚稳结构，因为亚稳结构在这种热力学环境中会比稳定结构更容易成核。因此，金刚石结构的硅团簇可以通过一系列晶面滑移、扭曲等方式转变为闪锌矿结构的硅团簇并最终成核。

最后，随着等离子体羽在液体环境束缚作用下迅速淬灭，生成的亚稳相闪锌矿结构晶核会被迅速"冻结"，以及完成晶体生长。由于淬灭及晶核生长过程非常短暂（我们实验中约 20 ns），因此亚稳相闪锌矿结构硅纳米方块会被保留在最终产物中。需要注意的是，在 LAL 合成过程中，无机盐离子都在等离子体羽内部影响着立方体状的晶粒形貌的生成。在亚稳相形核过程中，由于晶面生长存在各向异性，因此等离子体羽内部的无机盐离子会选择晶体能量最低的晶面进行附着，从而形成一种导向媒介的作用。这种优势吸附的导向作用能够降低所附着晶面的表面能，从而阻碍垂直于该晶面方向的晶体生长，最终会影响生长晶粒的形貌[18]。例如，在 LAL 合成硅纳米颗粒过程中，具有亚稳相闪锌矿结构的硅纳米方块和稳定相金刚石结构的硅纳米球都会在合成过程中生成，具体的合成机制及过程示意如图 6-6 所示。

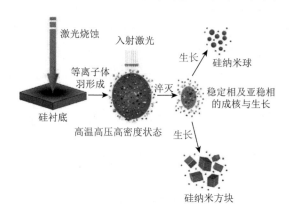

图 6-6　LAL 合成硅纳米方块和硅纳米球的机制及过程示意图

这里需要说明一点的是，在 LAL 合成亚稳相闪锌矿结构硅纳米方块中，我们发现单晶硅靶上的非晶碳层在纳米方块的立方体形貌形成中起到很大的作用。然而，由于 LAL 过程相当复杂，因此目前还很难理解非晶碳成分在合成中的作用。相关的进一步研究将会告诉我们答案。

6.2 四方相锗纳米晶

硅和锗是半导体材料的"双子星"，在微电子和光电子领域发挥着巨大作用。研究人员发现，纳米尺度的锗晶体具有多种量子效应，例如，锗纳米晶在一定激发条件下能够发射可见光[19]。同时，锗能很好地和硅基器件匹配[20]，因此，纳米锗和纳米硅一样，多年来一直受到半导体材料研究人员的高度关注，并且与纳米硅相比[21]，纳米锗在纳米器件制备上有着更广泛的应用。例如，纳米锗有着比纳米硅更小的禁带宽度，同时又拥有更强的量子限域效应，这就使得纳米锗能比纳米硅更适合应用于光电子学[22, 23]。众所周知，锗具有普通的金刚石结构，但是，研究人员发现锗还存在一种不寻常的亚稳相四方结构，被称为 ST-12 结构。事实上，四方结构锗最早是在 1965 年由 Bates 等[24]发现的，当时他们通过对金刚石结构锗单晶施加超过 12 GPa 的压强合成了四方结构锗晶体。有趣的是，这种亚稳相四方结构锗晶体拥有许多新颖的物性，例如，理论研究表明，四方结构锗是一种直接带隙半导体材料[25]，这就使得它在半导体光电器件领域上有着非常大的应用潜力。因此，实现环境条件下亚稳相四方结构锗的制备将对锗半导体器件的应用和发展产生深远影响。可喜的是，已经有理论研究表明，亚稳相四方结构锗纳米晶（四方相锗纳米晶）有可能在常温常压下存在[25-27]。我们应用 LAL 方法，在环境条件下，在国际上首次合成出了四方相锗纳米晶[28]。

LAL 合成四方相锗纳米晶的技术路线同 LAL 合成纳米晶，这里，靶材为金刚石结构锗单晶，环境液体选择甲苯溶液，唯一不同之处就是给靶材施加了一个静电场。具体操作为将靶材用不锈钢固件倒置固定在石英槽上方并施加−47 V 直流电压，同时在靶材下方放置一单晶硅片并施加 47 V 直流电压，然后向石英槽内缓慢注入纯度大于 99.5%的甲苯溶液使整个反应装置都浸没在溶液中。合成产物如图 6-7 所示。我们可以看到，LAL 合成的锗纳米晶呈均匀球形，粒径在 10～40 nm，平均尺寸大约在 20 nm（图 6-7 插图）。

为了标定 LAL 合成的锗纳米晶的晶体结构，我们分别对合成产物（T-Ge）和单晶锗靶（C-Ge）进行了系统的 XRD 比较分析，结果如图 6-8 所示。通过对比分析，我们可以很清楚看到，单晶锗靶的 XRD 只有一个 27.36°的衍射峰来自于金刚石结构锗的（111）面，而在合成产物的 XRD 中，除了在 27.36°有一个来自于金刚石结构锗的衍射峰以外，出现了 7 个新的衍射峰。通过对所有新的衍射峰进行分析，我们发现 45.28°和 53.70°这两个新峰分别归属于金刚石结构锗的（220）和（311）面，其余 5 个衍射峰可以标定为亚稳相四方结构锗的（101）、（111）、（102）、（201）及（211）面（JCPDS 卡片文件号 180549）[24]。因此，通过 XRD 对比分析我们可以明确看到，合成产物中除了保留有少量的金刚石结构锗单晶成分外，主要是亚稳相四方结构锗纳米晶。

图 6-7　LAL 合成的四方相锗纳米晶的 SEM 图像

插图为基于 SEM 数据得到的纳米晶粒径分布图

我们进一步地采用 TEM 对 LAL 合成的锗纳米晶进行了微观晶体结构分析，如图 6-9 所示。我们可以看到，合成的纳米颗粒呈球状，相应 SAED 图像和尺寸分布的结果表明，合成的纳米晶具有四方结构，平均直径在 17~18 nm。同时，样品的 EDS 分析［图 6-9（b）］告诉我们，合成纳米晶的成分可看作是纯锗（EDS 中的 O、Cu、Cr、C 峰分别来自于锗纳米晶外围的稀薄氧化层、铜网、TEM 样品、碳支持膜）。为了更进一步地分析合成的锗纳米晶的微观结构，我们对样品进行了 HRTEM 结构分析［图 6-9（c）］，给出了锗纳米晶的二维晶格条纹相。通过仔细测量后，我们得到两种晶格条纹相的面间距分别是 0.357 nm 和 0.272 nm，这两个结果与亚稳相四方结构锗中的（111）面和（201）面的面间距十分吻合。此外，我

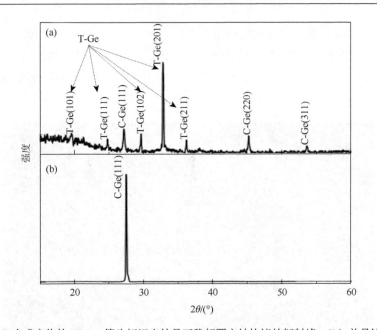

图 6-8 （a）合成产物的 XRD，箭头标识出的是亚稳相四方结构锗的衍射峰；（b）单晶锗靶的 XRD

们还测量了两种晶格条纹相的晶面夹角，$\theta = 69.1°$。根据晶体学中的晶面夹角公式，我们可以计算出四方结构锗晶体的（111）面和（201）面的夹角为 $69.14°$，这一理论计算结果与我们测量得到的晶面夹角 $\theta = 69.1°$ 结果符合得很好。众所周知，不同结构的晶体其晶面夹角都是不同的，或者说，一种晶体结构，当其中的两个晶面指数确定的时候，两晶面的夹角是唯一的。因此，我们实验测量结果与理论计算结果的一致性更进一步证明了合成的锗纳米晶确实具有亚稳相四方结构。

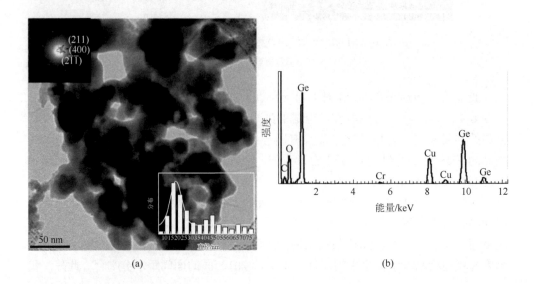

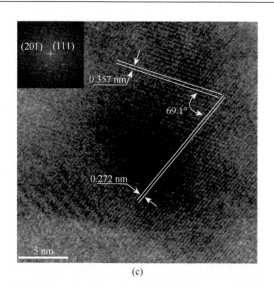

图 6-9 （a）LAL 合成产物的 TEM 图像，插图分别是相应 SAED 图像（左上）和颗粒尺寸分布（右下）；（b）合成产物的 EDS 分析；（c）纳米晶的 HRTEM 图像，插图是相应快速傅里叶变换（FFT）分析，标定了 T-Ge 结构的（201）和（111）晶面

在 TEM 表征中，我们还对合成产物中一些尺寸较大的纳米晶颗粒进行了分析，结果发现对于尺寸大于 30 nm 的锗纳米颗粒，它们基本上都具有金刚石结构，而对于尺寸小于 30 nm 的锗纳米颗粒，普遍都是亚稳相四方结构。这些结果说明了纳米尺寸效应在亚稳相四方锗纳米晶的生长中发挥了作用，使得它们能够在常温常压下保存下来。

除了 XRD 和 TEM 表征以外，我们还利用微区激光拉曼光谱技术对合成的锗纳米晶与单晶锗靶进行了对比分析，结果如图 6-10 所示。通过比较我们可以清楚看到，锗纳米晶拉曼光谱中出现了 4 个特征峰，其中 520.2 cm^{-1} 峰是来自单晶硅衬底（样品涂敷在单晶硅衬底上进行表征），299.5 cm^{-1} 峰来自金刚石结构锗靶的碎片，而 248 cm^{-1} 和 279 cm^{-1} 被标定为四方结构锗纳米晶的特征峰。1972 年，Kobliska 等[29]曾经报道了四方结构锗纳米晶会在 246±3（Γ_5）cm^{-1} 和 273±3（Γ_3）cm^{-1} 两个位置出现特征峰。所以，我们样品中的 248 cm^{-1} 和 279 cm^{-1} 特征峰应该是来自亚稳相四方结构锗纳米晶。需要注意，我们的实验测量结果均比 Kobliska 所报道的要大一些，我们推断这是由于锗纳米晶表面存在氧化层从而使得拉曼光谱产生了微弱红移。所以，样品的拉曼光谱表明合成的锗纳米晶具有四方结构。

亚稳相四方结构锗纳米晶的 LAL 合成机制与前文 LAL 合成亚稳纳米晶基本一样，只是这里使用了有机液体甲苯并且在靶材上施加了一个静电场。所以，我们重点讨论有机液体的选择和静电场的作用。在我们的合成中，使用甲苯作为环境液体是因为它对入射脉冲激光有着良好透明性和在静电场下的稳定性。甲苯的低介电常数（约 2.38）使其很容易形成内部静电场的精细结构，这应该是静电场作用于反应团簇的直接因素。另外，施加在靶材上的静电场有助于激光诱导的等离子体羽保持高能状态，并且在纳米颗粒上绝缘的电荷可能成为亚稳态的瞬态能量保护。

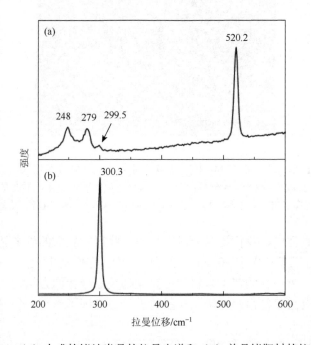

图 6-10 （a）合成的锗纳米晶的拉曼光谱和（b）单晶锗靶材的拉曼光谱

综上所述，我们应用 LAL 方法，在环境条件下捕获了锗的亚稳相四方结构并合成了四方相锗纳米晶，这一结果证明了 LAL 方法在探索新颖高压纳米相材料方面的独特优势。

6.3 双层六角密堆积铁纳米晶

铁是自然界中含量第二的金属元素，也是地核的组成元素。铁的氧化物种类繁多，各种结构的铁的氧化物及合金在物性上表现出非常丰富的多样性。因此，铁和相关氧化物及合金广泛应用于工业生产、医疗卫生和日常生活中[30-39]。随着纳米科学的发展，纳米尺度下的铁及其氧化物不断地通过各种材料制备手段合成。例如，通过化学溶液法合成各种形貌的铁及其氧化物纳米颗粒[30, 32, 34, 38]，利用分子束外延生长铁的某些亚稳结构如面心立方（face-centered cubic，FCC）结构的多层铁薄膜等[40-43]。单质铁除了立方结构之外还有六角密堆积(hexagonal close-packed,HCP)结构,HCP 铁是一种高压相,环境条件下无法存在,被认为是地核的主要成分[44,45]。1995 年,Saxena 等[44]利用金刚石砧室(diamond anvil cell,DAC)压力系统和连续激光加热系统测试块体单质铁,并通过同步 XRD 系统原位观测铁的相变过程,发现当压强为 35~40 GPa,铁转变为一种新的结构(β 相),这种结构被命名为双层六角密堆积(double-layer hexagonal close-packed,DHCP)结构。但是,实验表明,DHCP 铁只能存在于 35~40 GPa 的高压条件下,一旦高压撤除,它会自动变成六角或立方结构[44]。所以,人们无法在环境条件下观察到 DHCP 铁的存在。我们知道,纳米尺寸效应诱导产生的内压有可能使铁的某些亚稳相在环境条件下稳定存在,已知的有 FCC 铁和 HCP 铁等[30-43]。我们应用 LAL 方法,在环境条件下,成功合成了 DHCP 铁纳米晶,并且实验证明纳米尺寸效应能够使得 DHCP 铁纳米晶可以较长时间存在于大气环境中[46]。

LAL 合成 DHCP 铁纳米晶的技术路线如同前述,环境液体选择去离子水,99.97% 的铁作为靶材。合成产物的 XRD 谱如图 6-11 所示。我们可以看到一系列尖锐的衍射峰。对照粉末 X 射线衍射数据[44],这一系列衍射峰分别与 DHCP 铁的(100)、(101)、(004)、(102)、(103)和(110)等晶面的衍射峰非常吻合,这表明合成产物是 DHCP 铁。另外,图 6-11 中还有两个衍射峰应该来自于 γ-Fe_2O_3(JCPDS 卡片文件号 50-1275)。

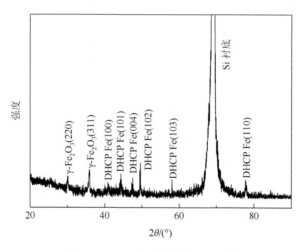

图 6-11 LAL 合成产物的 XRD 谱

为了进一步确定样品中氧化物的成分,我们使用微区激光拉曼光谱技术进行了表征,如图 6-12 所示。很清楚,我们可以看到,在 500 cm^{-1} 附近的特征峰应该是来自硅衬底,而另外三个分别位于 700 cm^{-1}、1360 cm^{-1}、1580 cm^{-1} 的散射峰,

与 γ-Fe$_2$O$_3$ 的拉曼散射谱非常一致[33]。因此，这些结果证明了样品中除了 DHCP 铁之外，还存在铁的氧化物 γ-Fe$_2$O$_3$。我们知道，单质铁作为金属，不存在拉曼散射，因此在我们的拉曼光谱中没有相应的信号。

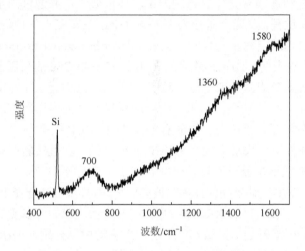

图 6-12　LAL 合成产物的拉曼光谱

非常有趣的是，LAL 合成 DHCP 铁纳米晶样品在干燥室温条件下放置 6 个月后，再次进行 XRD 测试（图 6-13），结果与第一次（图 6-11）完全不同。

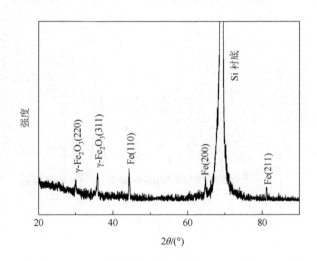

图 6-13　LAL 合成产物室温下放置 6 个月后的 XRD 谱

除了之前观察到的两个 γ-Fe$_2$O$_3$ 的衍射峰依然存在之外，其余 DHCP 铁的信号全部消失，取而代之的是稳定相的体心立方（body-centered cubic，BCC）铁相应的 XRD 信号[47]。这也从另一个方面验证了亚稳相材料的结构特点。尽管 DHCP 铁纳米晶在常温常压环境中可以存在一段时间，但是自身的晶格缺陷和周边环境（如温度、湿度、氧化等）所带来的微扰动在一个相对长时间的积累之下，会引起晶格弛豫使晶格向着能量更低和更加稳定的状态转变，也就是回到稳定相。因此，亚稳相铁纳米晶在存放了 6 个月之后，自发地从亚稳相 DHCP 铁变成为稳定相 BCC 铁。

接下来，我们应用 TEM 技术对 LAL 合成产物中的 DHCP 铁和 $\gamma\text{-}Fe_2O_3$ 进行了系统表征，如图 6-14 所示。图 6-14（a）低倍 TEM 照片显示样品为不规则纳米颗粒，主要尺寸分布在 5~20 nm。EDS 表明，样品的含氧量较高，与铁的原子个数比值大约为 1∶1，这说明样品中有大量的铁被氧化，C 和 Cu 信号则是来自承载样品的覆盖了碳支持膜的铜网。由于样品中纳米颗粒的尺寸普遍较小，用透射电子束难以测量到单个颗粒的 SAED 图像，因此我们使用快速傅里叶变换将 HRTEM 照像通过数学转换生成相对应的 SAED 图像，以便纳米晶面的标定。图 6-15（b）和图 6-15（c）分别代表 DHCP 铁和 $\gamma\text{-}Fe_2O_3$ 纳米晶的 HRTEM 图像和相应的经过 FFT 得到的 SAED 图像。DHCP 铁纳米晶的直径大约为 20 nm，并且较为清楚地呈现出＜101＞方向上的晶格层间结构和相应的 FFT 所得衍射点，另一个方向受 TEM 样品台转角限制只得到了较为模糊的晶格层间结构和相应的衍射点，SAED 图像中衍射点对应 DHCP 铁的（103）晶面。$\gamma\text{-}Fe_2O_3$ 纳米晶的尺寸要比 DHCP 铁小，直径约为 5 nm，二维的晶格层间结构和对应的 SAED 图像分别代表了 $\gamma\text{-}Fe_2O_3$ 的（311）和（222）晶面。

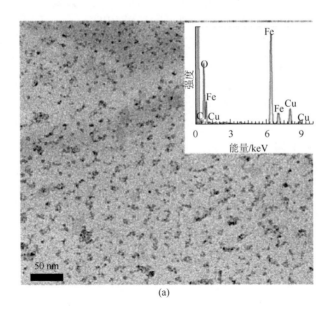

(a)

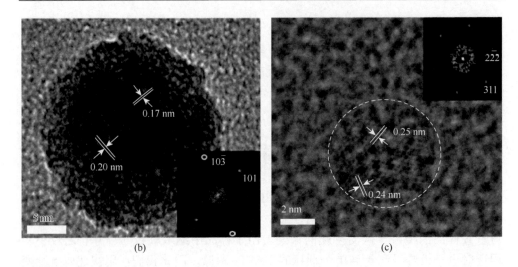

图 6-14 （a）LAL 合成 DHCP 铁纳米晶低倍 TEM 图像，插图是相应的 EDS；（b）DHCP 铁纳米晶的 HRTEM 照片，插图是经过 FFT 生成的 SAED 图像；（c）γ-Fe_2O_3 纳米晶的 HRTEM 照片，插图是经过 FFT 生成的 SAED 图像

DHCP 铁纳米晶是通过最简单的 LAL 方法即在水中激光烧蚀铁靶合成的，所以合成机制这里不再赘述。DHCP 铁被认为可能是地核的组成成分之一，但是，截至目前，除了在高压金刚石砧室系统的 XRD 原位观测中检测到 DHCP 铁的存在，尚未有其他实验能够在常温常压下合成这种只存在于高压下的特殊结构的亚稳相。然而，应用 LAL 方法，我们在常温常压下合成了 DHCP 铁纳米晶。而且纳米尺寸效应所产生的强大内压影响了物质的相图，改变了相变点[48,49]，所以 DHCP 铁纳米晶在环境条件下可以存在较长时间。但是，自然环境下存在着各种能量微扰动，如氧化、温度、湿度等，它们对有着很小表面积与体积之比的宏观块体材料的影响通常微乎其微，但是对于有着极大表面积与体积之比的纳米材料的影响则不可忽略。这些微扰动对纳米材料施加额外能量，会使纳米材料在小范围内偏离平衡态，当微扰动时间足够长、累积的能量足够大时，就可能驱动位于亚稳能级的纳米材料越过亚稳能级与稳态能级之间的势垒，以相变的形式回到稳态能级，这就是亚稳相的物理本质。所以，DHCP 铁的 LAL 合成有助于人们更深入地探索地球内部的组分和结构，研究铁的各种结构之间的热力学关系和物性的变化，也体现出 LAL 在合成亚稳纳米相方面表现出的其他热力学方法所不能替代的特点。

6.4 立方氮化硼纳米晶

科学家对立方氮化硼（cubic boron nitride，CBN 或 c-BN）材料的兴趣源于其独特的物理和化学性质,如高物理硬度(仅次于金刚石的第二硬材料)、高导热性、光学透明度等[50],更重要的是,与金刚石相比,c-BN 作为超硬材料在加工含铁材料时化学稳定好,没有任何热诱导的氧化反应[51]。自从 20 世纪 50 年代人们首次采用高温高压法合成立方氮化硼以来,已经发展了许多制备立方氮化硼材料的技术[52]。例如,PLD 技术作为一种重要的制备方法用于 c-BN 薄膜的制备[53-56]。众所周知,氮化硼的稳定相是六方氮化硼(hexagonal boron nitride,h-BN),而立方氮化硼是它的亚稳相。所以常规的基于平衡热力学的材料制备技术在一般情况下无法生长出立方氮化硼。我们应用 LAL 方法,在环境条件下,在国际上首次合成出了立方氮化硼纳米晶[57]。

LAL 合成立方氮化硼纳米晶的技术路线同前述 LAL 纳米晶合成。这里,我们选择丙酮(纯度 96%)为环境液体,六方氮化硼(纯度 99.7%)为靶材,合成产物的 XRD 谱如图 6-15 所示。我们可以看到,在 43.16°、74.16°、90.08°和 136.1°处有 4 个衍射峰,它们分别被标定为 c-BN 的(111)、(220)、(311)和(331)面。由于所获得的样品尚未被纯化处理,因此我们仍然可以在合成样品中观察到 h-BN 的衍射峰。因此,XRD 分析表明,在 LAL 合成产物中存在有氮化硼的立方相。同时,我们采用傅里叶变换红外光谱技术表征了 LAL 合成产物,如图 6-16 所示,因为 FTIR 光谱(400~4000 cm^{-1})可以通过探测最近邻键的性质来测量氮化硼的横光学声子频率。显然,我们可以看到 12 个红外吸收带,即 548 cm^{-1}、649 cm^{-1}、699 cm^{-1}、783 cm^{-1}、927 cm^{-1}、1027 cm^{-1}、1104 cm^{-1}、1195 cm^{-1}、1252 cm^{-1}、1435 cm^{-1}、2260 cm^{-1} 和 3218 cm^{-1}。研究人员曾在 c-BN 薄膜中观察到横光学声子频率[58-60],并且进行了理论计算[61],结果表明横光学声子频率位于 1004 cm^{-1}。

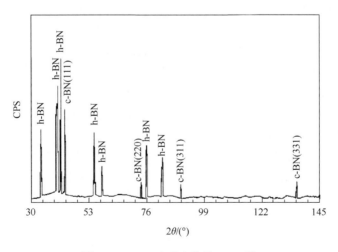

图 6-15 LAL 合成产物的 XRD 谱

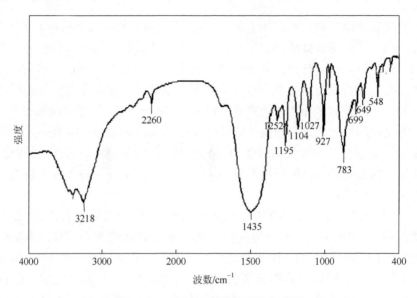

图 6-16 LAL 合成产物的 FTIR 光谱

考虑到 c-BN 的应变，所以它的红外吸收峰通常会向高频移动，在 1050 cm^{-1}～1100 cm^{-1}[62]。此外，Gielisse 等[63]和 Chrenko[64]测量了 c-BN 单晶在 400～3000 cm^{-1} 的红外光谱，分别在 650 cm^{-1}、700 cm^{-1}、1370 cm^{-1}、1580 cm^{-1}、1785 cm^{-1}、1830 cm^{-1}、2000 cm^{-1}、2140 cm^{-1}、2230 cm^{-1}、2465 cm^{-1}、2560 cm^{-1}、2700 cm^{-1} 和 2910 cm^{-1} 处观察到吸收峰。所以，与已报道的研究结果相比，这里 LAL 合成产物的绝大部分红外吸收峰都可以标定为氮化硼的相应相结构，649 cm^{-1}、699 cm^{-1}、1027 cm^{-1}、1104 cm^{-1}、1195 cm^{-1}、2260 cm^{-1} 属于 c-BN；783 cm^{-1}、1435 cm^{-1} 属于 h-BN；927 cm^{-1} 和 1252 cm^{-1} 属于爆炸相氮化硼（E-BN）[65]。但是，548 cm^{-1} 和 3218 cm^{-1} 的吸收带我们目前还并不清楚。因此，FTIR 光谱表征确认 XRD 的结果，LAL 合成了亚稳相立方氮化硼。

进一步，我们采用 TEM 技术对合成产物形貌进行表征，结果如图 6-17 所示。我们看到，合成产物为准球形纳米颗粒，其直径分布在 30～80 nm，类似于一种多晶形态，这种形态通常在金刚石及相关材料纳米晶中发现[66,67]。所以，LAL 合成产物为亚稳相立方氮化硼纳米晶。

无论是在技术路线还是合成机制，LAL 合成亚稳相立方氮化硼纳米晶都与 LAL 合成纳米金刚石颇为类似。当脉冲激光烧蚀固体靶 h-BN 时，所产生的激光诱导等离子体羽包含大量来自靶材的具有 sp^2 键的 B、N、B-N 及其离子等基团。随后，由于环境液体的束缚效应将激光诱导的等离子体羽驱动到一个高温高压高密度的极端热力学状态。根据氮化硼的热力学平衡相图，在此热力学状态下，sp^3 键的 c-BN 相是稳定相，而 sp^2 键的 h-BN 相则是亚稳相。因此，在等离子体羽发

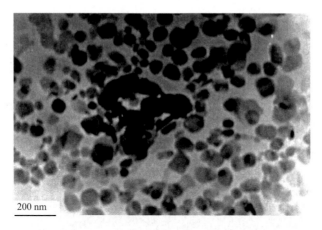

图 6-17　LAL 合成产物的 TEM 形貌图

生化学反应时，c-BN 晶核比 h-BN 晶核更易于生成。同时，等离子体羽中来自环境液体的如 OH^-、H^+ 及其离子等基团会参与到 c-BN 的成核反应，并增强 c-BN 晶核的形成和生长。例如，这些 OH^- 和 H^+ 可以通过抑制 sp^2 键的存在来促进 sp^2 向 sp^3 的转化[68]。最后，c-BN 晶核在等离子体羽快速淬灭中生长成纳米晶。因此，LAL 显然为合成亚稳相立方氮化硼纳米晶提供了一种有效的方法。

6.5　立方碳氮纳米晶

受四面体固体体积模量的经验模型的启发[69]，Liu 和 Cohen[70]通过第一性原理赝势能总能计算来研究 $\beta\text{-}C_3N_4$（C_3N_4，在 Si_3N_4 中用 C 替代 Si）的结构和物性，研究结果表明共价 C-N 固体的体积模量可以与金刚石相当，是超硬材料的良好候选者。这些理论研究推动了全球范围的 $\beta\text{-}C_3N_4$ 实验合成与表征和理论计算与模拟研究[71-81]。Teter 和 Hemly[80]从理论上发现了 C_3N_4 的 5 种形式，即 $\alpha\text{-}C_3N_4$、$\beta\text{-}C_3N_4$、赝立方相 C_3N_4、立方相 C_3N_4 和石墨相 C_3N_4，其中，$\alpha\text{-}C_3N_4$ 和石墨相 C_3N_4 在能量上优于 $\beta\text{-}C_3N_4$，而立方相 C_3N_4 可能具有超过金刚石的零压力模量，并且在环境条件下是亚稳相。目前，大多数用于制备这些碳氮材料的技术都是基于气相法（物理气相沉积和化学气相沉积），然而，这些技术所制备材料的结构表征表明，它们基本呈现出镶嵌有 $\beta\text{-}C_3N_4$ 小晶粒的非晶态。近年来，探索立方相 C_3N_4 的实验工作已经开始[81]。应用 LAL 方法，我们在国际上首次合成出了亚稳相立方结构 C_3N_4 纳米晶[67]。

在亚稳相立方结构 C_3N_4 纳米晶的 LAL 合成中，我们选择氨溶液（化学纯，氨浓度为 25%）为环境液体，高纯（≥99.99%）石墨为固体靶。我们采用 HRTEM 技术对合成样品的结构进行了分析，并采用能量色散 X 射线谱（EDS）对样品组分进行了分析。合成产物的 HRTEM 图像如图 6-18 所示,显然,我们可以清楚地观察到均匀的柱状晶体,相应的 SAED 图像如图 6-19 所示。表 6-2 列出了样品晶体结构的标定结果。显然,任何已知的碳同素异形体(金刚石、石墨、石墨等)都无法与 D_{exp} 匹配,而立方相 C_3N_4 的理论值[80]则与我们的实验值 D_{exp} 十分吻合。另外,样品 EDS 显示合成纳米晶体的组分为 C:45%(标准大气压)和 N:55%(标准大气压),并且 N/C≈1.2,接近 N/C = 4/3(1%的测量误差)。所以,我们合成的 C_3N_4 纳米晶的结构被标定为亚稳相立方结构。

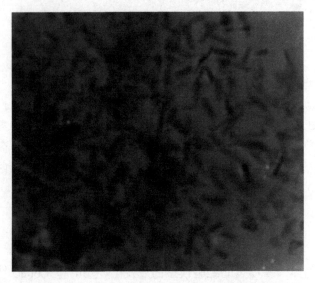

图 6-18 LAL 合成的 C_3N_4 纳米晶的 HRTEM 图像

纳米晶呈柱状多面体形状,平均尺寸(长度)约 50 nm

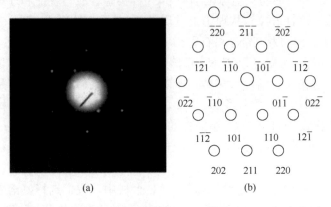

图 6-19 (a)合成纳米晶相应 SAED 图像和(b)标定结果

其中基轴 b = [111]

表 6-2　立方相 C_3N_4 纳米晶的透射电子衍射谱标定结果

(hkl)	D_{exp}/nm	D_{cala}/nm	强度
101	0.3802	0.3806	s
211	0.2202	0.2203	s
220	0.1911	0.1908	w
222	0.1555	0.1558	s
321	0.1440	0.1442	m
400	0.1345	0.1349	s
330	0.1269	0.1272	w
332	0.1155	0.1151	m
521	0.0983	0.0985	vw

D_{exp} 为实验值，D_{cala} 为理论值；s、m、w、vw 分别表示强、中等、弱、很弱。

基于上述实验结果，我们能够确信 LAL 合成了亚稳相立方结构 C_3N_4 纳米晶。同时，我们发现，合成产物中的石墨成分比例较高，C_3N_4 纳米晶占比相对较低。然而，需要注意的是，合成产物中的石墨呈现片状和球形，而 C_3N_4 纳米晶呈柱状，因此，我们可以很容易区分石墨和 C_3N_4 纳米晶。

LAL 合成亚稳相立方结构 C_3N_4 纳米晶的物理和化学过程可以分为三个阶段[66]：①激光诱导的等离子体羽的产生及高温高压高密度极端热力学状态的形成；②由于等离子体羽与环境液体的相互作用导致在界面处产生一个主要成分来自环境液体的由已有等离子体羽诱导的新等离子体羽，它一经产生就融入激光诱导的等离子体羽中；③来自固体靶和环境液体的活性基团分别在等离子体羽中和与环境液体之间的界面处发生化学反应，形成亚稳相纳米晶核。因此，随着等离子体羽在环境液体束缚中快速淬灭，亚稳相纳米晶核长大。由于生长时间非常短，所以在反应过程中形成的所有亚稳相都可以保留在最终产物中。此外，该 LAL 合成具有三个显著特征：第一，合成是在高温高压高密度状态下进行的，在该状态下优先形成亚稳相。第二，合成的化合物的组成分别来自固体靶和环境液体。第三，最终合成的材料在常温常压下通常具有亚稳结构。因此，LAL 为合成新颖亚稳相纳米晶体提供了一种新的方法。

6.6　立方氮化镓纳米晶

氮化镓（GaN）是第三代半导体即宽禁带半导体中的重要一员，由于其具有良好的蓝光发射能力而在光电纳米器件中有着巨大的应用潜力。众所周知，

GaN 晶体具有两种结构，一种是稳定相六方结构（h-GaN），另一种是亚稳相立方结构（c-GaN）。在常规的 GaN 材料合成与制备中人们能够得到的都是稳定相六方结构，而亚稳相立方 GaN 却很难在热力学平衡条件下制备出来。然而，与六方 GaN 相比，立方 GaN 具有一些更优越的物理性能[82]。因此，对立方 GaN 的研究，特别是对稳定相六方结构向亚稳相立方结构的转化研究，一直以来都是 GaN 材料研究领域里的热点[83-85]。我们应用 LAL 方法，通过脉冲激光辐照六方 GaN 粉末的悬浮液，成功实现 GaN 的六方相向立方相的转化，合成了立方 GaN 纳米晶[86]。

我们首先将纯度为 98%的六方 GaN 晶体在玛瑙研钵中研磨成均匀的细小粉末，然后与纯度大于 99.7%的无水乙醇充分混合配置成 GaN 粉末悬浊液，最后将脉冲激光聚焦于悬浊液并通过不断振荡悬浊液以便激光能够辐照而使反应均匀进行。合成产物的 XRD 谱如图 6-20 所示。我们从六方 GaN 靶材 XRD 衍射峰中可以清楚地看到，除了一个微弱的 O-Ga 特征峰外，其余的衍射峰都属于六方 GaN，这就表明靶材是纯净的六方 GaN 晶体。而通过对比合成产物与靶材的 XRD 衍射峰，我们可以发现，合成产物的衍射峰中除了原有的六方 GaN 衍射峰外，还出现了 5 个新的衍射峰。这 5 个峰除了一个已经被标定是 O-Ga 的特征峰、一个为 Si 特征峰（来自玛瑙研钵）外，其他 3 个新峰（如图 6-20 中箭头所示）被明确标定为立方 GaN 的(111)、(200)和(220)面的衍射峰。因此，XRD 谱表明合成产物中有立方 GaN 存在。

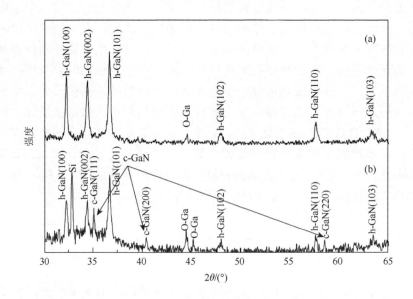

图 6-20　（a）六方 GaN 粉末 XRD 谱和（b）LAL 合成产物 XRD 谱

为了准确地获得合成产物的微观结构信息，我们对样品进行了细致的 TEM 分析，如图 6-21 所示。我们看到，在合成产物中，小尺寸颗粒普遍呈类球状形貌，尺寸范围大约在 50～150 nm。通过对单个类球状纳米晶 SAED 图像的分析，我们明确标定出所有衍射点分别隶属于立方 GaN 晶体的{111}、{220}、{131}和{311}晶面族。而其他小颗粒的电子衍射分析也证明了合成产物中的多数类球状纳米颗粒都具有立方结构，这也证明了 LAL 合成产物中存在立方 GaN 纳米晶。

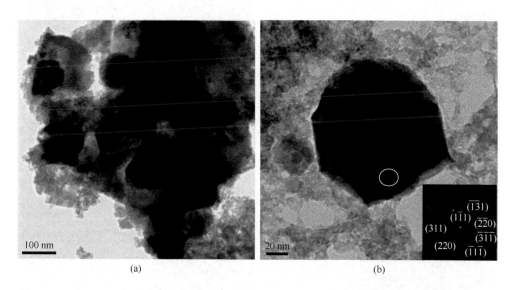

图 6-21　（a）LAL 合成纳米晶低倍 TEM 图像；（b）单个纳米晶高倍 TEM 明场图像，插图是相应的 SAED 图像，显示该 GaN 纳米晶具立方结构

为了进一步地表征合成纳米晶的相变情况，我们对产物中的纳米晶进行了 HRTEM 分析，如图 6-22 所示。我们不仅能够明确观察到立方 GaN 的晶格条纹相［图 6-22（a）］，0.258 nm 的晶格条纹相属于立方 GaN 晶体的(111)晶面，而且在纳米晶的不同区域中观测到了 GaN 的六方相和立方相的孪晶及层错混合结构［图 6-22（a）中画圈区域］，相应 SAED 图像明确显示出具有两种晶格相［图 6-22（b）］。因此，这种六方相和立方相共生的结果直观地反映了一部分 GaN 纳米晶中存在不完全的相变状态，同时也证明了 LAL 合成 GaN 纳米晶过程中，立方 GaN 纳米晶是由六方 GaN 晶体通过固-固相变而得到的。

同时，我们专门对其他数量较少、尺寸较大的有棱角 GaN 微晶进行了 TEM、SAED 和 HRTEM 分析，结果如图 6-22（c）和图 6-22（d）所示。我们发现这些有棱角的非球状颗粒普遍尺度较大（大于 200 nm），而且相应 SAED 分析表明

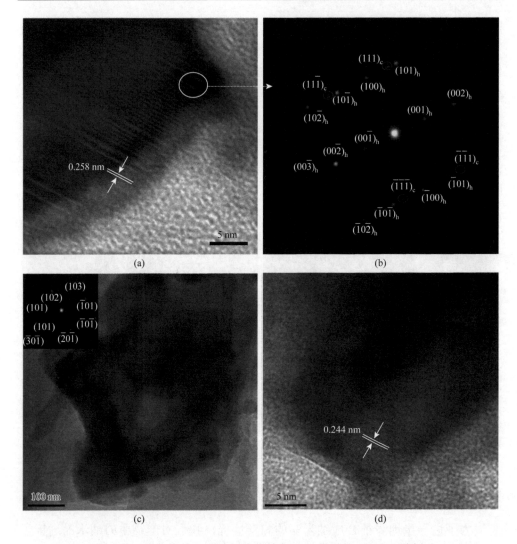

图 6-22 类球状 GaN 纳米晶和有棱角 GaN 微晶的 HRTEM 分析

(a) 类球状 GaN 纳米晶的高分辨图像发现其中存在层错区域；(b) 层错区域的 SAED 图像，其中画圈衍射点属于立方 GaN；(c) 有棱角 GaN 微晶的 TEM 明场图像及对应 SAED 图像，显示其结构属于 GaN 晶体的六方相；(d) 一个棱角 GaN 微晶的 HRTEM

它们都具有六方结构[图 6-22(c)中 SAED 图像可明确标定属于六方结构的{101}、{102}和{103}晶面族]。同时，HRTEM 分析也明确显示其中晶格条纹相的面间距为 0.244 nm，与六方 GaN 晶体（101）晶面的面间距吻合得很好。所以，这些颗粒都是六方 GaN 晶体。这些结果说明在 LAL 过程中，部分大尺寸六方 GaN 没有完全相变为立方 GaN。

下面我们给出 LAL 过程中 GaN 从稳定相六方结构转化为亚稳相立方结构的物理机制。如前所述，LAL 过程是一个非常短暂并且远离热力学平衡的过程，因此，许多在激光诱导等离子体羽的产生、膨胀、淬灭中生成的亚稳相会被保留在最终产物中，特别是具有中间态的亚稳相纳米材料。在 LAL 合成立方 GaN 纳米晶过程中，六方 GaN 相变成立方 GaN 纳米晶同样经历了三个阶段。首先，当脉冲激光辐照六方 GaN 微晶悬浊液的时候，会烧蚀微晶颗粒产生等离子体羽，其中包含了大量来自六方 GaN 的成分。然后，环境液体的束缚效应会将等离子体羽驱动到高温高压高密度的极端热力学状态，而在这个状态中亚稳相立方结构会比稳定相六方结构更加稳定。因此，许多具有六方结构的各种基团或团簇在高温高压作用下转化成立方结构。具体而言，六方 GaN 的（0002）基面在高温高压作用下能够发生压缩滑移，同时在（0002）基面邻近的原子会被迫调整它们的位置。通过理论计算，我们证明在这种相变状态下的 Ga 原子滑移距离只有原来原子间距的 0.7%，也就是说这种滑移相变在高温高压下的等离子体羽内部是很容易实现的。最后，六方 GaN 晶体的（0002）基面在经过滑移变化后转化成为立方 GaN 晶体的（111）晶面，并最终保留在合成产物中[87, 88]，生成亚稳相立方 GaN 纳米晶。图 6-23 给出了相应的 LAL 过程六方结构向立方转结构变的固-固相变机制。

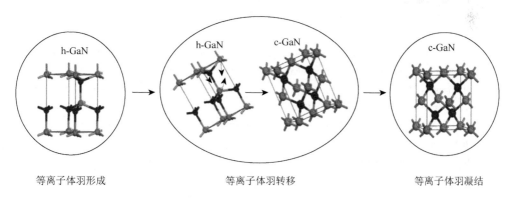

等离子体羽形成　　　　　　等离子体羽转移　　　　　　等离子体羽凝结

图 6-23　LAL 过程六方结构向立方结构转变的固-固相变机制

此外，我们运用第一性原理方法计算了六方 GaN 和立方 GaN 晶体的能量与晶体体积的变化关系，如图 6-24 所示。我们可以看到，这两种结构的总能差是非常小的。这个重要理论说明了两个问题，其一是这么小的总能差通过远离热力学平衡的 LAL 方法可以很容易地克服[89, 90]，从而实现稳定相六方结构向亚稳相立方结构的转变；其二是因为这两种结构总能差非常小，所以立方相在与六方相竞争生长中的优势不是太大，这就导致在最终产物中存在两相共存的现象。

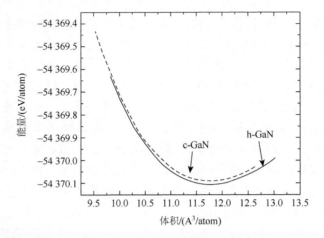

图 6-24　立方 GaN 和六方 GaN 晶体的能量随体积的变化

参 考 文 献

[1] Molas G, De Salvo B, Mariolle D, et al. Single electron charging and discharging phenomena at room temperature in a silicon nanocrystal memory[J]. Solid-State Electronics, 2003, 47 (10): 1645-1649.

[2] Takahashi Y, Ono Y, Fujiwara A, et al. Development of silicon single-electron devices[J]. Physica E: Low-dimensional Systems and Nanostructures, 2003, 19 (1-2): 95-101.

[3] Emiroglu E G, Hasko D G, Williams D A. Single-electron polarization of an isolated double quantum dot in silicon[J]. Microelectronic Engineering, 2004, 73-74: 701-706.

[4] Ota T, Hatano T, Ono K, et al. Single electron spectroscopy in a single pair of weakly coupled self-assembled InAs quantum dots[J]. Physica E: Low-dimensional Systems and Nanostructures, 2004, 22 (1-3): 510-513.

[5] Heitmann J, Schmidt M, Zacharias M, et al. Fabrication and photoluminescence properties of erbium doped size-controlled silicon nanocrystals[J]. Materials Science and Engineering: B, 2003, 105 (1-3): 214-220.

[6] Larsson M, Elfving A, Holtz P O, et al. Photoluminescence study of Si/Ge quantum dots[J]. Surface Science, 2003, 532-535: 832-836.

[7] Švrček V, Sasaki T, Shimizu Y, et al. Blue luminescent silicon nanocrystals prepared by ns pulsed laser ablation in water[J]. Applied Physics Letters, 2006, 89 (21): 213113.

[8] Mazen F, Baron T, Papon A M, et al. A two steps CVD process for the growth of silicon nano-crystals[J]. Applied Surface Science, 2003, 214 (1-4): 359-363.

[9] Baron T, Martin F, Mur P, et al. Silicon quantum dot nucleation on Si_3N_4, SiO_2 and SiO_xN_y substrates for nanoelectronic devices[J]. Journal of Crystal Growth, 2000, 209 (4): 1004-1008.

[10] Dubček P, Desnica U V, Desnica-Franković I D, et al. GISAXS study of shape and size of CDS nanocrystals formed in monocrystalline silicon by ion implantation[J]. Nuclear Instruments and Methods in Physics Research Section B: Beam Interactions with Materials and Atoms, 2003, 200: 138-141.

[11] Liu P, Cao Y L, Cui H, et al. Micro-and nanocubes of silicon with zinc-blende structure[J]. Chemistry of Materials, 2008, 20 (2): 494-502.

[12] Haro E, Balkanski M, Wallis R F, et al. Theory of the anharmonic damping and shift of the Raman mode in

silicon[J]. Physical Review B, 1986, 34 (8): 5358.

[13] Zhang S L, Hou Y T, Ho K S, et al. Raman investigation with excitation of various wavelength lasers on porous silicon[J]. Journal of Applied Physics, 1992, 72 (9): 4469-4471.

[14] Poborchii V, Tada T, Kanayama T. Giant heating of Si nanoparticles by weak laser light: Optical microspectroscopic study and application to particle modification[J]. Journal of Applied Physics, 2005, 97 (10): 104323.

[15] Wiley B J, Im S H, Li Z Y, et al. Maneuvering the surface plasmon resonance of silver nanostructures through shape-controlled synthesis[J]. The Journal of Physical Chemistry B, 2006, 110 (32): 15666-15675.

[16] Yeh C Y, Lu Z W, Froyen S, et al. Zinc-blende-wurtzite polytypism in semiconductors[J]. Physical Review B, 1992, 46 (16): 10086.

[17] Wang Z L. Book reprint: Characterization of nanophase materials[J]. Particle & Particle Systems Characterization, 2001, 18 (3): 142-165.

[18] Yang S W, Gao L. Controlled synthesis and self-assembly of CeO_2 nanocubes[J]. Journal of the American Chemical Society, 2006, 128 (29): 9330-9331.

[19] Xia J S, Nemoto K, Ikegami Y, et al. Silicon-based light emitters fabricated by embedding Ge self-assembled quantum dots in microdisks[J]. Applied Physics Letters, 2007, 91 (1): 011104.

[20] Lin C W, Lin S Y, Lee S C, et al. Structural and optical properties of germanium nanoparticles[J]. Journal of Applied Physics, 2002, 91 (3): 1525-1528.

[21] Jurbergs D, Rogojina E, Mangolini L, et al. Silicon nanocrystals with ensemble quantum yields exceeding 60%[J]. Applied Physics Letters, 2006, 88 (23): 233116.

[22] Wilcoxon J P, Provencio P P, Samara G A. Synthesis and optical properties of colloidal germanium nanocrystals[J]. Physical Review B, 2001, 64 (3): 035417.

[23] Nesher G, Kronik L, Chelikowsky J R. Ab initio absorption spectra of Ge nanocrystals[J]. Physical Review B, 2005, 71 (3): 035344.

[24] Bates C H, Dachille F, Roy R. High-pressure transitions of germanium and a new high-pressure form of germanium[J]. Science, 1965, 147 (3660): 860-862.

[25] Joannopoulos J D, Cohen M L. Electronic properties of complex crystalline and amorphous phases of Ge and Si. I. Density of states and band structures[J]. Physical Review B, 1973, 7 (6): 2644.

[26] Sato S, Nozaki S, Morisaki H, et al. Tetragonal germanium flims deposited by the cluster-beam evaporation technique[J]. Applied Physics Letters, 1995, 66 (23): 3176-3178.

[27] Nozaki S, Sato S, Rath S, et al. Optical properties of tetragonal germanium nanocrystals deposited by the cluster-beam evaporation technique: New light emitting material for future[J]. Bulletin of Materials Science, 1999, 22: 377-381.

[28] Liu P, Cao Y L, Chen X Y, et al. Trapping high-pressure nanophase of Ge upon laser ablation in liquid[J]. Crystal Growth & Design, 2009, 9 (3): 1390-1393.

[29] Kobliska R J, Solin S A, Selders M, et al. Raman scattering from phonons in polymorphs of Si and Ge[J]. Physical Review Letters, 1972, 29 (11): 725.

[30] Cozzoli P D, Snoeck E, Garcia M A, et al. Colloidal synthesis and characterization of tetrapod-shaped magnetic nanocrystals[J]. Nano Letters, 2006, 6 (9): 1966-1972.

[31] Sardan O, Yalcinkaya A D, Alaca B E. Self-assembly-based batch fabrication of nickel-iron nanowires by electroplating[J]. Nanotechnology, 2006, 17 (9): 2227.

[32] Yang D D, Ni X M, Zhang D G, et al. Preparation and characterization of hcp Co-coated Fe nanoparticles[J]. Journal of Crystal Growth, 2006, 286 (1): 152-155.

[33] Han Q, Liu Z H, Xu Y Y, et al. Growth and properties of single-crystalline γ-Fe_2O_3 nanowires[J]. The Journal of Physical Chemistry C, 2007, 111 (13): 5034-5038.

[34] Xiong Y, Ye J, Gu X Y, et al. Synthesis and assembly of magnetite nanocubes into flux-closure rings[J]. The Journal of Physical Chemistry C, 2007, 111 (19): 6998-7003.

[35] Han Y C, Cha H G, Kim C W, et al. Synthesis of highly magnetized iron nanoparticles by a solventless thermal decomposition method[J]. The Journal of Physical Chemistry C, 2007, 111 (17): 6275-6280.

[36] Yan Q F, Purkayastha A, Kim T, et al. Synthesis and assembly of monodisperse high-coercivity silica-capped FePt nanomagnets of tunable size, composition, and thermal stability from microemulsions[J]. Advanced Materials, 2006, 18 (19): 2569-2573.

[37] Patolsky F, Zheng G F, Lieber C M. Nanowire sensors for medicine and the life sciences[J]. Nanomedicine, 2006, 1 (1): 51.

[38] Dumestre F, Chaudret B, Amiens C, et al. Superlattices of iron nanocubes synthesized from $Fe[N(SiMe_3)_2]_2$[J]. Science, 2004, 303 (5659): 821-823.

[39] Zhang L C, Calin M, Paturaud F, et al. Deformation-induced nanoscale high-temperature phase separation in Co-Fe alloys at room temperature[J]. Applied Physics Letters, 2007, 90 (20): 201908.

[40] Pan F, Zhang M, Ding M, et al. Metastable rhombohedral Fe phase formed in Fe/Sb multilayers and its magnetic properties[J]. Physical Review B, 1999, 59 (17): 11458.

[41] Rueff J P, Krisch M, Cai Y Q, et al. Magnetic and structural α-ε phase transition in Fe monitored by X-ray emission spectroscopy[J]. Physical Review B, 1999, 60 (21): 14510.

[42] Wu Y Z, Won C, Scholl A, et al. Magnetic stripe domains in coupled magnetic sandwiches[J]. Physical Review Letters, 2004, 93 (11): 117205.

[43] Song C, Wei X X, Geng K W, et al. Magnetic-moment enhancement and sharp positive magnetoresistance in Co/Ru multilayers[J]. Physical Review B, 2005, 72 (18): 184412.

[44] Saxena S K, Dubrovinsky L S, Häggkvist P, et al. Synchrotron X-ray study of iron at high pressure and temperature[J]. Science, 1995, 269 (5231): 1703-1704.

[45] Andrault D, Fiquet G, Kunz M, et al. The orthorhombic structure of iron: An in situ study at high-temperature and high-pressure[J]. Science, 1997, 278 (5339): 831-834.

[46] Chen X Y, Cui H, Liu P, et al. Double-layer hexagonal Fe nanocrystals and magnetism[J]. Chemistry of Materials, 2008, 20 (5): 2035-2038.

[47] Varadwaja K S, Panigrahi M K, Ghose J. Effect of capping and particle size on Raman laser-induced degradation of γ-Fe_2O_3 nanoparticles[J]. Journal of Solid State Chemistry, 2004, 177 (11): 4286-4293.

[48] Wang C X, Yang G W. Thermodynamics of metastable phase nucleation at the nanoscale[J]. Materials Science and Engineering: R: Reports, 2005, 49 (6): 157-202.

[49] Kong L T, Liu B X. Correlation of magnetic moment versus spacing distance of metastable fcc structured iron[J]. Applied Physics Letters, 2004, 84 (18): 3627-3629.

[50] Davis R F. III-V nitrides for electronic and optoelectronic applications[J]. Proceedings of the IEEE, 1991, 79 (5): 702-712.

[51] Ichinose Y, Saitoh H, Hirotsu Y. Synthesis of cubic BN from the gas phase by a new plasma chemical vapour deposition method using r.f. waves and a tungsten filament[J]. Surface and Coatings Technology, 1990, 43-44 (1):

116-127.

[52] Wentorf Jr R H. Cubic form of boron nitride[J]. The Journal of Chemical Physics, 1957, 26 (4): 956.

[53] Miller J C. Laser ablation: Principles and applications[M]. Heidelberg: Springer-Verlag, 1994.

[54] Medlin D L, Friedmann T A, Mirkarimi P B, et al. Microstructure of cubic boron nitride thin films grown by ion-assisted pulsed laser deposition[J]. Journal of Applied Physics, 1994, 76 (1): 295-303.

[55] Zhang C Y, Zhong X L, Wang J B, et al. Room-temperature growth of cubic nitride boron film by RF plasma enhanced pulsed laser deposition[J]. Chemical Physics Letters, 2003, 370 (3-4): 522-527.

[56] Francis A A, Marcus H L. Laser synthesis of boron nitride powder[J]. International Journal of Powder Metallurgy (1986), 2001, 37 (6): 67-72.

[57] Wang J B, Yang G W, Zhang C Y, et al. Cubic-BN nanocrystals synthesis by pulsed laser induced liquid-solid interfacial reaction[J]. Chemical Physics Letters, 2003, 367 (1-2): 10-14.

[58] Sanjurjo J A, López-Cruz E, Vogl P, et al. Dependence on volume of the phonon frequencies and the ir effective charges of several Ⅲ-V semiconductors[J]. Physical Review B, 1983, 28 (8): 4579.

[59] Saitoh H, Yarbrough W A. Preparation and characterization of nanocrystalline cubic boron nitride by microwave plasma-enhanced chemical vapor deposition[J]. Applied Physics Letters, 1991, 58 (20): 2228-2230.

[60] Mineta S, Kohata M, Yasunaga N, et al. Preparation of cubic boron nitride film by CO_2 laser physical vapour deposition with simultaneous nitrogen ion supply[J]. Thin Solid Films, 1990, 189 (1): 125-138.

[61] Wentzcovitch R M, Chang K J, Cohen M L. Electronic and structural properties of BN and BP[J]. Physical Review B, 1986, 34 (2): 1071.

[62] Zhao Y N, Zou G T, Wang B, et al. Studies on the stress in cubic boron nitride films by infrared spectroscopy[J]. Chemical Journal of Chinese Universities, 1998, 19 (7): 1136-1139.

[63] Gielisse P J, Mitra S S, Plendl J N, et al. Lattice infrared spectra of boron nitride and boron monophosphide[J]. Physical Review Online Archive, 1967, 155 (3): 1039.

[64] Chrenko R M. Ultraviolet and infrared spectra of cubic boron nitride[J]. Solid State Communications, 1974, 14 (6): 511-515.

[65] Olszyna A, Konwerska-Hrabowska J, Lisicki M. Molecular structure of E-BN[J]. Diamond and Related Materials, 1997, 6 (5-7): 617-620.

[66] Yang G W, Wang J B, Liu Q X. Preparation of nano-crystalline diamonds using pulsed laser induced reactive quenching[J]. Journal of Physics: Condensed Matter, 1998, 10 (35): 7923.

[67] Yang G W, Wang J B. Carbon nitride nanocrystals having cubic structure using pulsed laser induced liquid-solid interfacial reaction[J]. Applied Physics A, 2000, 71: 343-344.

[68] Xiao R F. Growing diamond films from an organic liquid[J]. Applied Physics Letters, 1995, 67 (21): 3117-3119.

[69] Cohen M L. Calculation of bulk moduli of diamond and zinc-blende solids[J]. Physical Review B, 1985, 32 (12): 7988.

[70] Liu A Y, Cohen M L. Prediction of new low compressibility solids[J]. Science, 1989, 245 (4920): 841-842.

[71] Corkill J L, Cohen M L. Calculated quasiparticle band gap of β-C_3N_4[J]. Physical Review B, 1993, 48 (23): 17622.

[72] Yao H Y, Ching W Y. Optical properties of β-C_3N_4 and its pressure dependence[J]. Physical Review B, 1994, 50 (15): 11231.

[73] Guo Y L, Godard W A. Is carbon nitride harder than diamond? No, but its girth increases when stretched (negative Poisson ratio) [J]. Chemical Physics Letters, 1995, 237 (1-2): 72-76.

[74] Han H X, Feldman B J. Structural and optical properties of amorphous carbon nitride[J]. Solid State Communications, 1988, 65 (9): 921-923.

[75] Sekine T, Kanda H, Bando Y, et al. A graphitic carbon nitride[J]. Journal of Materials Science Letters, 1990, 9: 1376-1378.

[76] Niu C M, Lu Y Z, Lieber C M. Experimental realization of the covalent solid carbon nitride[J]. Science, 1993, 261 (5119): 334-337.

[77] Marton D, Boyd K J, Al-Bayati A H, et al. Carbon nitride deposited using energetic species: A two-phase system[J]. Physical Review Letters, 1994, 73 (1): 118.

[78] Li D, Chu X, Cheng S C, et al. Synthesis of superhard carbon nitride composite coatings[J]. Applied Physics Letters, 1995, 67 (2): 203-205.

[79] Sjöström H, Stafström S, Boman M, et al. Superhard and elastic carbon nitride thin films having fullerenelike microstructure[J]. Physical Review Letters, 1995, 75 (7): 1336.

[80] Teter D M, Hemley R J. Low-compressibility carbon nitrides[J]. Science, 1996, 271: 53-55.

[81] Peng Y G, Ishigaki T, Horiuchi S. Cubic C_3N_4 particles prepared in an induction thermal plasma[J]. Applied Physics Letters, 1998, 73 (25): 3671-3673.

[82] Wang W Y, Xu Y P, Zhang D F, et al. Synthesis and dielectric properties of cubic GaN nanoparticles[J]. Materials Research Bulletin, 2001, 36 (12): 2155-2162.

[83] Dhara S, Datta A, Wu C T, et al. Hexagonal-to-cubic phase transformation in GaN nanowires by Ga^+ implantation[J]. Applied Physics Letters, 2004, 84 (26): 5473-5475.

[84] Purdy A P. Ammonothermal synthesis of cubic gallium nitride[J]. Chemistry of Materials, 1999, 11(7): 1648-1651.

[85] Al-Sharif A I. Structural phase transformation of GaN under high-pressure: An exact exchange study[J]. Solid State Communications, 2005, 135 (8): 515-518.

[86] Liu P, Cao Y L, Cui H, et al. Synthesis of GaN nanocrystals through phase transition from hexagonal to cubic structures upon laser ablation in liquid[J]. Crystal Growth & Design, 2008, 8 (2): 559-563.

[87] Yang G W, Wang J B. Pulsed-laser-induced transformation path of graphite to diamond via an intermediate rhombohedral graphite[J]. Applied Physics A, 2001, 72: 475-479.

[88] Wang C X, Liu P, Cui H, et al. Nucleation and growth kinetics of nanocrystals formed upon pulsed-laser ablation in liquid[J]. Applied Physics Letters, 2005, 87 (20): 201913.

[89] Wang C X, Yang Y H, Liu Q X, et al. Phase stability of diamond nanocrystals upon pulsed-laser-induced liquid-solid interfacial reaction: Experiments and ab initio calculations[J]. Applied Physics Letters, 2004, 84 (9): 1471-1473.

[90] Liu Q X, Yang G W, Zhang J X. Phase transition between cubic-BN and hexagonal BN upon pulsed laser induced liquid-solid interfacial reaction[J]. Chemical Physics Letters, 2003, 373 (1-2): 57-61.

第 7 章 LAL 纳米制备技术

根据对 LAL 纳米制备技术所涉及的基本物理和化学过程的描述，我们不难发现，LAL 在纳米晶合成和纳米结构组装等若干方面具有显著的优势。例如，LAL 过程中脉冲激光诱导的等离子体羽极短的淬灭时间为合成小尺寸纳米晶及超小纳米晶或亚纳米晶提供了动力学通道；由于 LAL 纳米制备技术基本无须复杂的化学前驱体，因此所制备的纳米晶具有超洁净的表面；环境液体的束缚效应将脉冲激光诱导的等离子体羽驱动到一个局域高温高压高密度的远离热力学平衡的极端状态，从而使得亚稳相的形成成为可能。这些独特的优势使得 LAL 在纳米制备中拥有巨大的应用潜力[1]。本章，我们介绍如何通过外场（电场、磁场、温度场、电化学等）辅助来拓展 LAL 在纳米材料制备和加工中的应用，进而发展一系列 LAL 纳米制备技术。

7.1 电场辅助 LAL 用于金属氧化物纳米晶形貌控制

7.1.1 多种形貌金属氧化物半导体纳米晶

GeO_2 是一种高 k 电介质材料，具有良好的结构稳定性，GeO_2 纳米材料及其结构已经广泛用于光电子学领域[1, 2]。通常具有可控形状的 GeO_2 纳米材料可以通过溶胶-凝胶反应合成。然而，溶胶-凝胶反应会有大量的封端剂附着在 GeO_2 纳米材料的表面，这会极大地降低其性能如 k 值等[3, 4]。为了解决这个问题，我们发展了一种独特的纳米制备技术即电场辅助 LAL（EFLAL），无须任何催化剂或有机添加剂，用于可控组装 GeO_2 微米和立方体纳米晶及具有高折射率面的纺锤体纳米晶[5]。EFLAL 和施加于等离子体羽的外部电场对等离子体羽的作用示意图分别如图 7-1（a）和图 7-1（b）所示。我们可以看到，常规 LAL 和 EFLAL 的区别在于，EFLAL 是在固体靶两侧施加来自两个平行电极的可调电压的直流（DC）电场。EFLAL 制备的样品表现出两种不同的形态，由（1011）高折射率面组成的立方体纳米晶 [图 7-1（c）] 和纺锤体纳米晶 [图 7-1（d）]。不同的形貌与所施加的电场有关，立方体纳米晶在 14.5 V 的电压下获得，而纺锤体纳米晶在 32 V 的电压下形成。有趣的是，当样品形貌从立方体演变为纺锤体时，可以观察到其发光波长的红移，也就是它们的发光表现出强烈的形貌依赖 [图 7-1（e）]。这些结果充分表明了在 EFLAL 纳米制备中电场对纳米材料形貌的调控。

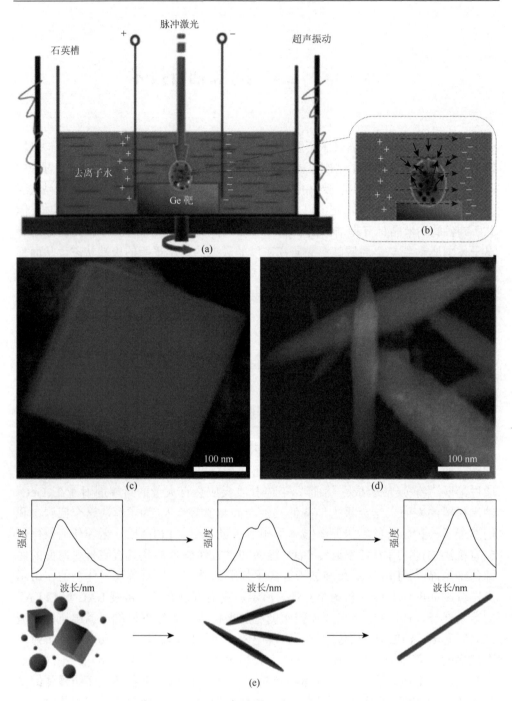

图 7-1 （a）EFLAL 示意图；（b）外部电场对等离子体羽的作用示意图；不同外部电场条件下制备的（c）立方体纳米晶和（d）纺锤体纳米晶的高倍 SEM 图像；（e）具有不同形貌 GeO_2 纳米结构的发光红移

7.1.2 电场对纳米晶形貌形成的影响

为了比较 LAL 和 EFLAL，我们在无外部电场的相同条件下进行了重复实验，以证明所施加的电场对制备纳米材料形貌的控制作用。我们发现不加外部电场的对照实验皆得到了球形纳米颗粒，这与相关报道是一致的[6, 7]。所以，我们可以得到结论，在 EFLAL 过程中，外部电场对高晶面指数的立方体纳米晶和纺锤体纳米晶的形成具有显著影响。另外，在 LAL 过程中，制备的球形纳米颗粒是 Ge 而不是 GeO_2。因此，这个重要结果表明环境液体水在 EFLAL 过程中可能被分解，而这有利于金属氧化物的形成。

GeO_2 立方体纳米晶和纺锤体纳米晶的 EFLAL 合成可以归因于激光与固体靶、环境液体、外部电场的相互作用。在 EFLAL 过程中，首先，激光烧蚀固体靶会产生脉冲激光诱导的 Ge 等离子体羽。然后，等离子体羽的高温会使界面处的水分解并产生 O^{2-} 和 OH^- 活性基团。最后，这些活性基团会与等离子体羽反应，生成 GeO_2 纳米颗粒。我们的研究发现，样品的 X 射线衍射（XRD）和选区电子衍射（SAED）图像均表明了 GeO_2 的（1011）面占主导地位。所以，这些结果表明，施加的外部电场有助于稳定的（1011）面的生长，并会阻碍垂直于该面的气态晶面的生长。在这种情况下，6 个（1011）面很容易形成立方体的形态。随着施加的外部电场增强，（1011）面的生长将沿 [0001] 方向被拉长，从而形成纺锤体的形貌。这就是施加不同的外部电场会获得不同形貌的 GeO_2 纳米结构的原因。因此，这些研究证明，EFLAL 纳米制备技术不仅在化学上是清洁的，而且所合成纳米材料的形貌可以通过辅助电场来调控。所以，该技术可以用于合成其他类型的具有可控形貌和功能的金属氧化物半导体纳米材料。

然而，一些研究人员在不施加外部电场的 LAL 中也制备了类似的纳米结构[8, 9]。这就提出了一个问题，在 EFLAL 过程中，外部电场是否在所制备的纳米材料形成中起着决定性的作用。为此，我们比较了其他研究人员制备的类似纳米结构及制备条件，总结了这些纳米结构形成的机制。Khan 等[9]在 LAL 纳米制备技术中使用波长为 1070 nm 的高功率、高亮度连续光纤激光合成 NiO 纳米颗粒。他们发现，制备的纳米颗粒形貌受表面活性剂 SDS 的影响，随着 SDS 浓度的提高，纳米颗粒形貌从球形变为四方形，如 SDS 浓度为 0.1 mol/L 时，大多数纳米颗粒为四方晶系。根据选择性吸附理论，表面活性剂或表面活性剂混合物会选择性地吸附在生长晶体的不同晶面上，这就改变了各个晶面的自由能并导致所得晶体具有各向异性的形貌。所以，在这类研究中，不同形貌纳米晶的获得是通过表面活性剂对生长晶面自由能的调节来调控纳米结构形貌的。Wang 等[8]报道了 YVO_4：Eu^{3+} 多晶的卵形纳米结构，该纳米结构是由许多小纳米颗粒聚集而成的。实际上，这是一种常见的由小尺寸纳米颗粒聚集成大尺寸纳米结构的定向附着机制，其物理根源就是聚集体自由能的最小化。而在 EFLAL 制备 GeO_2 研究中，我们证明了外部电场对立方体纳米晶和纺锤体纳米晶形成的作用，在没有外部电场的情况下只能得到球形纳米颗粒，这意味着电场在控制纳米颗粒形貌方面起着重要作用。并且我们在 EFLAL 中没有添加任何表面活性剂，只有高纯 Ge 靶和去离子水。由此可见，在 EFLAL 中，外部电场确实对纳米材料形貌生成有重要影响。

7.2 电场辅助 LAL 用于纳米结构组装

7.2.1 金属氧化物功能纳米结构

我们不仅可以通过外部电场来控制 LAL 过程中纳米晶的形貌，还可以通过电场辅助把 LAL 合成的小尺寸纳米晶组装成功能纳米结构[10]。前文我们讨论了附加电场对晶体生长方向的影响，本节，我们探讨如何通过电场将纳米晶组装成有序的纳米结构。我们以 CuO 纺锤体纳米晶的组装作为实例介绍 EFLAL 在纳米结构组装中的应用。在 EFLAL 中，当激光烧蚀去离子水中的 Cu 靶时，我们在反应池的两侧施加了电压可调的直流电场。需要注意的是，在 7.1 节中，外部电场是施加在固体靶的两边，而在这里，外部电场是施加在反应池的两侧。研究结果表明，EFLAL 合成的小纳米晶自发组装成了纺锤体纳米结构，这些纳米结构由棒状和球形小纳米颗粒组成[图 7-2（a）和图 7-2（b）]。重要的是，这些纺锤体纳米结构的光学吸收峰取决于它们的形态，如图 7-2（c）所示。显然，在没有施加外部电场的情况下，EFLAL 合成的样品仅能观察到位于 400 nm 处的一个宽峰，与之前的研究报道类似[11-13]。但是，在施加 40 V、80 V 和 120 V 电压时，EFLAL 合成的样品的吸收光谱中出现了两个宽峰。具体而言，这些吸收峰可以分为两组，包括吸收峰波长相对较长的和相对较短的，可以观察到的两个峰，其纳米材料光吸收取决于它的偏振和能带结构，而这些物性与其形貌和尺寸密切相关。因此，两个峰的出现表明 CuO 纺锤体纳米结构的尺寸和形态在不同的外加电场下发生了变化。

(a)

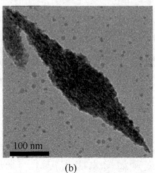

(b)

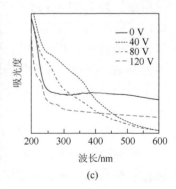

(c)

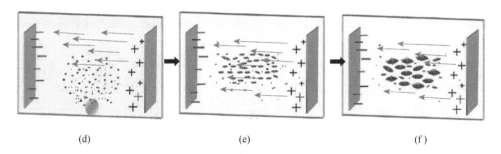

图 7-2　80 V 电场辅助 LAL 组装 CuO 纺锤体纳米结构的（a）低倍和（b）高倍 TEM 图像；（c）CuO 纺锤体纳米结构的尺寸依赖的紫外可见吸收光谱；（d）纳米晶的合成；（e）由附加电场引起的纳米晶的聚集；（f）由附加电场引起的纺锤体纳米结构的组装

7.2.2　纳米结构组装中的定向附着机制

通常来说，在 EFLAL 过程中 CuO 纺锤体纳米结构的形成归因于纳米颗粒的定向附着和聚集[14, 15]。在这种机制中，大尺寸纳米颗粒是由最初的小尺寸纳米颗粒通过定向附着生长形成的。在生长过程中，具有共同的晶体取向的相邻小纳米颗粒会在平面界面处结合，降低系统的整体自由能，这便是自组装过程[10, 16-20]。这一机制可以通过透射电子显微镜（TEM）观察到，我们发现所组装的 CuO 纺锤体纳米结构包含许多沿特定方向排列的结构单元。所以，在 EFLAL 组装 CuO 纺锤体纳米结构过程中，首先 LAL 合成小尺寸的 CuO 纳米晶[图 7-2（d）]，然后这些小纳米晶通过定向附着机制聚集形成棒状 CuO 纳米结构[图 7-2（e）]，最后在外部电场的辅助下，CuO 纳米晶和纳米棒会作为构筑砖块（building blocks）进一步聚集组装成纺锤体纳米结构[图 7-2（f）]。这表明，外部电场在纺锤体纳米结构的自组装中起着至关重要的作用。我们注意到，Wang 等[8]在不施加外部电场的情况下通过 LAL 制备了具有相似形貌的 $YVO_4:Eu^{3+}$ 多晶纳米结构。这意味着对于某些特定的金属氧化物纳米结构，通过定向附着机制来组装纺锤形纳米结构是可行的。

在 EFLAL 组装 CuO 纺锤体纳米结构中，当施加的外部电场强度增加时，纳米结构的形状会发生变化，特别是纺锤体纳米结构的长度与直径的平均比会发生变化。为了量化 CuO 纳米结构在外部电场作用下的形貌变化，我们采用紫外可见吸收光谱来分析纳米结构光学性质随外部电场的变化，因为纳米结构的形状和尺寸强烈影响其光学性质。换句话说，LAL 过程中纳米材料生长发生了定向附着，而电场会增强这一过程的发生。之所以能够实现这种定向组装，是因为 CuO 是一种极性金属氧化物，在外部电场作用下很容易发生极化。这样的结果就是，由于电场引起偶极相互作用，CuO 在施加电场方向上定向排列，这是纳米晶定向聚集的驱动力之一[18]。所以，EFLAL 组装纳米结构的方法可以用于纳米构筑单元组装功能纳米结构。

7.3 温度场辅助 LAL 用于纳米结构组装

7.3.1 一维和二维纳米结构

在没有外场辅助的情况下，LAL 合成的纳米材料总是具有球形形态，包括贵金属、氧化物、硫化物和硒化物等纳米颗粒[21-24]，这是因为在脉冲激光诱导的等离子体羽淬灭过程中（相变和凝结），相同体积下球形形态的表面积最小或表面自由能最小。研究人员提出，通过在一定温度下的熟化（ripening）过程，LAL 合成的球形纳米晶会自发组装成各种纳米结构，例如，一维纳米线和二维纳米片等[25, 26]。

Xiao 等[25, 26]发展了用于功能纳米结构组装的温度场辅助 LAL 技术。我们以温度场辅助 LAL 组装 MnOOH 纳米线为例介绍这项简单实用的纳米组装技术。首先是通过常规 LAL 合成纳米晶，即在去离子水中用纳秒激光烧蚀固体锰（Mn）靶合成尺寸均匀（10 nm）的 Mn_3O_4 纳米晶胶体溶液，如图 7-3（a）和图 7-3（b）所示。然后，该纳米晶胶体溶液在室温下原位静置 5 d，可以观察到这些球形 Mn_3O_4 纳米晶已经自发组装成 MnOOH 纳米线。SEM 图像显示，MnOOH 纳米线的直径在 20～50 nm，长度可达数微米，纵横比高达 100 左右［图 7-3（c）］。此外，SAED 图像显示 MnOOH 纳米线是单晶，具有沿 [101] 方向的择优取向［图 7-3（d）］。同样地，$H_2WO_4·H_2O$ 纳米片亦可用上述温度场辅助 LAL 方法组装。首先通过 LAL 合成 10 nm 的多晶水合钨纳米颗粒的胶体溶液，如图 7-3（e）和图 7-3（f）所示，然后该胶体溶液在室温下静置 3 d，可以观察到 $H_2WO_4·H_2O$ 纳米片的生成。TEM 观察表明，10 nm 的纳米颗粒团簇会自发转变为平面尺寸为 1.5～3.5 μm 的叶状单晶纳米片［图 7-3（g）］。从相应的 SAED 图像可以看到，每个纳米片都是单晶［图 7-3（h）］。这些研究结果表明，纳米晶胶体溶液的室温静置过程实际上就是在一定温度下的熟化过程，在这一过程中，LAL 合成的纳米颗粒可以从小变大，从不规则形态变为规则形貌，从结晶不良如非晶、多晶等变为良好单晶。

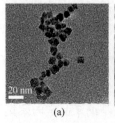 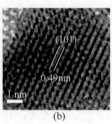 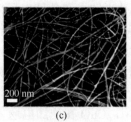 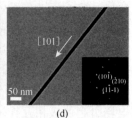

(a)　　　　　　　　(b)　　　　　　　　(c)　　　　　　　　(d)

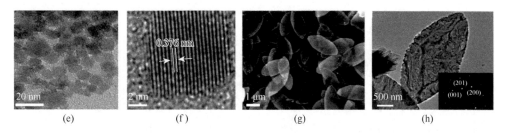

图 7-3 在去离子水中用纳秒激光烧蚀固体锰靶合成的 Mn_3O_4 纳米晶的(a)TEM 和(b)HRTEM 图像;(c)纳米晶胶体溶液在室温下原位静置 5 d,纳米晶自发组装成 MnOOH 纳米线;(d)单根 MnOOH 纳米线及其 SAED 图像;在去离子水中用纳秒激光烧蚀固体钨靶合成的 $H_2WO_4·H_2O$ 纳米颗粒的(e)TEM 和(f)HRTEM 图像;(g)纳米晶胶体溶液在室温下静置 3 d,纳米晶自发组装成 $H_2WO_4·H_2O$ 纳米片;(h)单个 $H_2WO_4·H_2O$ 纳米片及其 SAED 图像

7.3.2 纳米结构组装中的 Ostwald 熟化机制

我们把温度场辅助 LAL 组装纳米结构的物理机制归因于奥斯特瓦尔德(Ostwald)熟化过程。众所周知,Ostwald 熟化过程是指晶体在溶液生长过程中,尺寸较小的晶体溶解度较高,而尺寸较大的晶体则溶解度较低。这样的话,在晶体生长中,较小的晶体会溶解至最终消失,其质量转移到较大的晶体上使其生长得更大,这种现象在晶体生长中很常见[17, 18]。这是以热力学驱动的自发过程,因为尺寸较大的晶体比较小的自由能更低。换句话说,大的纳米颗粒会变得更大,而充当大纳米颗粒生长原料的小纳米颗粒会变得更小,这是因为纳米颗粒表面的分子的能量不如颗粒内部有序堆积的分子稳定[27, 28]。

在温度场辅助 LAL 的一维纳米线组装中,由于这些纳米颗粒是在水中形成而不暴露于空气,因此它们保留了洁净的表面,并最大限度地减少最高能量表面的暴露,重新排列形成共同的晶体取向,这意味着这些纳米颗粒会连接形成链状结构即纳米线。与此同时,在 Ostwald 熟化的帮助下形成的纳米线的直径会进一步增加。因此,纳米线的生长可以分为两个步骤。首先,通过定向附着机制,无序的纳米颗粒通过头尾连接自发组装成纳米线。其次,遵循 Ostwald 熟化机制,较小的纳米颗粒会被不断地消耗,而纳米线的直径则会不断增加,直至小纳米颗粒被消耗殆尽。通常情况下,Ostwald 熟化和定向附着会共同影响晶体生长[29]。这一机制也适用于温度场辅助 LAL 中 $H_2WO_4·H_2O$ 纳米片的形成。实际上,LAL 合成的纳米颗粒并不稳定并且容易聚集。在纳米颗粒聚集和生长过程中,生长速度决定了产物的形态。在这个实例中,对于单斜晶系 $H_2WO_4·H_2O$,$d(100)>d(001)$,因此,由于不同平面的生长速度不同,从而导致了 $H_2WO_4·H_2O$ 纳米片的形成。由于其独特的平面形态,与 LAL 合成的多晶纳米颗粒不同,这些单晶纳米片在 XRD 谱中显示出明显的择优取向,如图 7-4 所示。

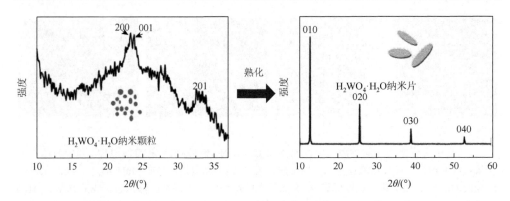

图 7-4　熟化处理后从多晶 $H_2WO_4 \cdot H_2O$ 纳米颗粒到单晶 $H_2WO_4 \cdot H_2O$ 纳米片演变的 XRD 谱

Liu 等[30]报道了单质 Ge 纳米颗粒的 Ostwald 熟化过程。尽管这种机制已经被广泛研究，但关于单元素纳米颗粒生长的 Ostwald 熟化报道很少。有趣的是，遵循 Ostwald 熟化规则，Ge 纳米颗粒经历了三个阶段的相变，从无定形结构到四方结构再到立方结构，如图 7-5 所示。这些相变的热力学驱动力是系统的形成能 E_f，即 $E_{f(非晶)} > E_{f(四方)} > E_{f(立方)}$。为了最小化 E_f，初始非晶态 Ge 纳米颗粒会自发生长[30]。

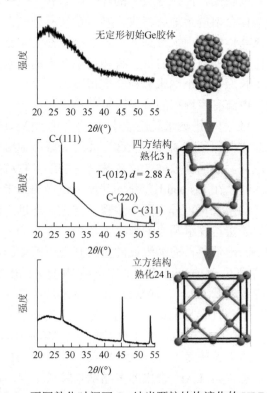

图 7-5　不同熟化时间下 Ge 纳米颗粒结构演化的 XRD 谱

还有一种晶体生长模型,我们称之为聚结。Ostwald 熟化和聚结的区别在于,Ostwald 熟化与纳米颗粒尺寸依赖的溶解度密切相关,这由 Gibbs-Thomson 关系式决定,而聚结则不然[31]。事实上,聚结通常是由纳米颗粒旋转控制的[32],根据该机制,相邻晶粒的旋转导致了晶粒-晶粒界面的联结,进而共同晶界的消除导致了相邻晶粒的聚结,从而形成单个较大的晶粒。Schaumberg 等[33] 发表了小纳米颗粒合并形成较大纳米颗粒的情况。他们发现,LAL 合成的小纳米颗粒会在持续的激光烧蚀过程中聚结。当聚结体吸收足够量的激光能量会完全熔化,然后通过聚结过程形成更大的纳米颗粒。这些结果说明,当初始的小纳米颗粒聚结时,它们被激光熔化并结晶成比初始纳米颗粒更大的次级纳米颗粒。Jendrzej 等[34]在报道中提到,熟化和聚结都有助于纳米颗粒生长。他们发现 LAL 合成的纳米晶胶体溶液早期阶段的生长机制是由无障碍的聚结动力学控制的。例如,在熟化过程中,经过第一天的聚结之后,在 1~50 d,原子簇数量减少所导致的慢生长过程受到聚结和 Ostwald 熟化二者共同影响。

综上所述,我们可以清楚地看到,$H_2WO_4 \cdot H_2O$ 纳米片和 Ge 纳米颗粒的 LAL 合成过程都经历了 Ostwald 熟化。这两个例子具有相似的晶体生长特征。首先,LAL 合成小纳米颗粒后,初始纳米颗粒的结晶度非常差。而熟化后,它们变得比初始纳米颗粒大得多。在最终产物中,由于溶液中较小纳米颗粒的溶解度和表面自由能较高,所以它们会重新溶解,通过物质输运促进较大纳米颗粒生长。因此,我们认为这二者皆遵循 Ostwald 熟化机制。另外,MnOOH 纳米线的生成遵循定向附着机制。相关研究表明,这些纳米颗粒大多排列整齐并按特定取向熔合在一起,它们通过重排共享共同的晶体取向,组装成链状纳米结构[35],但是,初始链状纳米结构的尺寸与初始纳米颗粒的尺寸相当,这与 $H_2WO_4 \cdot H_2O$ 和 Ge 的情况不同。因此,MnOOH 纳米线的合成主要受定向附着机制控制。

7.4 电化学辅助 LAL 用于复杂纳米结构组装

7.4.1 多金属氧酸盐纳米结构

在外场对 LAL 纳米制备作用的讨论中,我们知道,EFLAL 是控制纳米晶生长和纳米结构组装的有效方法。值得注意的是,EFLAL 中使用的固体靶和电极不会发生反应。那么,如果我们选择在 EFLAL 装置中可以发生相互反应的靶材和电极材料时会发生什么呢?基于这种考虑,Liu 等[36]发展了电化学辅助

LAL 即 ECLAL，通过选择不同的固体靶和电极材料组装出系列多金属氧酸盐（POM）纳米结构。这些 POM 纳米结构由至少两种（通常是三种或更多）过渡金属氧阴离子组成，它们通过共享的氧原子连接在一起，形成一个大的、封闭的三维框架。如今，ECLAL 已经用于组装各种微米和纳米团簇，该方法主要基于二元金属（如 Mo、W、V 或 Ta）和另一种过渡金属（如 Cu、Fe、Co 或 Ni）的氧化物[37, 38]。

众所周知，POM 纳米结构因其在炼油中的高催化活性、良好的排放特性及适合储能应用等而引起了大量的关注[39-42]，因此，研究人员开发了许多合成 POM 纳米结构的技术[43-45]。然而，这些方法都存在一些缺陷，例如，需要高温、需要引入各种模板或添加剂、最终产物中出现多杂质等。与这些方法相比，ECLAL 具有三个优点：①ECLAL 是一种化学"干净"的复杂纳米结构组装技术，因为其起始材料简单且形成的副产物有限；②ECLAL 工作在室温和环境压强下，无须极端的温度和压强条件；③研究人员可以通过组合特定的靶材、电极材料和液体种类灵活地设计各种组装的纳米结构。

例如，Liu 等[36]选取钼作为固体靶、铜作为电极、去离子水作为环境液体，通过 ECLAL 一步法合成 POM。其中两个铜电极放置于反应池的两侧，由直流电源产生电压为 20 V 的稳定电场，实验设置如图 7-6（a）所示。合成产物的 XRD 谱表明，所组装的纳米结构是纯单斜晶系的结晶锂辉石[$Cu_3(OH)_2(MoO_4)_2$]，并且未发现非晶钼、铜化合物或其他氧化物相。这说明，ECLAL 组装的纳米结构样品具有高纯度和高结晶性。此外，将样品在 500℃下退火 5 h 诱导了从钼铜矿到二钼酸三铜（$Cu_3Mo_2O_9$）的相变。

如上所述，ECLAL 组装 POM 纳米结构的优点就是扩展了研究人员对固体靶的选择范围并用电化学的方法来合成 POM。例如，我们可以变换 ECLAL 中靶材和电极材料的组合，从而组装出多种多样的复杂纳米结构。Liang 等使用了三种不同的组合，钒（靶材）+铜（电极）[46]、钒（靶材）+银（电极）[47]和钼（靶材）+锌（电极）[48]，分别组装出了 $ZnMoO_4$ 纳米板和纳米棒［图 7-6（b）～图 7-6（e）］、$Cu_3(OH)_2V_2O_7·2H_2O$ 花状纳米结构［图 7-6（f）和图 7-6（g）］和 $Ag_2V_4O_{11}$ 纳米刷［图 7-6（h）和图 7-6（i）］。重要的是，这些组装的纳米结构表现出了令人印象深刻的功能特性。例如，$ZnMoO_4$ 纳米棒在退火后可以发出绿光[48]；$Ag_2V_4O_{11}$ 纳米刷表现出了优异的传感器性能，尤其是在低工作温度下依然对 10～600 mg/L 的乙醇有良好响应[47]等。

上述例子表明 ECLAL 是制造简单 POM 纳米结构的有效通用策略。该技术是一种简单、绿色、无催化剂的方法，可在一般环境中进行。除了这些优点之外，ECLAL 还为研究人员提供了更广泛的固体靶选择范围，以制造 POM 用于基础研究和应用。

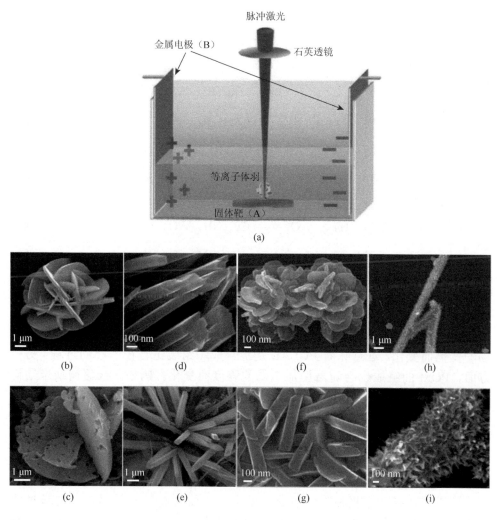

图 7-6 (a)ECLAL 装置示意图；(b)~(e)ZnMoO$_4$ 纳米板和纳米棒；(f)和(g)Cu$_3$(OH)$_2$V$_2$O$_7$·2H$_2$O 花状纳米结构；(h) 和 (i) Ag$_2$V$_4$O$_{11}$ 纳米刷

7.4.2 纳米结构组装中的化学反应

一般来说，ECLAL 过程中纳米结构的形成可分为三个过程：①当激光烧蚀固体靶如 V 或 Mo 时，在环境液体和固体靶之间的界面处会产生包含靶材粒子（V、Mo）在内的多种物质粒子的脉冲激光诱导的等离子体羽[49]。由于等离子体羽处于高温高压高密度状态，等离子体羽与环境液体界面处的去离子水会被电离成 H$^+$、OH$^-$ 和 O^{2-} 等活性基团[50,51]，这些活性基团可以与等离子体羽中的靶材粒子如 V 或 Mo 发生化学反应，生成氧化钒或氧化钼。②电化学反应。

当 ECLAL 中阳极（由 Cu、Zn 或 Ag 组成）在适当的电压下发生电解时，相应的离子（Cu^{2+}、Zn^{2+} 或 Ag^+）很容易溶解在环境液体中，并通过与水反应生成氢氧根离子[46-48]。

$$Cu + 2H_2O \longrightarrow Cu(OH)_2 + H_2 \quad (7\text{-}1)$$

$$Zn + 2H_2O \longrightarrow Zn(OH)_2 + H_2 \quad (7\text{-}2)$$

$$2Ag + 2H_2O \longrightarrow 2AgOH + H_2 \quad (7\text{-}3)$$

③在纳米结构组装中，电极材料粒子（Cu^{2+}、Zn^{2+} 或 Ag^+）会通过电泳效应移动到等离子体羽中，并与等离子体羽中的高活性物质发生反应。

$$3Cu(OH)_2 + V_2O_5 \longrightarrow Cu_3(OH)_2V_2O_7 + 2H_2O \quad (7\text{-}4)$$

$$2MoO_3 + 5Zn(OH)_2 \longrightarrow Zn_5Mo_2O_{11} \cdot 5H_2O \quad (7\text{-}5)$$

$$2AgOH + 2V_2O_5 \longrightarrow Ag_2V_4O_{11} + H_2O \quad (7\text{-}6)$$

另外，在整个 ECLAL 纳米结构组装过程中，施加电场的强度对 POM 纳米结构的形成起着重要作用。一方面，组装的纳米结构的形貌会随着施加电场的变化而变化，这意味着 ECLAL 是将合成和组装结合在一起的过程。另一方面，POM 的组成也会随着施加电场的强度而变化。例如，钒酸铜纳米结构只在高于 5 V 的电压下形成。因此，ECLAL 中的化学反应是一种组合反应，而不是简单的电化学或传统的 LAL 反应，因为 LAL 是一个极快的、远离平衡的过程。ECLAL 中的电场强度非常大，大约为 2.7×10^3 V/m[48]。因此，ECLAL 中等离子体羽处于高温高压高密度及强外电场状态下，这对形成 POM 纳米结构非常有效。

关于固体靶和电极材料的组合，原则上，如果选择 Mo、W 或 V 等过渡金属作为靶材，那么由于它们的高活性则极有可能发生化学反应。由于钒酸盐、钼酸盐和钨酸盐是最常见的 POM，且易于合成，因此 Mo、W 和 V 等是靶材的理想选择。ECLAL 对电极材料没有严格要求，只要能转变为相应的氧化物和氢氧化物即可。因此，我们期待 ECLAL 用于更多功能 POM 纳米结构的组装。

7.5 磁场辅助 LAL 用于磁性纳米链组装

7.5.1 磁性纳米颗粒一维链束

磁性纳米颗粒一维链束也称磁性纳米链，是一种重要的功能材料，在许多不同领域如光子晶体、磁记忆材料和 DNA 分离等有着重要应用[52-56]。但是，组装磁性纳米链，也就是实现各向同性纳米颗粒的各向异性一维组装，仍然是一个挑战[57]。目前有两种常见的磁性纳米颗粒一维链束组装技术，直接相互作用方法和需要模板或外场的间接作用方法[58-60]。组装过程通常可分为两步：第一步是纳米颗粒的合成，第二步是磁性纳米链的组装。我们发展的磁场辅助 LAL（MFLAL）技术可以一步组装出磁性纳米链[61]，并且采用 MFLAL 组装出了铁碳复合材料的微米纤维[62]、亚微米 Co_3C 颗粒链[63]和铁基双金属合金纳米颗粒的一维链束[64]。这些研究表明，MFLAL 是一种很有前途的磁性纳米结构组装技术，可以让研究人员通过选择不同的固体靶和环境液体来组装所需要的有序磁性纳米结构。

下面我们介绍所发展的 MFLAL 纳米组装技术。由亚微米碳化铁（Fe_3C）球构成的微米纤维是第一个通过 MFLAL 组装的一维链束[62]。我们采用铁作为靶材，乙醇作为环境液体，并施加了一个强度为 9 T 的稳定、均匀磁场用于铁靶的激光烧蚀。图 7-7（a）为 MFLAL 实验装置的照片，图 7-7（b）为 MFLAL 组装过程中等离子体羽的形成示意图。图 7-8（a）和图 7-8（b）显示了 MFLAL 组装的样品是由一维链束以有序方式紧密捆绑在一起构成的有序微米纤维，相应的 XRD 谱 [图 7-8（c）] 证实合成产物主要是 Fe_3C。重要的是，这些组装的一维链束的饱和磁化强度在室温下为 261 emu/g [图 7-8（d）]，是报道的 Fe_3C 纳米颗粒的（47 emu/g）[65]5.6 倍，并且超过了块体 Fe 的（212 emu/g）[66]。我们把这种磁性纳米颗粒的一维链束异常大的饱和磁化强度归因于其独特的结构和形状。显然，一维链束中磁性纳米球之间及纳米链之间的交换耦合对于增强微米纤维的磁能起着重要作用。

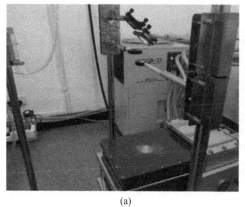 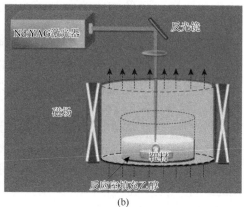

(a) (b)

图 7-7 MFLAL 的（a）实验装置及（b）组装过程中等离子体羽的形成示意图

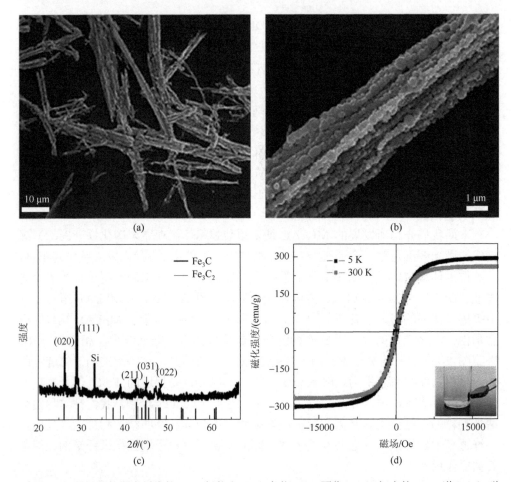

图 7-8 LAL 组装的微米纤维的（a）低倍和（b）高倍 SEM 图像；（c）相应的 XRD 谱；（d）磁性表征，插图表示样品在几秒之内受磁性作用快速向磁铁运动

采用相同的 FMLAL 技术，Liang 等通过用钴代替铁，成功组装了亚微米碳化钴（Co_3C）球构成的一维链束[67-69]，从相应的 XRD 谱中观察到的衍射峰都可对应于 Co_3C 的斜方结构的特征峰。此外，组装的一维链束的饱和磁化强度在 300 K 和 5 K 下分别为 232 emu/g 和 261 emu/g，是截至 2015 年获得的钴基磁性纳米材料的最高值。在此基础上，除了简单的单元素靶材外，$Fe_{21}Ni_{79}$、$Fe_{65}Co_{35}$ 和 $Fe_{52}Pt_{48}$ 等合金靶材也被用于 MFLAL 一维链束的组装[64]，其链的形态和相应 EDS 表征如图 7-9（a）～图 7-9（c）所示。上述实验结果表明，MFLAL 是一种简便的、绿色的组装一维链束的通用技术。值得注意的是，虽然磁性纳米链在 MFLAL 中是通过磁场辅助进行组装的，但是，Barcikowski 等[70]提出了一种无须磁场的方法将磁性纳米线组装为高纵横比的纳米链，并将其掺入聚合物中。

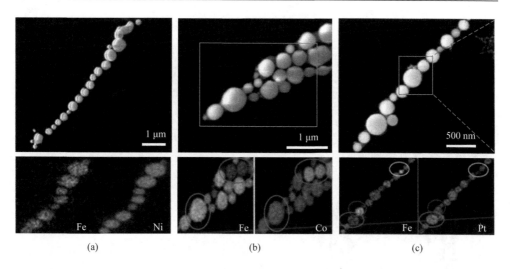

图 7-9 （a）FeNi 链的形态（上）和相应的 EDS 表征（下）；（b）FeCo 链的形态（上）和相应的 EDS 表征（下）；（c）FePt 链的形态（上）和相应的 EDS 表征（下）

7.5.2 磁场感应定向附着机制

MFLAL 组装磁性纳米颗粒一维链束的物理机制。一般来说，分散在溶液中的磁性纳米颗粒在外部磁场下受到三种力：范德瓦耳斯力 F_{vdw}、偶极相互作用力 F_{dd} 和偶极场力 F_m。其中，F_{vdw} 来自所有原子、分子和块体材料内正电荷和负电荷不断运动所产生的电磁波，F_{dd} 来自两个磁性粒子的相互作用，而 F_m 来自 $\nabla(m \cdot H)$，并且可以改变系统的能量分布。

当两个磁性纳米颗粒足够接近时，F_{vdw} 会使它们保持分开，并且是各向同性的；F_{dd} 会使两个磁性纳米颗粒保持相吸或相斥，并且是各向异性的；F_m 会使每个磁性纳米颗粒的磁矩与外部场保持一致。这样，我们就可以将这三种相互作用表示为势 U_{vdw}、偶极间能量 U_{dd} 和静磁能 U_m。同时，施加的外部磁场对磁性纳米颗粒的组装有两个主要作用：①在磁性纳米颗粒中诱导偶极矩；②根据磁场方向磁化纳米颗粒。U_{dd} 是将两个具有磁矩 m_1 和 m_2 的任意偶极子从无穷远拉近至有限的距离 $\hat{r}$ 所需要的功[71]：

$$U_{dd} = \frac{m_1 \cdot m_2 - 3(m_1 \cdot \hat{r})(m_2 \cdot \hat{r})}{4\pi\mu_o r^3} \tag{7-7}$$

与 U_m 和 U_{dd} 不同，U_{vdw} 是各向同性的。对于两个半径为 a 且中心距为 r 的球体，可以通过哈马克（Hamaker）积分近似得到以下公式[71]：

$$U_{vdw} = \frac{A}{3}\left\{\frac{a^2}{r^2-(2a)^2} + \frac{a^2}{r^2} + \frac{1}{2}\ln\left[\frac{r^2-(2a)^2}{r^2}\right]\right\} \tag{7-8}$$

式中，A 是铁氧体颗粒的 Hamaker 常数（10^{-19} J）。显然，U_{vdw} 仅取决于 r/a 的值。为了简化计算，r/a 设置为 2.1，这样计算出的 U_{vdw} 为 4.93×10^{-20} J。

外场 H 中偶极粒子的 U_m 由 $U_m = -m\cdot H$ 给出。我们以 9 T 外部磁场下的单个线性链系统为例，计算了上述三个相互作用能，如图 7-10（a）所示，其中图 7-10（b）提供了针对不同纳米颗粒尺寸计算的 U_{dd} 和 U_m。我们可以看出，对于任何尺寸的纳米颗粒，U_m 主导它们的相互作用，使磁性纳米颗粒一维链束容易发生自组装现象。

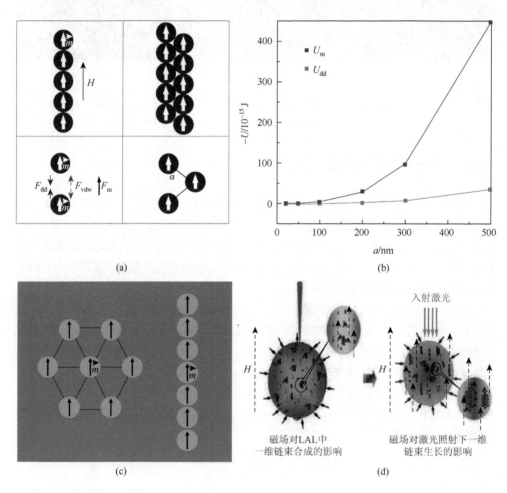

图 7-10 （a）磁性纳米颗粒一维链束的物理模型及在外部磁场作用下两个纳米球之间的力的分析；（b）不同尺寸磁性粒子的静磁能和偶极间能量；（c）分散的球形磁性颗粒（左）和一维链束（右）在磁场作用下的两种典型构型；（d）MFLAL 磁性纳米颗粒一维链束组装示意图

此外,我们还通过计算来比较图 7-10(c)中两种情况的能量:①具有六边形结构的分散颗粒(左侧),其纳米颗粒处于非链状聚集的状态;②纳米颗粒沿磁场方向链状排列(右侧)。这两种情况都是基于两个基本假设:其一,颗粒尺寸相同;其二,颗粒在空间上被磁矩均匀磁化。将方程(7-7)和方程(7-8)相加并对所有粒子求和,我们得到总能量

$$U = \sum_i U_m^{(i)} + \frac{1}{2}\sum_i \sum_{j \neq i} U_{dd}^{(i,j)} + \frac{1}{2}\sum_i \sum_{j \neq i} U_{vdw}^{(i,j)} \tag{7-9}$$

计算得到六边形的能量为 U_1,线性的能量为 U_2。显然,我们可以发现 $U_1 \gg U_2$,这意味着当组装这些磁性纳米颗粒时,链束结构比无序结构在能量上更稳定。

Liang 等提出了磁性纳米颗粒一维链束 MFLAL 组装的一般机制。首先,MFLAL 合成的磁性纳米球通过外部磁场磁化,然后,这些纳米球相互吸引并倾向于沿着磁力线形成线性链。与此同时,这些磁性纳米球会因为激光辐照提供的能量而进一步生长。由于 U_m 远大于排斥势,因此磁性纳米颗粒一维链束的形成在能量上是优先的[图 7-10(d)]。

本章,我们全面介绍了外场辅助 LAL 纳米材料合成和纳米结构组装,包括了 EFLAL、温度场辅助 LAL、ECLAL、MFLAL 等,表 7-1 总结了不同环境条件对 LAL 纳米组装技术的影响。虽然大多数外场辅助 LAL 装置都是使用纳秒激光,但是,其他类型脉冲激光(例如,毫秒、皮秒和飞秒激光等)在外场辅助 LAL 中的应用也是值得深入探讨的。对于温度场辅助 LAL 而言,纳秒、皮秒和飞秒激光应该都有类似的作用,即首先合成纳米晶胶体溶液,然后是设定温度下的熟化过程,这是一个两步法技术。然而,毫秒激光比纳秒、皮秒和飞秒激光产生更多的热量,因此 ms-LAL 过程可能直接发生熟化。对于 EFLAL 和 ECLAL,固体靶和电极材料的组合及附加电场的强度都会影响到最终的产物。在这里,脉冲激光的脉宽所起的作用不大。例如,可以使用纳秒、皮秒和飞秒激光烧蚀水中的 V 和 Mo 靶来合成 V_2O_5 和 MoO_3。由于 V 和 Mo 是高熔点金属,毫秒激光无法将它们变成金属液滴,因此,毫秒激光不适合 EFLAL。MFLAL 与 EFLAL 类似。施加磁场强度是影响磁性纳米颗粒一维链束组装的主要因素。所以,我们可以看到,纳秒、皮秒和飞秒激光应用于外场辅助 LAL 具有相似的效果,而毫秒激光仅对一些低熔点金属起作用。

表 7-1 不同环境条件的外场辅助 LAL 方法总结

外场类型	外场强度	固体靶	环境液体	外场作用
温度场	室温	W,Mn	去离子水	在室温下熟化后,原始球形纳米颗粒转变为一维或二维纳米结构。

续表

外场类型	外场强度	固体靶	环境液体	外场作用
磁场	0~9 T	Fe, Co, Ni, $Fe_{21}Ni_{79}$, $Fe_{65}Co_{35}$, $Fe_{52}Pt_{48}$	乙醇	随着磁场强度的增加,磁性纳米颗粒可以相互吸引,并倾向于沿着磁力线形成线性链,最终产生一维链束。
电场/电化学	0~200 V	Ge, Cu	去离子水	对于极性金属氧化物,如 CuO,它们很容易在电场下极化,由于电场诱导的偶极相互作用,它们将在施加的场方向上排列。它们的形状取决于电场的强度。例如,较大的电场导致主轴形态具有较大的长径比。
		V(靶材)+Cu(电极); V(靶材)+Ag(电极); Mo(靶材)+Zn(电极)		对于 ECLAL,多金属氧酸盐纳米结构的形态和组成在很大程度上取决于电场强度。例如,如果电场强度增加,则会相应地增加更多的电极材料成分,这会影响多金属氧酸盐纳米结构的最终组成。

参 考 文 献

[1] Hoffman M, Veinot J G C. Understanding the formation of elemental germanium by thermolysis of sol-gel derived organogermanium oxide polymers[J]. Chemistry of Materials, 2012, 24(7): 1283-1291.

[2] Castellani C E S, Kelleher E J R, Popa D, et al. Mode-locking by nanotubes of a Raman laser based on a highly doped GeO_2 fiber[C]//CLEO: Science and Innovations. Optica Publishing Group, 2012: CTu1I. 4.

[3] Wu H P, Liu J F, Ge M Y, et al. Preparation of monodisperse GeO_2 nanocubes in a reverse micelle system[J]. Chemistry of Materials, 2006, 18(7): 1817-1820.

[4] Davis T M, Snyder M A, Tsapatsis M. Germania nanoparticles and nanocrystals at room temperature in water and aqueous lysine sols[J]. Langmuir, 2007, 23(25): 12469-12472.

[5] Liu P, Wang C X, Chen X Y, et al. Controllable fabrication and cathodoluminescence performance of high-index facets GeO_2 micro-and nanocubes and spindles upon electrical-field-assisted laser ablation in liquid[J]. The Journal of Physical Chemistry C, 2008, 112(35): 13450-13456.

[6] Musaev O R, Sutter E A, Wrobel J M, et al. Au, Ge, and AuGe nanoparticles fabricated by laser ablation[J]. Journal of Nanoparticle Research, 2012, 14: 654.

[7] Vadavalli S, Valligatla S, Neelamraju B, et al. Optical properties of germanium nanoparticles synthesized by pulsed laser ablation in acetone[J]. Frontiers in Physics, 2014, 2: 57.

[8] Wang H H, Odawara O, Wada H. Facile and chemically pure preparation of YVO$_4$: Eu^{3+} colloid with novel nanostructure via laser ablation in water[J]. Scientific Reports, 2016, 6: 20507.

[9] Khan S Z, Yuan Y D, Abdolvand A, et al. Generation and characterization of NiO nanoparticles by continuous wave fiber laser ablation in liquid[J]. Journal of Nanoparticle Research, 2009, 11: 1421-1427.

[10] Lin X Z, Liu P, Yu J M, et al. Synthesis of CuO nanocrystals and sequential assembly of nanostructures with shape-dependent optical absorption upon laser ablation in liquid[J]. The Journal of Physical Chemistry C, 2009, 113 (40): 17543-17547.

[11] Chen X Y, Cui H, Liu P, et al. Shape-induced ultraviolet absorption of CuO shuttlelike nanoparticles[J]. Applied Physics Letters, 2007, 90 (18): 183118.

[12] Xu X D, Zhang M, Feng J, et al. Shape-controlled synthesis of single-crystalline cupric oxide by microwave heating using an ionic liquid[J]. Materials Letters, 2008, 62 (17-18): 2787-2790.

[13] Zhang Y G, Wang S T, Li X B, et al. CuO shuttle-like nanocrystals synthesized by oriented attachment[J]. Journal of Crystal Growth, 2006, 291 (1): 196-201.

[14] Lorazo P, Lewis L J, Meunier M. Short-pulse laser ablation of solids: From phase explosion to fragmentation[J]. Physical Review Letters, 2003, 91 (22): 225502.

[15] Shih C Y, Wu C P, Shugaev M V, et al. Atomistic modeling of nanoparticle generation in short pulse laser ablation of thin metal films in water[J]. Journal of Colloid and Interface Science, 2017, 489: 3-17.

[16] Penn R L, Banfield J F. Imperfect oriented attachment: Dislocation generation in defect-free nanocrystals[J]. Science, 1998, 281 (5379): 969-971.

[17] Banfield J F, Welch S A, Zhang H Z, et al. Aggregation-based crystal growth and microstructure development in natural iron oxyhydroxide biomineralization products[J]. Science, 2000, 289 (5480): 751-754.

[18] Tang Z Y, Kotov N A, Giersig M. Spontaneous organization of single CdTe nanoparticles into luminescent nanowires[J]. Science, 2002, 297 (5579): 237-240.

[19] Pradhan N, Xu H F, Peng X G. Colloidal CdSe quantum wires by oriented attachment[J]. Nano Letters, 2006, 6 (4): 720-724.

[20] Xu H L, Wang W Z, Zhu W, et al. Hierarchical-oriented attachment: From one-dimensional Cu(OH)$_2$ nanowires to two-dimensional CuO nanoleaves[J]. Crystal Growth & Design, 2007, 7 (12): 2720-2724.

[21] Kabashin A V, Meunier M. Synthesis of colloidal nanoparticles during femtosecond laser ablation of gold in water[J]. Journal of Applied Physics, 2003, 94 (12): 7941-7943.

[22] Franzel L, Bertino M F, Huba Z J, et al. Synthesis of magnetic nanoparticles by pulsed laser ablation[J]. Applied Surface Science, 2012, 261: 332-336.

[23] Anikin K V, Melnik N N, Simakin A V, et al. Formation of ZnSe and CdS quantum dots via laser ablation in liquids[J]. Chemical Physics Letters, 2002, 366 (3-4): 357-360.

[24] Gondal M A, Saleh T A, Drmosh Q A. Synthesis of nickel oxide nanoparticles using pulsed laser ablation in liquids and their optical characterization[J]. Applied Surface Science, 2012, 258 (18): 6982-6986.

[25] Xiao J, Liu P, Liang Y, et al. High aspect ratio β-MnO$_2$ nanowires and sensor performance for explosive gases[J]. Journal of Applied Physics, 2013, 114 (7): 073513.

[26] Xiao J, Liu P, Liang Y, et al. Porous tungsten oxide nanoflakes for highly alcohol sensitive performance[J]. Nanoscale, 2012, 4 (22): 7078-7083.

[27] Voorhees P W. The theory of Ostwald ripening[J]. Journal of Statistical Physics, 1985, 38: 231-252.

[28] Kabalnov A. Ostwald ripening and related phenomena[J]. Journal of Dispersion Science and Technology, 2001,

22（1）：1-12.

[29] Lin M, Fu Z Y, Tan H R, et al. Hydrothermal synthesis of CeO$_2$ nanocrystals: Ostwald ripening or oriented attachment? [J]. Crystal Growth & Design, 2012, 12（6）：3296-3303.

[30] Liu J, Liang C H, Tian Z F, et al. Spontaneous growth and chemical reduction ability of Ge nanoparticles[J]. Scientific Reports, 2013, 3: 1741.

[31] Thanh N T K, Maclean N, Mahiddine S. Mechanisms of nucleation and growth of nanoparticles in solution[J]. Chemical Reviews, 2014, 114（15）：7610-7630.

[32] Leite E R, Giraldi T R, Pontes F M, et al. Crystal growth in colloidal tin oxide nanocrystals induced by coalescence at room temperature[J]. Applied Physics Letters, 2003, 83（8）：1566-1568.

[33] Schaumberg C A, Wollgarten M, Rademann K. Metallic copper colloids by reductive laser ablation of nonmetallic copper precursor suspensions[J]. The Journal of Physical Chemistry A, 2014, 118（37）：8329-8337.

[34] Jendrzej S, Gökce B, Amendola V, et al. Barrierless growth of precursor-free, ultrafast laser-fragmented noble metal nanoparticles by colloidal atom clusters: A kinetic in situ study[J]. Journal of Colloid and Interface Science, 2016, 463: 299-307.

[35] Zhang H M, Liang C H, Tian Z F, et al. Organization of Mn$_3$O$_4$ nanoparticles into γ-MnOOH nanowires via hydrothermal treatment of the colloids induced by laser ablation in water[J]. CrystEngComm, 2011, 13（4）：1063-1066.

[36] Liu P, Liang Y, Lin X Z, et al. A general strategy to fabricate simple polyoxometalate nanostructures: Electrochemistry-assisted laser ablation in liquid[J]. ACS Nano, 2011, 5（6）：4748-4755.

[37] Coronado E, Gómez-García C J. Polyoxometalate-based molecular materials[J]. Chemical Reviews, 1998, 98（1）：273-296.

[38] Long D L, Burkholder E, Cronin L. Polyoxometalate clusters, nanostructures and materials: From self assembly to designer materials and devices[J]. Chemical Society Reviews, 2007, 36（1）：105-121.

[39] Breysse M, Geantet C, Afanasiev P, et al. Recent studies on the preparation, activation and design of active phases and supports of hydrotreating catalysts[J]. Catalysis Today, 2008, 130（1）：3-13.

[40] Ivanov K, Dimitrov D, Boyanov B. Optimization of the methanol oxidation over iron: Molybdate catalysts[J]. Chemical Engineering Journal, 2009, 154（1-3）：189-195.

[41] Kim S S, Ogura S, Ikuta H, et al. Reaction mechanisms of MnMoO$_4$ for high capacity anode material of Li secondary battery[J]. Solid State Ionics, 2002, 146（3-4）：249-256.

[42] Song C S, Ma X L. New design approaches to ultra-clean diesel fuels by deep desulfurization and deep dearomatization[J]. Applied Catalysis B: Environmental, 2003, 41（1-2）：207-238.

[43] Chu W G, Wang H F, Guo Y J, et al. Catalyst-free growth of quasi-aligned nanorods of single crystal Cu$_3$Mo$_2$O$_9$ and their catalytic properties[J]. Inorganic Chemistry, 2009, 48（3）：1243-1249.

[44] Wang Q M, Yan B. Hydrothermal mild synthesis of microrod crystalline Y$_x$Gd$_{2-x}$(MoO$_4$)$_3$: Eu^{3+} phosphors derived from facile co-precipitation precursors[J]. Materials Chemistry and Physics, 2005, 94（2-3）：241-244.

[45] Yu S, Lin Z B, Zhang L Z, et al. Preparation of monodispersed Eu^{3+} : CaMoO$_4$ nanocrystals with single quasihexagon[J]. Crystal Growth & Design, 2007, 7（12）：2397-2399.

[46] Liang Y, Liu P, Li H B, et al. Synthesis and characterization of copper vanadate nanostructures via electrochemistry assisted laser ablation in liquid and the optical multi-absorptions performance[J]. CrystEngComm, 2012, 14（9）：3291-3296.

[47] Liang Y, Zhu L F, Liu P, et al. Ag$_2$V$_4$O$_{11}$ nanostructures for highly ethanol sensitive performance[J].

CrystEngComm, 2013, 15 (31): 6131-6135.

[48] Liang Y, Liu P, Li H B, et al. ZnMoO$_4$ micro-and nanostructures synthesized by electrochemistry-assisted laser ablation in liquids and their optical properties[J]. Crystal Growth & Design, 2012, 12 (9): 4487-4493.

[49] Yang G W. Laser ablation in liquids: Applications in the synthesis of nanocrystals[J]. Progress in Materials Science, 2007, 52 (4): 648-698.

[50] Liang C H, Sasaki T, Shimizu Y, et al. Pulsed-laser ablation of Mg in liquids: Surfactant-directing nanoparticle assembly for magnesium hydroxide nanostructures[J]. Chemical Physics Letters, 2004, 389 (1-3): 58-63.

[51] Sakka T, Iwanaga S, Ogata Y H, et al. Laser ablation at solid-liquid interfaces: An approach from optical emission spectra[J]. The Journal of Chemical Physics, 2000, 112 (19): 8645-8653.

[52] Zeng H, Li J, Liu J P, et al. Exchange-coupled nanocomposite magnets by nanoparticle self-assembly[J]. Nature, 2002, 420: 395-398.

[53] Koenig A, Hébraud P, Gosse C, et al. Magnetic force probe for nanoscale biomolecules[J]. Physical Review Letters, 2005, 95 (12): 128301.

[54] Kinsella J M, Ivanisevic A. Magnetotransport of one-dimensional chains of CoFe$_2$O$_4$ nanoparticles ordered along DNA[J]. The Journal of Physical Chemistry C, 2008, 112 (9): 3191-3193.

[55] He L, Wang M S, Ge J P, et al. Magnetic assembly route to colloidal responsive photonic nanostructures[J]. Accounts of Chemical Research, 2012, 45 (9): 1431-1440.

[56] Chen J, Dong A G, Cai J, et al. Collective dipolar interactions in self-assembled magnetic binary nanocrystal superlattice membranes[J]. Nano Letters, 2010, 10 (12): 5103-5108.

[57] Ku J Y, Aruguete D M, Alivisatos A P, et al. Self-assembly of magnetic nanoparticles in evaporating solution[J]. Journal of the American Chemical Society, 2011, 133 (4): 838-848.

[58] Butter K, Bomans P H H, Frederik P M, et al. Direct observation of dipolar chains in iron ferrofluids by cryogenic electron microscopy[J]. Nature Materials, 2003, 2: 88-91.

[59] Pileni M P. Nanocrystal self-assemblies: Fabrication and collective properties[J]. The Journal of Physical Chemistry B, 2001, 105 (17): 3358-3371.

[60] Tang Z Y, Ozturk B, Wang Y, et al. Simple preparation strategy and one-dimensional energy transfer in CdTe nanoparticle chains[J]. The Journal of Physical Chemistry B, 2004, 108 (22): 6927-6931.

[61] Wang H Q, Pyatenko A, Kawaguchi K, et al. Selective pulsed heating for the synthesis of semiconductor and metal submicrometer spheres[J]. Angewandte Chemie International Edition, 2010, 49 (36): 6361-6364.

[62] Liang Y, Liu P, Xiao J, et al. A microfibre assembly of an iron-carbon composite with giant magnetisation[J]. Scientific Reports, 2013, 3: 3051.

[63] Liang Y, Liu P, Xiao J, et al. A general strategy for one-step fabrication of one-dimensional magnetic nanoparticle chains based on laser ablation in liquid[J]. Laser Physics Letters, 2014, 11 (5): 056001.

[64] Liang Y, Liu P, Yang G W. Fabrication of one-dimensional chain of iron-based bimetallic alloying nanoparticles with unique magnetizations[J]. Crystal Growth & Design, 2014, 14 (11): 5847-5855.

[65] Giordano C, Kraupner A, Wimbush S C, et al. Iron carbide: An ancient advanced material[J]. Small, 2010, 6 (17): 1859-1862.

[66] Gao C, Doyle W D, Shamsuzzoha M. Quantitative correlation of phase structure with the magnetic moment in rf sputtered Fe-N films[J]. Journal of Applied Physics, 1993, 73 (10): 6579-6581.

[67] Dutta P, Seehra M S, Thota S, et al. A comparative study of the magnetic properties of bulk and nanocrystalline Co$_3$O$_4$[J]. Journal of Physics: Condensed Matter, 2007, 20 (1): 015218.

[68] Park J B, Jeong S H, Jeong M S, et al. Synthesis of carbon-encapsulated magnetic nanoparticles by pulsed laser irradiation of solution[J]. Carbon, 2008, 46 (11): 1369-1377.

[69] Lukanov P, Anuganti V K, Krupskaya Y, et al. CCVD synthesis of carbon-encapsulated cobalt nanoparticles for biomedical applications[J]. Advanced Functional Materials, 2011, 21 (18): 3583-3588.

[70] Barcikowski S, Baranowski T, Durmus Y, et al. Solid solution magnetic FeNi nanostrand: Polymer composites by connecting-coarsening assembly[J]. Journal of Materials Chemistry C, 2015, 3 (41): 10699-10704.

[71] Bishop K J M, Wilmer C E, Soh S, et al. Nanoscale forces and their uses in self-assembly[J]. Small, 2009, 5 (14): 1600-1630.

第 8 章 液体环境中纳米图案 LAL 组装

8.1 液体环境中脉冲激光沉积用于纳米图案组装

8.1.1 液体环境中的脉冲激光沉积

将孤立的纳米颗粒图案组装为功能纳米结构是一项极具挑战性的工作。为此，我们发展了液体环境中的脉冲激光沉积（也叫液相 PLD）用于透明衬底表面纳米颗粒图案组装[1]，液相 PLD 的装置示意图和纳米颗粒图案组装原理示意图如图 8-1 所示。具体而言，当脉冲激光穿过透明衬底和环境液体辐照到固体靶表面时，相继在靶材表面产生等离子体羽和空泡。由于等离子体羽的反冲效应，它会在透明衬底的背表面冷却、凝结，最终形成图案化的纳米结构。与真空或气体环境中的传统 PLD 相比，液相 PLD 的等离子体羽受到环境液体的束缚，使得微纳图案的组装更加可控。相比之下，真空或气体环境中脉冲激光沉积的薄膜通常较为疏松[2]。受环境液体的束缚，激光诱导的等离子体羽的产生、膨胀和凝聚与在真空情况下不同，因此其产物往往具有亚稳相纳米结构。所以，只要选择合适的液体种类，我们就可以应用液相 PLD 组装具有复杂成分和图案的功能纳米结构，而这些在真空或气体环境中的 PLD 是难以实现的。

8.1.2 在透明基片上组装纳米颗粒图案

我们以在玻璃基片表面组装的 Ag 纳米颗粒图案作为表面增强拉曼散射（SERS）衬底为例来阐明液相 PLD 关于功能纳米图案组装的应用[1]。实际上，液相 PLD 的工作原理与常规 LAL 类似。如图 8-1 所示，当脉冲激光聚焦在固体 Ag 靶表面上时，靶材中的物质会被高能激光轰击出等离子体羽，然后反冲效应会导致这个等离子体羽喷射到玻璃基片的背表面（面向靶材的那一面）并冷却、凝结，形成由纳米颗粒构筑的图案。在沉积过程结束后，我们可以在玻璃基片的背表面看到轮廓分明的 Ag 纳米颗粒图案，如图 8-2（a）所示，每个 Ag 纳米颗粒图案的尺寸约为 50 μm，相应的原子力显微镜（AFM）图像表明 Ag 纳米颗粒图案由尺寸约为 50 nm 的 Ag 纳米晶构成［图 8-2（b）］。考虑到 SERS 衬底表面粗糙[3]，因此我们将组装的 Ag 纳米颗粒图案作为 SERS 衬底进行了表征，如图 8-2（c）所示，Ag 纳米颗粒图案对罗丹明 6 G（R6G）溶液具有高达 10^9 的增强因子。这些研究结果表明，普通玻璃或其他透明基片都可以用于纳米颗粒图案的液相 PLD 组装。此外，液相 PLD 的组装面积被限制在几十微米的范围内，这比真空或气体环境中的传统 PLD 小得多。考虑到其便捷、可控和清洁的特性，液相 PLD 是一项极具潜力的纳米颗粒图案组装技术，可以用于各种领域如 SERS 衬底和其他纳米生物传感芯片的组装。

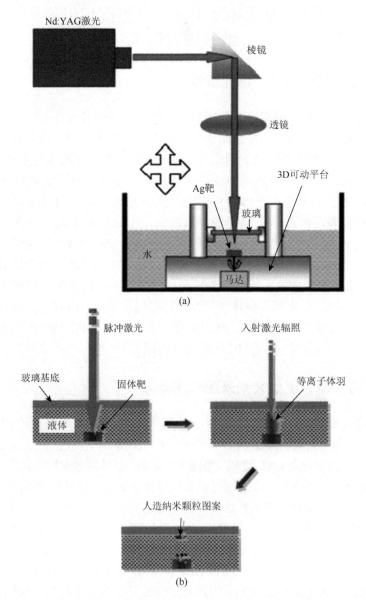

图 8-1　(a) 液相 PLD 装置示意图；(b) 纳米颗粒图案组装原理示意图

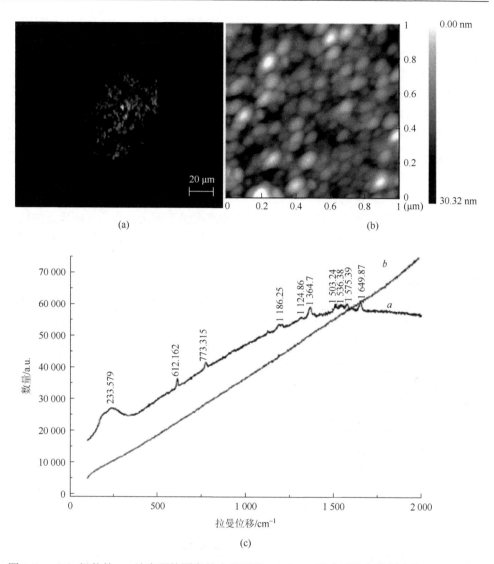

图 8-2 （a）组装的 Ag 纳米颗粒图案的光学图像；（b）Ag 纳米颗粒组装图案的 AFM 图像；（c）5.85×10^{-5} R6G 溶液的 SERS 光谱，a：Ag 纳米颗粒图案基片，b：纯玻璃

8.2 液体环境中功能纳米结构图案的激光直写

8.2.1 异质结构纳米图案的组装

随着器件尺寸的减小，微米和亚微米尺度的材料加工在微电子和光电子领域变得越来越重要。因此，开发新的微纳加工技术如刻蚀和图案化技术等一直是研

究人员追逐的目标。LAL作为一种功能强大的纳米制备技术,也可以用于液相中固体表面纳米图案的组装。如上所述,在LAL中,激光烧蚀在固-液界面处会产生一个局域极端的热力学和动力学环境,与其在真空和气体环境中有很大不同,为LAL在纳米图案化领域的应用提供了更大的发展空间。为此,基于LAL,我们发展了液体环境中功能纳米结构图案的激光直写技术,如图8-3所示[4]。我们以一个具体的异质结构纳米图案的组装为例来介绍这种独特的激光直写技术,通过对环境液体中的非晶碳膜进行激光直写,组装一种纳米金刚石周期性嵌入的独特纳米结构。首先我们在单晶硅衬底上沉积一层非晶碳膜;然后将该衬底放置在液体环境中的激光直写装置的顶部,该装置可以二维移动,从而实现在非晶碳膜表面均匀写入设定的图案;最后将脉冲激光聚焦在非晶碳膜表面进行直写,环境液体就是简单的去离子水。图8-4(a)和图8-4(b)显示了非晶碳膜表面上组装的均匀点阵,其中每个点的直径约为2~2.5 μm,TEM图像[图8-4(c)~图8-4(f)]证实了非晶碳膜上组装的点阵中每一个点都是由镶嵌在非晶碳基质中的大量金刚石纳米晶构筑。考虑到非晶碳膜是一种优异的场发射冷阴极材料,所以我们将组装的金刚石-非晶碳异质结构纳米图案应用于场发射冷阴极表征,如图8-5所示。可喜的是,与本征非晶碳膜相比,这种具有周期性的金刚石-非晶碳异质结构纳米图案可以显著改善场发射性能,展现了这种激光直写技术在纳米结构图案组装方面的有效性和适用性。

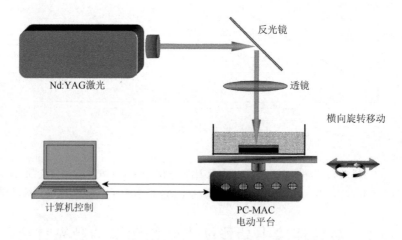

图8-3 液体环境中功能纳米结构图案的激光直写装置示意图

8.2.2 液体环境中激光诱导的相变

通过液体环境中激光直写组装异质结构纳米图案,最核心的过程就是从非晶碳到金刚石的相变。我们通过对非晶碳膜的结构表征知道,使用的非晶碳膜是含

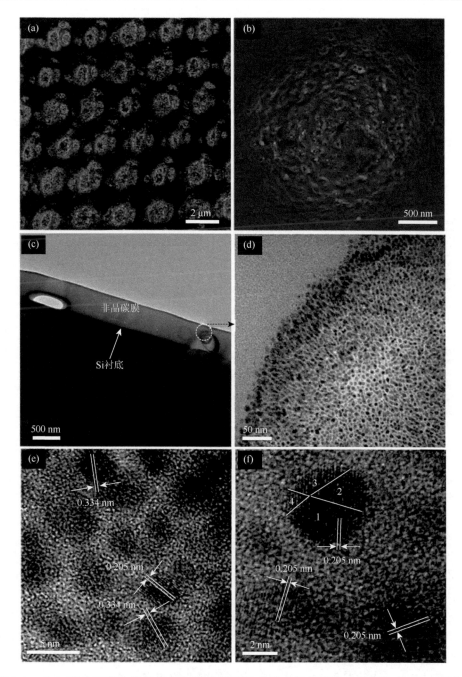

图 8-4 （a）液体环境中脉冲激光辐照非晶碳膜组装空间周期性点阵图案的 SEM 图像；（b）点阵中单个点位的典型 SEM 图像；（c）样品横截面的低分辨 TEM 图像；（d）凝聚纳米晶的 TEM 图像；（e）金刚石纳米晶的 HRTEM 图像；（f）三种金刚石纳米晶的高倍 TEM 图像

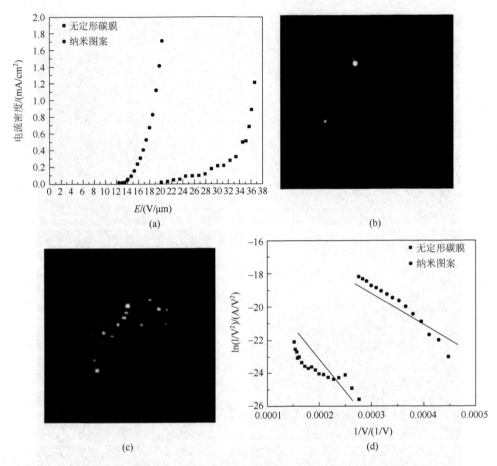

图 8-5 （a）本征非晶碳膜和金刚石-非晶碳异质结构纳米图案的场发射电流密度；（b）本征非晶碳膜发射位点；（c）金刚石-非晶碳异质结构纳米图案；（d）金刚石-非晶碳异质结构纳米图案场发射电流密度的 Fowler-Nordheim 图

有 sp^3 和 sp^2 成分的亚稳相。所以，在 LAL 过程中，非晶碳和金刚石之间的具体相变有两个通道。第一个通道是非晶碳向金刚石转变发生在脉冲激光诱导等离子体羽的相变过程。我们知道，受环境液体的束缚作用，脉冲激光诱导等离子体羽会被推到一个局域高温高压高密度的极端热力学状态，从而使得等离子体羽中的 sp^2 基团转化为 sp^3 基团。与此同时，由于等离子体羽的高温，界面处的水分子会分解成氢原子和氧原子等活性粒子并融入等离子体羽中，而氢原子的存在有利于金刚石的成核和生长[5,6]，进而引发等离子体羽中的相变。第二个通道就是在 LAL 过程中，在固体靶与环境液体界面处产生的脉冲激光诱导的等离子体羽的高温高压会直接导致从非晶碳到金刚石的相变，也就是说，等离子体羽会导致靶材表面的部分 sp^2 直接转变为 sp^3，进而引发金刚石的成核和生长[7]。

参 考 文 献

[1] Cui H, Liu P, Yang G W. Noble metal nanoparticle patterning deposition using pulsed-laser deposition in liquid for surface-enhanced Raman scattering[J]. Applied Physics Letters, 2006, 89 (15): 153124.

[2] Lowndes D H, Geohegan D B, Puretzky A A, et al. Synthesis of novel thin-film materials by pulsed laser deposition[J]. Science, 1996, 273 (5277): 898-903.

[3] Tian Z Q, Ren B, Wu D Y. Surface-enhanced Raman scattering: From noble to transition metals and from rough surfaces to ordered nanostructures[J]. The Journal of Physical Chemistry B, 2002, 106 (37): 9463-9483.

[4] Liu P, Wang C X, Chen J, et al. Localized nanodiamond crystallization and field emission performance improvement of amorphous carbon upon laser irradiation in liquid[J]. The Journal of Physical Chemistry C, 2009, 113 (28): 12154-12161.

[5] Joeris P, Benndorf C, Kröger R. Investigations concerning the role of hydrogen in the deposition of diamond films[J]. Surface and Coatings Technology, 1993, 59 (1-3): 310-315.

[6] Hu X J, Chen X H, Ye J S. The roles of hydrogen in the diamond/amorphous carbon phase transitions of oxygen ion implanted ultrananocrystalline diamond films at different annealing temperatures[J]. AIP Advances, 2012, 2 (4): 042109.

[7] Yang G W. Laser ablation in liquids: Applications in the synthesis of nanocrystals[J]. Progress in Materials Science, 2007, 52 (4): 648-698.

第 9 章　LAL 制备的纳米材料

9.1　纳米颗粒-聚合物复合材料

众所周知，环境液体在 LAL 纳米制备中有着重要的作用，因为它们不仅直接参与相关的化学反应，而且由于对激光诱导的等离子体羽的束缚效应而影响等离子体羽的产生、相变和淬灭等过程。在 LAL 纳米制备技术研究早期，研究人员主要关注成分简单的环境液体，例如，水、乙醇、丙酮和一些简单的表面活性剂[1,2]，这些环境液体常常只是起到束缚作用和对纳米晶生长的保护作用，而基本不参与 LAL 过程中各种化学反应。后来，为了进一步拓展 LAL 的应用空间，人们开始使用一些复杂的环境液体来制备结构和功能多样的纳米材料。研究人员在聚合物有机溶液中进行 LAL 纳米材料合成[3,4]。我们知道，纳米颗粒-聚合物复合材料可以将聚合物基质的高机械强度和优异的生物相容性与纳米颗粒优异的催化性质和抗菌活性相结合[5-8]，从而极大提升这类生物医用材料的性能和应用领域。为此，将无机纳米颗粒掺入聚合物基质中不仅可以有效改善材料的性能，而且有可能制备出具有独特性能的新材料。

一般来说，将纳米颗粒掺入聚合物基质中需要聚合物和纳米颗粒具备一定的相容性，这通常是通过对纳米颗粒表面进行改性来实现的。若纳米颗粒表面改性不充分，便会导致纳米颗粒在聚合物中聚集成团块状而无法实现均匀分散。为此，人们付出了巨大的努力，试图通过转移和接枝等方法制备具有合适功能化表面的纳米颗粒[9-11]。但是，这些方法都很烦琐，并且往往导致纳米颗粒在聚合物基质中的分散效率下降。与传统方法相比，LAL 纳米制备技术是一种一步法制备纳米颗粒-聚合物复合材料的绿色、高效技术。LAL 纳米制备技术有两个明显的优势：①这种合成方法不需要化学前驱体，因此所制备的纳米颗粒-聚合物复合材料非常纯净，这确保了最终产物将保持材料原本的固有特性；②最重要的是 LAL 与传统方法的主要区别在于不需要基质黏合剂来连接纳米颗粒和聚合物。因此，大部分纳米颗粒-聚合物复合材料都可以采用 LAL 技术制备。具体而言，此类复合材料的 LAL 制备主要包括如下步骤：首先将块状固体靶浸入环境液体聚合物有机溶剂中，然后使用脉冲激光烧蚀固体靶，最终合成均匀分散在聚合物基质中的纳米颗粒。需要注意的是，该方法是在没有任何化学添加剂、基质黏合剂和稳定剂的情况下进行的。迄今为止，LAL 纳米制备技术已经合成了多种纳米颗粒-聚合材料,并且这些复合材料在生物医学、能源、催化和荧光领域显示出巨大的应用潜力[12-21]。

我们介绍几个 LAL 制备功能纳米颗粒-聚合物复合材料的实例。Makridis 等[19]采用 LAL 技术在含有聚合物聚甲基丙烯酸甲酯(PMMA)的均匀四氢呋喃溶液中使用固体 Mg 靶合成了 Mg 纳米颗粒-PMMA 复合材料,如图 9-1(a)所示。我们可以从相应的 XRD 谱中发现,所制备复合材料中仅存在单相的六方 Mg,这是由于在 LAL 过程中复合基质的保护使得 Mg 不被氧化。重要的是,与其他类型的纯 Mg 相比,该复合材料具有更快的吸氢速度(250℃条件下小于 20 min)和高储氢量(6%的质量分数),因此可以成为非常合适的储氢材料[19]。除了能源领域的应用之外,这类复合材料在荧光方面也显示出应用潜力[22]。例如,为了制备先进的荧光纳米复合材料,我们可以采用不同浓度的热塑性聚氨酯(TPU)作为环境液体,通过脉冲激光烧蚀固体 ZnO 靶,利用 TPU 在 ZnO 纳米颗粒周围形成聚合物壳,原位增强 ZnO 纳米颗粒的稳定性,从而合成了高效的荧光纳米复合材料[13]。

由于贵金属纳米颗粒如 Ag 等具有优异的灭菌能力,所以 Wagener 等[22]通过在有机溶剂中激光烧蚀 Ag 靶和 Cu 靶制备了 Ag 和 Cu 的纳米颗粒-聚合物复合材料,并且使用了带有氟化侧链的两亲性共聚物对纳米颗粒进行原位功能化。这些功能化的纳米颗粒可以均匀地分散到全氟化聚合物中,并且对大肠杆菌也表现出高抗菌活性。此外,他们还发现全氟化聚合物尺寸和纳米颗粒分散性很大程度上取决于聚合物在 LAL 前的环境液体浓度。选择最佳的浓度,纳米颗粒的分散性可以得到显著改善。注意,这类纳米复合材料注射成型的过程不影响纳米颗粒的分散性,如图 9-1(b)所示。

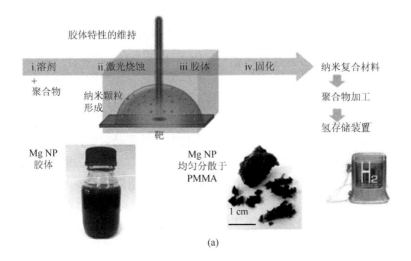

(a)

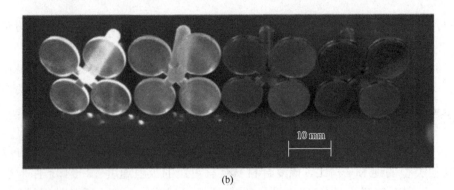

图 9-1　(a) PMMA 中 LAL 合成用于氢存储的 Mg 纳米颗粒-PMMA 复合材料；(b) Ag 浓度增加的注塑原型

Zhang 和 Barcikowski[23]总结了 LAL 制备纳米颗粒-聚合物复合材料的进展。他们不仅指出了 LAL 合成这类复合材料时的一些重要影响因素，其中包括聚合物特性、LAL 类型、激光参数和液体参数等，而且还给出了纳米颗粒-聚合物复合材料的八种不同应用，例如，抗菌、生物相容剂、离子释放、三维原位组装等。这些研究充分表明，LAL 纳米制备技术是一种有吸引力的纳米颗粒-聚合物复合材料制备方法，它可以通过简单、快速、绿色、直接基质偶联途径合成无机纳米颗粒-聚合物复合材料，而且不会产生污染，也不需要基质黏合剂[24]。同时，这些复合材料的聚合物外壳不仅可以有效防止纳米颗粒的聚集，而且赋予了复合材料用于临床诊断和能量存储等领域的功能。

9.2　掺杂半导体纳米晶

掺杂，即有目的地将杂质元素引入材料中来调制材料的本征物性，是调控半导体材料性能的基本方法。由于纳米晶中电子态的体积限制效应，在孤立纳米晶中的掺杂可能导致一些奇异的物理现象，而这些现象通常在相应的块体材料中是不存在的[25-27]，这就激发了研究人员极大的研究兴趣。目前人们已经取得了许多重要的进展，但是，还有一些问题阻碍了新型掺杂纳米晶材料的开发。其中最大的问题就是这类纳米材料通常都有"自清洁"特性，这是因为纳米晶具有自发地排出杂质元素的内在机制[28]。此外，即便能掺杂，掺杂浓度通常也非常低。因此，如何在半导体纳米晶晶格内引入杂质元素并且精准调控掺杂量已经成为该领域的核心科学问题。

2011 年，Liu 等[29]发表了关于在 LAL 过程中将前驱体材料作为掺杂剂引入半导体纳米晶的开创性研究，即将 Si 原子掺杂到赤铁矿（$\alpha\text{-}Fe_2O_3$）晶格中。他们没有使用化学离子进行掺杂，而是将在去离子水中脉冲激光烧蚀固体 Si 靶合成的

Si 胶体溶液作为掺杂剂。这种胶体溶液中含有分散性好、尺寸小且活性高的 Si 纳米颗粒，不含杂质或额外的化学配体。将上述 Si 胶体溶液和 $FeCl_3$ 溶液的混合物进行水热反应，可以合成出具有超晶格结构的 Si 掺杂 $\alpha\text{-}Fe_2O_3$ 纳米片，其中 Si 原子取代了 $\alpha\text{-}Fe_2O_3$ 中的部分 Fe 原子，产生了更大的晶格［图 9-2（a）和图 9-2（b）］。受这项研究工作的启发，Liu 等[30]又发展了一种普适的半导体纳米晶掺杂方法，将各种杂质元素（例如，Ge、Si、Mn、Ti 和 Sn 等）掺杂到单晶 $\alpha\text{-}Fe_2O_3$ 纳米结构的晶格中，成功合成了 Ge 掺杂的 $\alpha\text{-}Fe_2O_3$［图 9-2（c）］、Si 掺杂的 $\alpha\text{-}Fe_2O_3$［图 9-2（d）］、Mn 掺杂的 $\alpha\text{-}Fe_2O_3$［图 9-2（e）～图 9-2（g）］和 Sn 掺杂的 $\alpha\text{-}Fe_2O_3$［图 9-2（h）～图 9-2（j）］等。在这项技术中，采用 LAL 合成了固体靶的粒子团簇胶体溶液，例如，Ge、Si、TiO_x、SnO_x 和 MnO_x 的高反应性胶体团簇，这些团簇通常具有结合羟基自由基的表面。在高倍电子显微镜下，我们可以清楚地看到，掺杂剂原子要么形成超晶格结构（Ge 和 Si），要么以无序固溶体形式（Mn、Sn 和 Ti）分布在 $\alpha\text{-}Fe_2O_3$ 晶格内，其中掺杂剂的掺入过程基于 Fe^{3+} 与高反应性 LAL 衍生胶体团簇的独特相互作用。

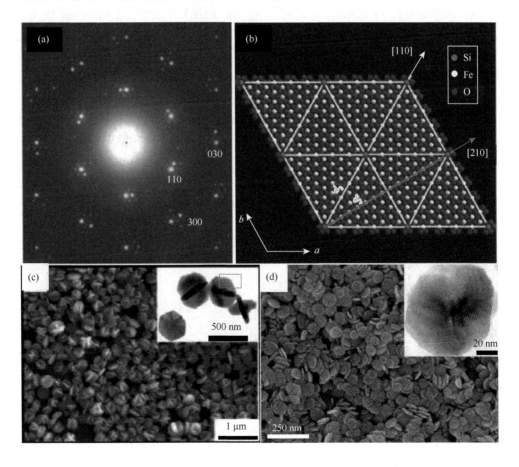

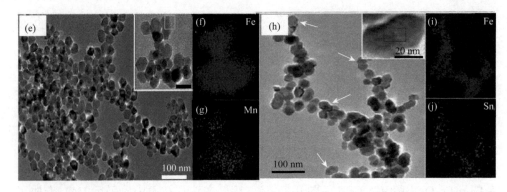

图 9-2 （a）具有超晶格结构的 Si 掺杂 α-Fe_2O_3 纳米片的 SAED 图像；（b）α-Fe_2O_3 晶格中掺杂的 Si 的二维结构模型；（c）Ge 掺杂的 α-Fe_2O_3；（d）Si 掺杂的 α-Fe_2O_3；（e）～（g）Mn 掺杂的 α-Fe_2O_3；（h）～（j）Sn 掺杂的 α-Fe_2O_3

许多研究报道了合成出不同 Ge 原子掺杂浓度的 α-Fe_2O_3 纳米结构，并且试图通过缩小带隙并改变氧化物的尺寸和形状来提高 α-Fe_2O_3 纳米结构的光电流密度[31, 32]。例如，Ge 掺杂的 α-Fe_2O_3 纳米片阵列在 1.23 V（vs. RHE）下表现出 1.4 mA/cm^2 的光电流密度，是未掺杂的 α-Fe_2O_3 纳米棒阵列的 50 倍以上[32]。Sn 掺杂的 α-Fe_2O_3 纳米晶也有类似的效果[33]。采用相同的 LAL 方法，研究人员成功合成了 Ag 簇掺杂的 TiO_2 纳米颗粒，并应用于光降解五氯苯酚[34]。还有一些掺杂半导体纳米晶展现了荧光应用的潜力。重要的是，这种 LAL 半导体纳米晶掺杂方法与其他掺杂方法不同，它为建立胶体化学掺杂模型提供了新的物理和化学认识。此外，这种使用胶体溶液作为掺杂剂的方法可以扩展到使用其杂质元素掺杂 α-Fe_2O_3 或将杂质元素掺入其他半导体纳米晶中。

9.3 亚微米球形颗粒

目前，人们已经发展了多种半导体、金属亚微米球形颗粒的合成技术，有力地推动了这类纳米材料在光子学、光电子学、等离子体羽等领域中的应用[35, 36]。我们很高兴地看到，Wang 等[37]提出的基于 LAL 的选择性脉冲激光加热法已经被证明为合成亚微米球形颗粒的有效方法，因为它可以抑制各向异性晶体生长、促进球形晶体的形成，这种方法被称为液相激光辐照（laser irradiation in liquids，LIL）。同时，该方法是合成宏观量亚微米球形颗粒的一种简便易行的方法。LIL 已经应用到金属、半导体、碳化物和氮化物的亚微米球形颗粒制备中[38-48]。图 9-3（a）～图 9-3（d）显示了 LIL 使用商业原料（包括 ZnO、WO_3、Cu 和 Fe 纳米粉体）合成相应亚微米球形颗粒的能力。LIL 纳米合成的物理机制可以概括为"加热—熔化—蒸发"过程，并且该过程可以解释 LIL 过程中亚微米球形颗粒的尺寸演变。在 LIL 的初始阶段，相对较小的初级颗粒会聚集；然后，由于激光加热效应，小颗粒继续生长，进入快速熔化和凝固过程，如此重复；最后，形成亚微米球形颗粒，其中球形颗粒的最大尺寸由激光功率密度决定[49, 50]。从纳米颗粒到亚微米颗粒演化的示意图如图 9-3（e）所示。

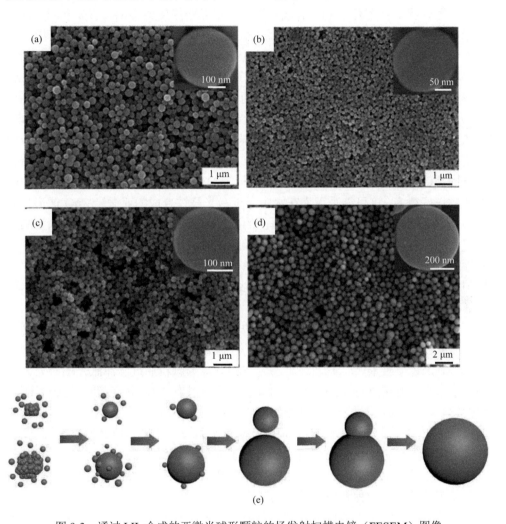

图 9-3 通过 LIL 合成的亚微米球形颗粒的场发射扫描电镜（FESEM）图像

分别使用了商用原料（a）ZnO、（b）WO$_3$、（c）Cu 和（d）Fe 纳米粉体，插图是相应的放大的 SEM 图像；（e）纳秒激光辐照下纳米颗粒向亚微米颗粒演化的示意图

在 LIL 制备技术中，我们可以通过对激光功率密度和激光烧蚀时间等参数简单地调整，灵活控制脉冲激光的选择性加热，从而合成不同尺寸的球形颗粒。这种方法合成的半导体和金属亚微米球形颗粒由于其高结晶度、清洁表面和不含任

何次级结构的球形形态，具有独特的电学、磁学和光电子特性。此外，LIL 合成亚微米球形颗粒的产率可以达到 100 mg/h，比 LAL 合成亚微米球形颗粒产率高约两个数量级[37]。因此，通过液相激光辐照合成亚微米球形颗粒的技术有助于推进亚微米球形颗粒的实际应用。

下面，我们介绍几个 LIL 合成亚微米球形颗粒应用的实例。首先介绍单晶亚微米 ZnO 球形颗粒构筑的紫外探测器[51]。我们可以通过 LIL 过程，采用商用 ZnO 纳米粉体作为原料，用去离子水作为环境液体，合成亚微米 ZnO 球形颗粒。由这种单晶亚微米 ZnO 球形颗粒构成的薄膜可以产生远高于目前所报道的多晶 ZnO 球形颗粒的光电流密度，因为球形颗粒的单晶性质和清洁表面促进了电子在相邻球形颗粒之间的界面传输，并消除了球形颗粒内电子传输的障碍。在研究中发现，亚微米 ZnO 球形颗粒可以选择性地捕获苯胺（苯胺是已知具有潜在的致癌性或致突变性的物质[52]），这是因为亚微米 ZnO 球形颗粒表面羟基的强配位作用。令人惊讶的是，经过紫外线/臭氧处理的亚微米 ZnO 球形颗粒的性能比未处理的 ZnO 纳米颗粒好得多，这可能是因为在水中通过激光烧蚀合成的亚微米 ZnO 球形颗粒去除了表面的碳，因此有利于 OH⁻基团的接枝。另外，亚微米 ZnO 球形颗粒具有与尺寸相关的消光系数并且可选择性地捕获苯胺。亚微米 ZnO 球形颗粒这些奇特的性质决定了它们在光学和传感设备等中的潜在应用。例如，Fujiwara 等[39]使用均匀亚微米尺寸的 ZnO 球形颗粒（包括黏附聚合物颗粒）开发了一种具有准单模和低激光阈值的随机激光器。

我们知道，在摩擦学领域，高硬度的球形颗粒很容易将滑动摩擦变为滚动摩擦，以减少磨损。研究人员以 LIL 合成的亚微米磁铁矿（Fe_3O_4）球形颗粒为例，研究了其摩擦学性能[43]。这里，亚微米 Fe_3O_4 球形颗粒通过脉冲激光辐照分散在去离子水中的商用立方 Fe_2O_3 纳米粉体而合成，具有较高的硬度和光滑的表面。他们发现，与 Fe_2O_3 纳米粉体和立方亚微米颗粒（不规则形状）相比，这些球形颗粒的摩擦和磨损分别降低了 40%和 20%。同时，对将这些球形颗粒作为润滑剂的抗磨损能力也进行了表征。图 9-4 比较了使用纯油和含有亚微米 Fe_3O_4 球形颗粒的油的磨损疤痕图像。我们可以清楚地看到，球形颗粒的存在导致疤痕直径更小，证明其可作为一种新型亚微米润滑剂。

实际上，LIL 不仅可以用于制备单晶亚微米实心球颗粒，还可以用于组装亚微米空心球颗粒[53]。LIL 组装空心球颗粒的物理机制是柯肯德尔效应，也就是在 LIL 过程中实心球颗粒表面熔化后不同组分的扩散速率不同而导致的空心化。由于独特的空心结构，这类亚微米空心球颗粒在许多领域都有重要的应用。例如，通过 LIL 组装的尺寸可控的亚微米 TiO_2 空心球颗粒可以在较宽的可见光范围内表现出可调谐的光散射 [图 9-5（a）和图 9-5（b）]，因此它们可以用作量子点敏化太阳电池（quantum dot sensitized solar cell，QDSSC）中的散射层材料 [图 9-5（c）和

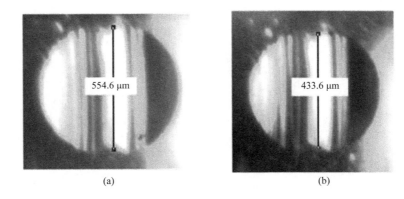

图 9-4 磨损疤痕测试

(a) 纯油作为润滑剂；(b) 含有 0.07%质量分数的亚微米 Fe_3O_4 球形颗粒的油作为润滑剂

图 9-5（d）]，使得太阳能-电能的转换效率提高 10%［图 9-5（e）和图 9-5（f）]。此外，这种 LIL 亚微米空心球颗粒组装技术也适用于其他金属和半导体小尺度空心球颗粒的组装，如 Fe、Ni、NiO 和 WO_3，这对于小尺度单晶无机空心球颗粒的相关研究有极大的促进作用[54]。

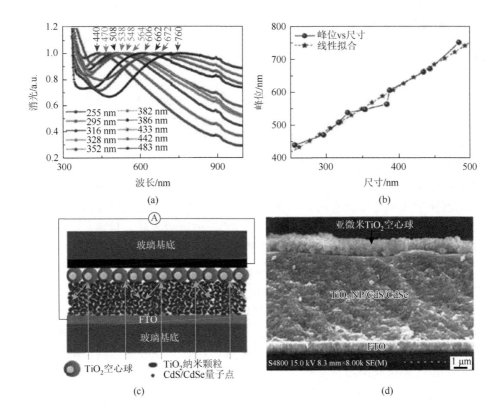

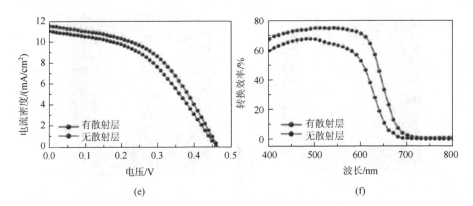

图 9-5 （a）不同激光能量密度辐照样品的归一化紫外可见消光光谱；（b）消光峰值与球形颗粒平均尺寸的关系；（c）QDSSC 的 TiO_2 空心球颗粒散射层示意图；（d）TiO_2 空心球颗粒覆盖的量子点敏化太阳电池介孔电极的横截面 SEM 图像；（e）电流密度与电压的关系；（f）有和没有 TiO_2 空心球颗粒散射层的 QDSSC 的入射光子-电子转换效率（后附彩图）

用脉冲激光辐照液体环境中的初始纳米颗粒来合成亚微米球形（空心球）颗粒是一种非常简单、绿色、高产额的通用技术，由于所合成的亚微米球形（空心球）颗粒具有光滑表面、高结晶度和尺寸可控的特性，在微电子学、光电子学、摩擦学和光学等领域显示出了巨大的潜在应用。

9.4 单分散胶体量子点

胶体量子点（colloidal quantum dot，CQD）具有可调的带隙和多激子生成等独特的物理和化学性质，在光电子学和生物医学中有着重要的应用[55-57]。但是，CQD 的性能与其尺寸和形貌有很大的关系，例如，要想获得理想的性能，CQD 的尺寸和形态分布都必须在一个非常窄的范围内[58]。以往，单分散 CQD 主要是利用胶体溶液化学技术来合成的，这种方法可以实现胶体量子点核的成核和生长[59, 60]。然而，胶体溶液化学技术通常成本高昂、反应前驱体不环保、工艺操作复杂。近年来，LAL 纳米制备技术已经被证明是一种合成 CQD 的有效方法。目前，研究人员已经采用 LAL 纳米制备技术合成了多种多样的 CQD 如 PbS、PbSe、Cd、立方氮化硼（c-BN）等 CQD[61-64]。这种简单、环保、高效的方法不仅简化了 CQD 的合成过程，而且合成出了以往的方法无法合成的新型 CQD。

开创性的工作是由 Yang 等[61]实现的，他们首次使用两步 LAL 方法合成了超级单分散 CQD。具体而言，首先他们使用毫秒激光对环境液体双对氯苯基三氯乙烷（滴滴涕，DDT）中的固体 Pb 靶进行烧蚀，合成了 PbS 纳米晶；然后他们使用了另一束非聚焦脉冲激光辐照 PbS 纳米晶溶液，从而合成了直径为 (5.5 ± 0.3)nm 且分散度仅为 5.5%的 CQD，如图 9-6（a）所示。我们可以看到，相应的 SAED

图像和高分辨 TEM 图像均证实了这些纳米晶是具有 FCC 结构的 PbS [图 9-6（b）和图 9-6（c）]。需要注意的是，这里的 PbS CQD 的表面覆盖了一层双对氯苯基三氯乙烷以防止 CQD 发生聚集 [图 9-6（d）]。进一步，他们把单分散 PbS CQD 的形成归因于强量子限制效应 [图 9-6（e）～图 9-6（h）]，该效应影响带隙宽度和低强度毫秒激光束对纳米晶的选择性蒸发。

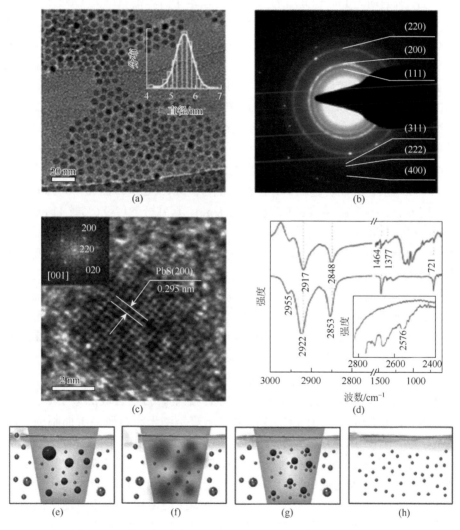

图 9-6　（a）激光辐照原始 PbS 纳米晶溶液 20 min 的 TEM 图像，插图为量子点的尺寸直方图和相应的高斯拟合；（b）PbS CQD 的 SAED 图像；（c）单个 CQD 的 HRTEM 图像，插图是相应的 FTT 图；（d）涂有 DDT（上）和未涂 DDT（下）的 PbS CQD 的 FTIR 光谱，插图为局部放大；（e）大尺寸纳米颗粒的选择性加热；（f）纳米颗粒被加热至蒸发；（g）蒸发的 PbS 原子气体的凝结，并且尺寸超过阈值的较小纳米颗粒被选择性地进一步加热；（h）单分散 PbS CQD 的形成

上面的工作显示了液相毫秒激光烧蚀/辐照合成单分散 PbS CQD 的情况，同样，纳秒激光也可以用于该技术。例如，Wu 等[63]发表了液相纳秒激光辐照天然方铅矿一步合成 CQD。他们使用未聚焦的纳秒激光辐照分散在环境液体油酸中的研磨方铅矿颗粒，一步合成单分散 CQD。图 9-7（a）和图 9-7（b）显示了方铅矿颗粒研磨前后的对比照片。在脉冲激光辐照后，可以获得尺寸为(5.5±0.5)nm、分散度为 11.1%的均匀 PbS CQD［图 9-7（c）］。这个过程的机制为 LaMer 成核。首先，方铅矿颗粒被纳秒激光裂解为活性单体（活性基团）并弥散在环境液体油酸中；然后，当油酸中的活性单体浓度在局域内达到过饱和时，LaMer 成核发生，也就是瞬间同时发生大量成核，也称"爆炸式成核"，接着这些晶核同时生长从而保证了尺寸均一性；最后，合成出尺寸均一、形状一致的单分散 PbS CQD。研究人员进行了毫秒激光和纳秒激光的对比实验，结果发现，由于毫秒激光很难产生高度过饱和的 PbS 单体，无法诱导 LaMer 成核发生，所以产物中只观察到不均匀的纳米颗粒。

这种 LAL 纳米制备技术除了可以合成单分散 PbS 和 PbSe 等半导体 CQD 以外，还可以用来合成非贵金属单分散纳米晶，例如，Cd 纳米晶（140~80 nm、尺寸分散度 6.5%~9.5%）和 Zn 纳米晶（15.8 nm、尺寸分散度 9.6%或 11 nm、尺寸分散度 10.2%）可以在环境液体环己烷和油酸中通过两步 LAL（烧蚀/辐照）方法合成[64]。具体而言，图 9-7（d）中的固体 Cd 靶在第一步激光烧蚀中转变为多分散的球形 Cd 纳米晶［图 9-7（e）］，随后进行第二步激光辐照，在这一过程中，多分散纳米晶转变成了单分散纳米晶。这一结果意味着，第二步的激光辐照过程在单分散纳米晶的合成中非常重要。图 9-7（f）和图 9-7（g）显示了使用不同激光功率密度获得的不同尺寸的 Cd 纳米晶，这些纳米晶相应的高分辨 TEM 和 SAED 图像证实了 Cd 纳米晶的生成［图 9-7（h）和图 9-7（i）］。

c-BN 和金刚石一样属于超宽禁带半导体材料，它们的 CQD 在光电子学、量子信息、生物医学等领域有着重要潜在应用。然而，目前合成的最小 c-BN 纳米晶尺寸为 14 nm，所以，c-BN CQD 的合成一直是纳米材料研究者面临的一项重大挑战。Du 等采用 LAL 纳米制备技术首次合成单分散 c-BN CQD[62]。他们采用的方法是用脉冲激光直接烧蚀环境液体来合成 CQD，使用氨的二噁烷溶液作为环境液体（同时也是液体靶材），通过自上而下（top-down）的光化学过程分解液体靶材为高活性基团，然后持续的激光辐照会使环境液体中高活性基团浓度达到高度的过饱和，从而诱导 LaMer 成核发生，最后合成出尺寸为 3.5 nm 的 c-BN CQD。

如上所述，利用 LAL 纳米制备技术合成单分散 CQD 的工艺极大地简化了高质量 CQD 的传统合成工艺，并且无须引入任何还原剂。因此，在高质量单分散半导体和金属 CQD 合成领域有着很大应用潜力。

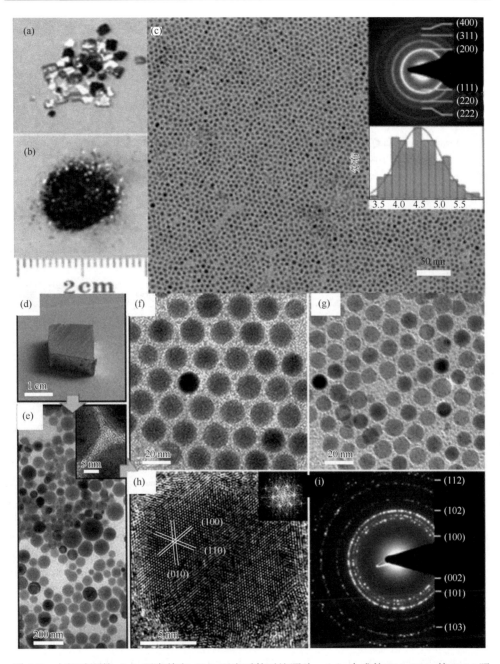

图 9-7 方铅矿颗粒（a）研磨前和（b）研磨后的对比照片；（c）合成的 PbS CQD 的 TEM 照片，插图是粒度分布和相应的高斯拟合（下）及 SAED 图像（上）；（d）Cd 靶的照片；（e）激光烧蚀 Cd 靶合成的多分散 Cd 纳米晶的 TEM 照片；（f）和（g）不同尺寸的单分散 Cd 纳米晶的 TEM 图像；（h）和（i）分别是（f）中 Cd 纳米晶的 HRTEM 和 SAED 图像

参 考 文 献

[1] Pyatenko A, Shimokawa K, Yamaguchi M, et al. Synthesis of silver nanoparticles by laser ablation in pure water[J]. Applied Physics A, 2004, 79: 803-806.

[1] Phuoc T X, Howard B H, Martello D V, et al. Synthesis of Mg(OH)$_2$, MgO, and Mg nanoparticles using laser ablation of magnesium in water and solvents[J]. Optics and Lasers in Engineering, 2008, 46 (11): 829-834.

[2] Amikura K, Kimura T, Hamada M, et al. Copper oxide particles produced by laser ablation in water[J]. Applied Surface Science, 2008, 254 (21): 6976-6982.

[3] Bärsch N, Jakobi J, Weiler S, et al. Pure colloidal metal and ceramic nanoparticles from high-power picosecond laser ablation in water and acetone[J]. Nanotechnology, 2009, 20 (44): 445603.

[4] Cioffi N, Torsi L, Ditaranto N, et al. Copper nanoparticle/polymer composites with antifungal and bacteriostatic properties[J]. Chemistry of Materials, 2005, 17 (21): 5255-5262.

[5] Balazs A C, Emrick T, Russell T P. Nanoparticle polymer composites: Where two small worlds meet[J]. Science, 2006, 314 (5802): 1107-1110.

[6] Crosby A J, Lee J Y. Polymer nanocomposites: The "nano" effect on mechanical properties[J]. Polymer Reviews, 2007, 47 (2): 217-229.

[7] Sambhy V, MacBride M M, Peterson B R, et al. Silver bromide nanoparticle/polymer composites: Dual action tunable antimicrobial materials[J]. Journal of the American Chemical Society, 2006, 128 (30): 9798-9808.

[8] Du F M, Scogna R C, Zhou W, et al. Nanotube networks in polymer nanocomposites: Rheology and electrical conductivity[J]. Macromolecules, 2004, 37 (24): 9048-9055.

[9] Huang Y Y, Ahir S V, Terentjev E M. Dispersion rheology of carbon nanotubes in a polymer matrix[J]. Physical Review B, 2006, 73 (12): 125422.

[10] Schaefer D W, Zhao J, Brown J M, et al. Morphology of dispersed carbon single-walled nanotubes[J]. Chemical Physics Letters, 2003, 375 (3-4): 369-375.

[11] Stelzig S H, Menneking C, Hoffmann M S, et al. Compatibilization of laser generated antibacterial Ag-and Cu-nanoparticles for perfluorinated implant materials[J]. European Polymer Journal, 2011, 47 (4): 662-667.

[12] Wagener P, Faramarzi S, Schwenke A, et al. Photoluminescent zinc oxide polymer nanocomposites fabricated using picosecond laser ablation in an organic solvent[J]. Applied Surface Science, 2011, 257 (16): 7231-7237.

[13] Sowa-Söhle E N, Schwenke A, Wagener P, et al. Antimicrobial efficacy, cytotoxicity, and ion release of mixed metal (Ag, Cu, Zn, Mg) nanoparticle polymer composite implant material[J]. BioNanoMaterials, 2013, 14(3-4): 217-227.

[14] Semaltianos N G, Perrie W, Romani S, et al. Polymer-nanoparticle composites composed of PEDOT：PSS and nanoparticles of Ag synthesised by laser ablation[J]. Colloid and Polymer Science, 2012, 290: 213-220.

[15] Schwenke A, Dalüge H, Kiyan R, et al. Non-agglomerated gold-PMMA nanocomposites by in situ-stabilized laser ablation in liquid monomer for optical applications[J]. Applied Physics A, 2013, 111: 451-457.

[16] Nachev P, van'T Zand D D, Coger V, et al. Synthesis of hybrid microgels by coupling of laser ablation and polymerization in aqueous medium[J]. Journal of Laser Applications, 2012, 24 (4): 042012.

[17] Mastrotto F, Caliceti P, Amendola V, et al. Polymer control of ligand display on gold nanoparticles for multimodal switchable cell targeting[J]. Chemical Communications, 2011, 47 (35): 9846-9848.

[18] Makridis S S, Gkanas E I, Panagakos G, et al. Polymer-stable magnesium nanocomposites prepared by laser ablation for efficient hydrogen storage[J]. International Journal of Hydrogen Energy, 2013, 38(26): 11530-11535.

[19] Kalyva M, Kumar S, Brescia R, et al. Electrical response from nanocomposite PDMS-Ag NPs generated by in situ laser ablation in solution[J]. Nanotechnology, 2012, 24 (3): 035707.

[20] Hahn A, Brandes G, Wagener P, et al. Metal ion release kinetics from nanoparticle silicone composites[J]. Journal of Controlled Release, 2011, 154 (2): 164-170.

[21] Wagener P, Brandes G, Schwenke A, et al. Impact of in situ polymer coating on particle dispersion into solid laser-generated nanocomposites[J]. Physical Chemistry Chemical Physics, 2011, 13 (11): 5120-5126.

[22] Zhang D S, Barcikowski S. Rapid nanoparticle-polymer composites prototyping by laser ablation in liquids[M]// Kobayashi S, Müllen K. Encyclopedia of polymeric nanomaterials. Berlin: Springer, 2014: 1-12.

[23] Zhang D S, Gökce B. Perspective of laser-prototyping nanoparticle-polymer composites[J]. Applied Surface Science, 2017, 392: 991-1003.

[24] Erwin S C, Zu L J, Haftel M I, et al. Doping semiconductor nanocrystals[J]. Nature, 2005, 436: 91-94.

[25] Jain P K, Manthiram K, Engel J H, et al. Doped nanocrystals as plasmonic probes of redox chemistry[J]. Angewandte Chemie International Edition, 2013, 52 (51): 13671-13675.

[26] Norris D J, Efros A L, Erwin S C. Doped nanocrystals[J]. Science, 2008, 319 (5871): 1776-1779.

[27] Dalpian G M, Chelikowsky J R. Self-purification in semiconductor nanocrystals[J]. Physical Review Letters, 2006, 96 (22): 226802.

[28] Liu J, Liang C H, Zhang H M, et al. Silicon-doped hematite nanosheets with superlattice structure[J]. Chemical Communications, 2011, 47 (28): 8040-8042.

[29] Liu J, Liang C H, Zhang H M, et al. General strategy for doping impurities (Ge, Si, Mn, Sn, Ti) in hematite nanocrystals[J]. The Journal of Physical Chemistry C, 2012, 116 (8): 4986-4992.

[30] Liu J, Cai Y Y, Tian Z F, et al. Highly oriented Ge-doped hematite nanosheet arrays for photoelectrochemical water oxidation[J]. Nano Energy, 2014, 9: 282-290.

[31] Liu J, Liang C H, Xu G P, et al. Ge-doped hematite nanosheets with tunable doping level, structure and improved photoelectrochemical performance[J]. Nano Energy, 2013, 2 (3): 328-336.

[32] Ruan G S, Wu S L, Wang P P, et al. Simultaneous doping and growth of Sn-doped hematite nanocrystalline films with improved photoelectrochemical performance[J]. RSC Advances, 2014, 4 (108): 63408-63413.

[33] Zhang H M, Liang C H, Liu J, et al. Defect-mediated formation of Ag cluster-doped TiO_2 nanoparticles for efficient photodegradation of pentachlorophenol[J]. Langmuir, 2012, 28 (8): 3938-3944.

[34] Wang H Q, Kawaguchi K, Pyatenko A, et al. General bottom-up construction of spherical particles by pulsed laser irradiation of colloidal nanoparticles: A case study on CuO[J]. Chemistry: A European Journal, 2012, 18 (1): 163-169.

[35] Li X Y, Pyatenko A, Shimizu Y, et al. Fabrication of crystalline silicon spheres by selective laser heating in liquid medium[J]. Langmuir, 2011, 27 (8): 5076-5080.

[36] Wang H Q, Pyatenko A, Kawaguchi K, et al. Selective pulsed heating for the synthesis of semiconductor and metal submicrometer spheres[J]. Angewandte Chemie International Edition, 2010, 49 (36): 6361-6364.

[37] Zhang D S, Lau M, Lu S W, et al. Germanium sub-microspheres synthesized by picosecond pulsed laser melting in liquids: Educt size effects[J]. Scientific Reports, 2017, 7: 40355.

[38] Fujiwara H, Niyuki R, Ishikawa Y, et al. Low-threshold and quasi-single-mode random laser within a submicrometer-sized ZnO spherical particle film[J]. Applied Physics Letters, 2013, 102 (6): 061110.

[39] Kawasoe K, Ishikawa Y, Koshizaki N, et al. Preparation of spherical particles by laser melting in liquid using TiN as a raw material[J]. Applied Physics B, 2015, 119: 475-483.

[40] Li X Y, Shimizu Y, Pyatenko A, et al. Tetragonal zirconia spheres fabricated by carbon-assisted selective laser heating in a liquid medium[J]. Nanotechnology, 2012, 23 (11): 115602.

[41] Li X Y, Shimizu Y, Pyatenko A, et al. Carbon-assisted fabrication of submicrometre spheres for low-optical-absorbance materials by selective laser heating in liquid[J]. Journal of Materials Chemistry, 2011, 21(38): 14406-14409.

[42] Song X Y, Qiu Z W, Yang X P, et al. Submicron-lubricant based on crystallized Fe_3O_4 spheres for enhanced tribology performance[J]. Chemistry of Materials, 2014, 26 (17): 5113-5119.

[43] Swiatkowska-Warkocka Z, Koga K, Kawaguchi K, et al. Pulsed laser irradiation of colloidal nanoparticles: A new synthesis route for the production of non-equilibrium bimetallic alloy submicrometer spheres[J]. RSC Advances, 2013, 3 (1): 79-83.

[44] Tsuji T, Higashi Y, Tsuji M, et al. Preparation of submicron-sized spherical particles of gold using laser-induced melting in liquids and low-toxic stabilizing reagent[J]. Applied Surface Science, 2015, 348: 10-15.

[45] Tsuji T, Yahata T, Yasutomo M, et al. Preparation and investigation of the formation mechanism of submicron-sized spherical particles of gold using laser ablation and laser irradiation in liquids[J]. Physical Chemistry Chemical Physics, 2013, 15 (9): 3099-3107.

[46] Wang H Q, Jia L C, Li L, et al. Photomediated assembly of single crystalline silver spherical particles with enhanced electrochemical performance[J]. Journal of materials chemistry A, 2013, 1 (3): 692-698.

[47] Ishikawa Y, Feng Q, Koshizaki N. Growth fusion of submicron spherical boron carbide particles by repetitive pulsed laser irradiation in liquid media[J]. Applied Physics A, 2010, 99: 797-803.

[48] Pyatenko A, Wang H Q, Koshizaki N, et al. Mechanism of pulse laser interaction with colloidal nanoparticles[J]. Laser & Photonics Reviews, 2013, 7 (4): 596-604.

[49] Pyatenko A, Wang H Q, Koshizaki N. Growth mechanism of monodisperse spherical particles under nanosecond pulsed laser irradiation[J]. The Journal of Physical Chemistry C, 2014, 118 (8): 4495-4500.

[50] Wang H, Pyatenko A, Koshizaki N, et al. Single-crystalline ZnO spherical particles by pulsed laser irradiation of colloidal nanoparticles for ultraviolet photodetection[J]. ACS Applied Materials & Interfaces, 2014, 6 (4): 2241-2247.

[51] Wang H Q, Koshizaki N, Li L, et al. Size-tailored ZnO submicrometer spheres: Bottom-up construction, size-related optical extinction, and selective aniline trapping[J]. Advanced Materials, 2011, 23 (16): 1865-1870.

[52] Wang H Q, Miyauchi M, Ishikawa Y, et al. Single-crystalline rutile TiO_2 hollow spheres: Room-temperature synthesis, tailored visible-light-extinction, and effective scattering layer for quantum dot-sensitized solar cells[J]. Journal of the American Chemical Society, 2011, 133 (47): 19102-19109.

[53] Wang Y L, Xia Y N. Bottom-up and top-down approaches to the synthesis of monodispersed spherical colloids of low melting-point metals[J]. Nano Letters, 2004, 4 (10): 2047-2050.

[54] Kim J Y, Voznyy O, Zhitomirsky D, et al. 25th anniversary article: Colloidal quantum dot materials and devices: A quarter-century of advances[J]. Advanced Materials, 2013, 25 (36): 4986-5010.

[55] Medintz I L, Uyeda H T, Goldman E R, et al. Quantum dot bioconjugates for imaging, labelling and sensing[J]. Nature Materials, 2005, 4: 435-446.

[56] Carey G H, Abdelhady A L, Ning Z J, et al. Colloidal quantum dot solar cells[J]. Chemical Reviews, 2015, 115 (23): 12732-12763.

[57] Rogach A L, Talapin D V, Shevchenko E V, et al. Organization of matter on different size scales: Monodisperse nanocrystals and their superstructures[J]. Advanced Functional Materials, 2002, 12 (10): 653-664.

[58] Weidman M C, Beck M E, Hoffman R S, et al. Monodisperse, air-stable PbS nanocrystals via precursor stoichiometry control[J]. ACS Nano, 2014, 8 (6): 6363-6371.

[59] Zhang J B, Crisp R W, Gao J B, et al. Synthetic conditions for high-accuracy size control of PbS quantum dots[J]. The Journal of Physical Chemistry Letters, 2015, 6 (10): 1830-1833.

[60] Yang J, Ling T, Wu W T, et al. A top-down strategy towards monodisperse colloidal lead sulphide quantum dots[J]. Nature Communications, 2013, 4: 1695.

[61] Liu H, Jin P, Xue Y M, et al. Photochemical synthesis of ultrafine cubic boron nitride nanoparticles under ambient conditions[J]. Angewandte Chemie International Edition, 2015, 54 (24): 7051-7054.

[62] Wu W T, Liu H, Dong C, et al. Gain high-quality colloidal quantum dots directly from natural minerals[J]. Langmuir, 2015, 31 (8): 2251-2255.

[63] Luo R C, Li C, Du X W, et al. Direct conversion of bulk metals to size-tailored, monodisperse spherical non-coinage-metal nanocrystals[J]. Angewandte Chemie International Edition, 2015, 54 (16): 4787-4791.

第 10 章　LAL 基纳米材料的应用

本章我们介绍采用 LAL 纳米制备技术制备的纳米材料在不同领域的应用。之所以要专门介绍 LAL 基纳米材料与纳米结构的应用，是因为这一类纳米材料通常在化学上表现出高活性，在物理上由于高结晶度和高清洁表面展现出其他方法制备的纳米材料所没有的奇异物性，因此，在不同应用领域都得到了极大关注。我们集中介绍 LAL 基纳米材料与纳米结构在光学、磁学、环境科学、能源和生物医学等领域的应用，从而希望在更多、更宽广的领域去发现它们新的应用。

10.1　光学功能纳米结构

10.1.1　荧光发射

众所周知，LAL 已经广泛地用于荧光纳米材料的合成[1-33]，并且研究人员已经证明，我们可以根据材料体系的不同调控 LAL 合成的荧光纳米材料的荧光来源。所以，根据目前已经取得的研究成果，我们可以根据荧光来源将这些 LAL 合成的荧光纳米材料分为三类：①本征荧光纳米材料；②缺陷诱导荧光纳米材料；③官能团诱导荧光纳米材料。

荧光纳米材料的本征荧光与其带隙内的带间跃迁密切相关。例如，在 LAL 合成的 SiC 纳米环荧光发射研究中［图 10-1（a）～图 10-1（d）][34]，人们发现，当 SiC 纳米环被波长大于 350 nm 的光激发时，表现出强烈的量子限域效应。然而，当激发波长小于 350 nm 时，SiC 纳米环则表现出异常的光谱红移，如图 10-1（e）和图 10-1（f）所示。这些奇异的行为可能是由 SiC 纳米环的表面结构引起的，这一推论得到了理论计算与模拟的支持[35]。

稀土掺杂纳米晶是除半导体量子点之外的一种典型的本征荧光纳米材料。众所周知，稀土掺杂材料的荧光与掺杂离子密切相关，但是，对于固定的掺杂浓度，材料的带隙也就相对固定了。因此，稀土掺杂材料可视作是一种本征荧光材料。稀土掺杂纳米晶因其高的发光稳定性、无闪烁和极窄的发射线而引起了人们的极大兴趣[36-38]。值得一提的是，镧系元素掺杂的氧化钆（Gd_2O_3）纳米颗粒具有优异的磁性和荧光特性，因而可以用于生物医学领域的双模态成像。例如，由于其独特的性能，Gd_2O_3 纳米颗粒可以充当磁共振成像（MRI）的造影剂，而掺杂

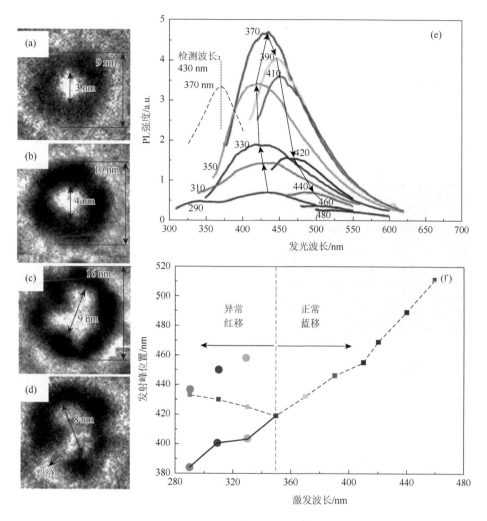

图 10-1 （a）～（d）各种形态的 SiC 纳米环；（e）不同波长激光激发下的 SiC 纳米环的光致发光光谱；（f）发射峰位置与激发波长的函数关系（后附彩图）

的镧系元素离子则可以作为荧光发射中心[39, 40]。并且，这些掺杂纳米颗粒既有 MRI 中高组织对比度，又有光学成像技术中高灵敏度的优点[38, 41]。然而，这些镧系元素掺杂的纳米晶一般是通过化学方法合成的，这就需要高纯度的化学试剂，并依赖于特定反应前驱体的适用性和苛刻的化学反应条件等[39]。因此，发展简单、绿色、高效的镧系元素掺杂氧化钆纳米材料的制备技术就成了研究人员的追求目标。

最近，Luo 等[42]发展了一种简便、绿色的方法，通过结合标准固态反应和 LAL 来组装 Tm^{3+}、Tb^{3+} 和 Eu^{3+} 掺杂的 Gd_2O_3 纳米颗粒。他们首先通过标准固态反应方法制备了镧系元素掺杂的 Gd_2O_3 作为固体靶，然后使用纳秒激光在去离子水中烧蚀制备的靶材，最后合成出无配体的 Tm^{3+}、Tb^{3+} 和 Eu^{3+} 掺杂的 Gd_2O_3 纳米颗粒。令人高兴的是，这些掺杂纳米颗粒发出强烈的蓝色、绿色和红色荧光，如图10-2（a）所示，图10-2（b）显示了掺杂纳米颗粒相应的发射光谱。显然，这些纳米颗粒的荧光源于光敏化剂 Gd^{3+} 和不同的激活剂（Tm^{3+}、Tb^{3+} 和 Eu^{3+}）之间的能量转移，如图10-2（c）所示。重要的是，细胞共焦显微镜图像（图10-3）表明这些纳米颗粒胶体是良好的荧光成像剂，这意味着它们可以用来进行细胞标记。

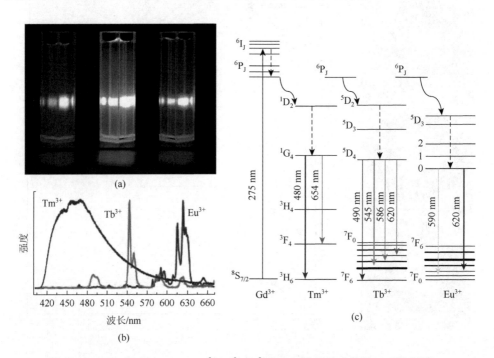

图 10-2　LAL 合成 Gd_2O_3：$Tm^{3+}/Tb^{3+}/Eu^{3+}$ 纳米颗粒的荧光特性（后附彩图）
（a）纳米颗粒的荧光图像，从左到右分别为 Gd_2O_3：$Tm^{3+}/Tb^{3+}/Eu^{3+}$，激光波长为 275 nm；（b）相应的发射光谱；（c）光敏化剂 Gd^{3+} 与不同激活剂 Tm^{3+}、Tb^{3+} 和 Eu^{3+} 之间的能量转移示意图

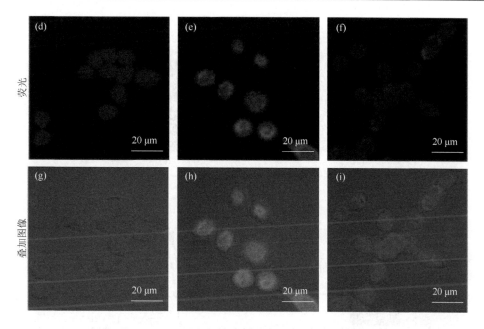

图 10-3　分别与 Gd_2O_3：$Tm^{3+}/Tb^{3+}/Eu^{3+}$ 纳米颗粒一起孵育的细胞的共焦显微镜图像
（后附彩图）

(a)～(c) 明场图像；(d)～(f) 405 nm 激光激发下的荧光图像；(g)～(i) 叠加图像

此外，基于相同的合成路线，Lu 等还合成了具有近红外（near-IR）到可见光发射和 near-IR 上转换（upconversion，UC）荧光的 Gd_2O_3：Yb^{3+}/Tm^{3+} 纳米颗粒。我们知道，对于生物医学应用而言，具有上转换荧光的纳米颗粒是理想的材料，因为生物组织对较长波长的光散射弱于较短波长的光，因此近红外可以有效地穿透生物组织，从而减少光损伤。所以，这一类 LAL 合成的荧光纳米颗粒在生物医学应用中引起了广泛关注[43-45]。LAL 合成的荧光纳米颗粒的近红外发射可以归因于 $^3F_4 \rightarrow {}^1G_4$ 跃迁比 $^3H_4 \rightarrow {}^1G_4$ 跃迁有着更小的能量失配，以及 $^1G_4 \rightarrow {}^3H_4$ 和 $^3F_4 \rightarrow {}^3H_4$ 跃迁之间的交叉弛豫。此外，通过将 Gd_2O_3：Yb^{3+} 纳米颗粒的 Yb^{3+} 含量从 0 增加到 15%，在 980 nm 波长激光激发下，可以实现连续颜色可调的上转换荧光，如图 10-4（a）所示。图 10-4（b）显示了 Gd_2O_3：Yb^{3+}（0～15%）胶体在 980 nm 激光激发下的荧光图像，以及相应的发射光谱 [图 10-4（c）]。

LAL 合成荧光纳米颗粒的第二种荧光来源是缺陷诱导的荧光。由于 LAL 过程可以产生远离热力学平衡的极端环境，即高温高压高密度高激发态的局部极端热力学环境，所以，LAL 合成的纳米材料总是含有高浓度的缺陷，这就为材料的缺陷发光提供了机会。例如，Zeng 等[46-51]通过大量研究阐明了 LAL 合成的 ZnO 纳米颗粒产生蓝色荧光的机制。他们通过系统的激发光谱、发射光谱和电

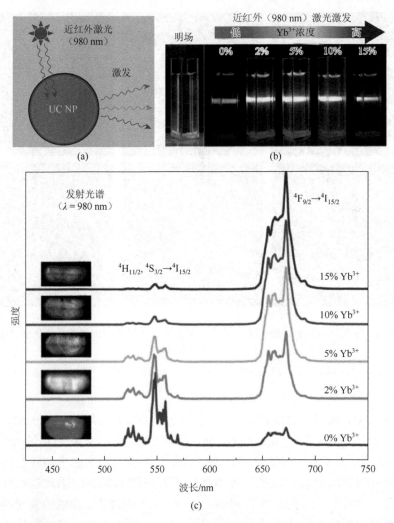

图 10-4 （a）近红外激光激发下纳米颗粒上转换过程的示意图；（b）Gd_2O_3：Yb^{3+}（0~15%）胶体在 980 nm 激光激发下的荧光图像；（c）相应的发射光谱（后附彩图）

子顺磁共振谱（EPRS）分析表明，初始状态位于 Zn 相关的缺陷能级的 ZnO 是产生蓝色荧光的原因[52]，如图 10-5 所示。这样，通过对缺陷的调控，他们实现了 ZnO 纳米颗粒可见光的可控发射。

LAL 合成的 Si 纳米颗粒也表现出明显的熟化增强的蓝光发射[53]。一些研究人员认为，Si 纳米材料发射的蓝光源于氧化物相关缺陷[54, 55]或量子尺寸限制效应[56-59]。Yang 等[60]通过监测 Si 纳米晶在各种处理过程中的光致发光（PL）演化，发现 Si 纳米晶的荧光不能简单地归因于上述两种效应。相反，最合理的发光机制是通过定向跃迁，首先，激子在 Si 纳米晶内部形成；随后，一些激子被 Si/SiO_2

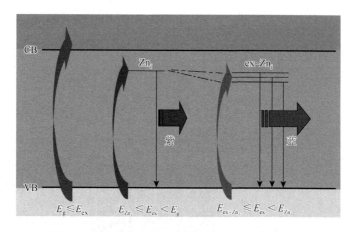

图 10-5　LAL 合成 ZnO 纳米颗粒的蓝光发射机制

界面的顺磁缺陷捕获，其他的激子则向附近的界面陷阱转移并重组从而发射蓝光[61]。然而，研究人员更多的是强调 S1 纳米结构的光致发光机制在很大程度上取决于其制备方法。

现在我们来讨论 LAL 合成荧光纳米材料的第三种荧光来源，即官能团诱导荧光。众所周知，块状碳材料由于缺乏适当的光学带隙而无法发光。然而，当它们的尺寸减小到小于 5 nm 时，可以观察到强烈的可见光发射，这就是所谓的碳量子点发光。Li 等[62]通过在普通有机溶剂中用非聚焦脉冲激光辐照碳纳米颗粒来合成碳量子点。他们发现，LAL 合成的碳量子点表现出可调谐的、稳定的可见光发射。实际上，金刚石纳米晶和其他碳纳米结构也可以观察到类似的发光现象[63-66]。然而，目前尚不清楚为什么不同的碳纳米结构表现出相似的荧光发射。究其根本，我们知道，石墨是导体而金刚石是绝缘体，他们在可见光范围内都不发光。但是，这些不同属性的碳材料的不同纳米结构则具有相似的荧光发射。因此，这些荧光的起源不应归因于它们的本体特性。事实上，根据大量实验研究，这些碳量子点必须经过有机溶剂钝化才能显示荧光现象，也就是需要必要的发光官能团的修饰后才能发光，这一点利用 XPS 得到了证实[62]。Xiao 等[64]系统研究了金刚石纳米晶的荧光发射，指出金刚石纳米晶的荧光源于其表面官能团，如—OH、酮基 C=O 和酯基 C=O 基团，不同荧光的产生是由官能团的协作和竞争造成的。官能团的协作体现在金刚石纳米晶的激发依赖性荧光上，而它们的竞争则体现在最佳激发波长和发射波长的变化上。这些研究结果证实了碳主链和化学基团的杂化对于荧光中心的形成很重要。更重要的是，这些结论并不局限于金刚石纳米晶的荧光发射，同样有助于理解其他基量子点的荧光发射，例如，石墨和石墨烯量子点[67-70]。图 10-6 总结了 LAL 合成的荧光纳米颗粒的三种荧光来源。

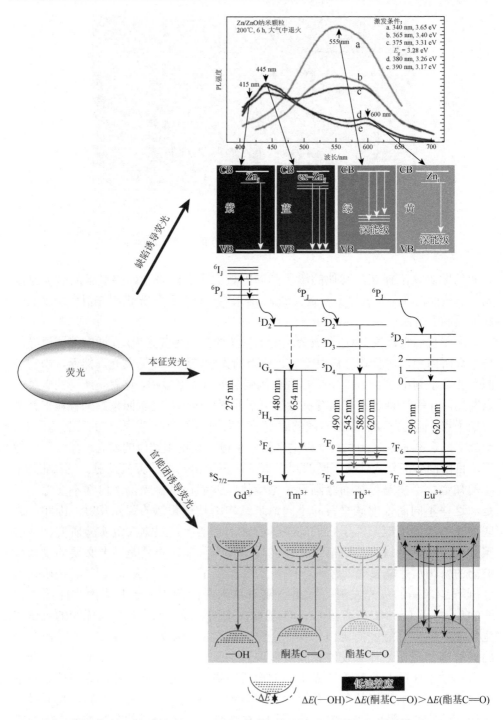

图 10-6　LAL 合成的荧光纳米颗粒的三种荧光发光来源（后附彩图）

10.1.2 可见光散射

贵金属等离子体纳米结构（plasma nanostructures）不仅在物理上展现出许多未曾预料的新奇物性，而且在纳米光子学器件应用上表现出传统器件无法比拟的优越性。但是，这类纳米结构有三个缺点：高光学损耗（尤其在可见光区）、资源稀缺价格昂贵、单一电偶极子共振模式，极大地限制了它们在器件中的实际应用。因此，人们开始探索具有低损耗、低成本及电磁共振模式丰富等优点的全介质纳米结构（all-dielectric nanostructures）作为贵金属等离子体纳米结构的互补、替代和超越者，并期待着能够看到更为奇妙的物理景象和更加优越的器件应用。最具代表性的全介质纳米结构就是硅纳米结构。由于在可见光区域的定向散射和低损耗特性，硅纳米颗粒被认为是优异的纳米尺度散射体，在定向纳米天线和可见光通信应用中优于贵金属纳米颗粒[71-73]。同时，硅纳米颗粒的强磁偶极响应在纳米光子学应用中发挥着至关重要的作用[74-76]。因此，制备高质量全介质硅纳米结构已经成为纳米光子学研究的重要方向。为此，研究人员制备了在近红外区域具有强磁偶极响应的亚微米硅纳米腔[76]，这些具有 380 nm 发射光的硅纳米颗粒是通过批量规模反应器在超临界正己烷中分解丙硅烷来制备的。然而，丙硅烷易燃且有毒，并且制备纳米颗粒的尺寸范围太大，无法作为有效的光散射剂。为了解决这个问题，Fu 等[72]通过飞秒激光技术制备的尺寸范围为 100～200 nm 的硅纳米颗粒表现出强烈的定向光散射。至此，开启了采用脉冲激光技术制备全介质硅纳米结构的研究。

Yan 等[77, 78]在国际上率先采用 fs-LAL 纳米制备技术制备出系列全介质纳米结构的材料，并且将其用于可见光散射，组装出若干原理性全介质纳米结构的光子学器件。接着，他们应用 fs-LAL 技术相继组装出了系列等离子体-全介质杂化纳米结构如 Si/Au、Si/Ag 和其他核-壳纳米结构等，这些新型纳米结构的光子学材料均显示出强烈的可见光散射模式[79-82]。Liu 等将 fs-LAL 制备的全介质纳米材料与纳米结构用于纳米光子学、新能源与环境科学、生物医学等多个领域[83-86]，充分展现了 LAL 基全介质纳米材料与纳米结构作为新型纳米结构的光子学材料的巨大应用潜力[77-86]。

我们知道，法诺（Fano）共振是一种在散射光谱中产生非对称线型的光学共振现象，源于量子力学离散态与连续态的相互干涉，在纳米光子学中有着广泛应用。目前，Fano 共振已经在光子晶体及贵金属等离子体纳米结构中实现。但是贵金属等离子体纳米结构须引入对称破缺来实现 Fano 共振。那么，能否取代传统光子晶体和贵金属等离子体纳米结构，通过简单的全介质纳米结构来实现 Fano 共振呢？这是全介质纳米光子学研究的一个重要科学问题。Yan 等[77]首次在 fs-LAL 组装的硅纳米球二聚体中观察到方向性 Fano 共振，同时揭示其背后的物理机制为硅纳米球二聚体结构具有独特的电磁耦合模式：两个相邻的硅纳米球可以在连接处产生宽谱电偶极子（ED）共振，而硅纳米球本身又拥有内在的窄带磁偶极子（MD）共振模式，这种电磁耦合导致了方向性 Fano 共振（前向散射和背向散射的峰谷对应），如图 10-7 所示。硅纳米结构 Fano 共振不同于传统贵金属如金等离子体纳米结构 Fano 共振：①它源于全介质纳米结构特有的磁偶极子和电偶极子共振，磁偶极

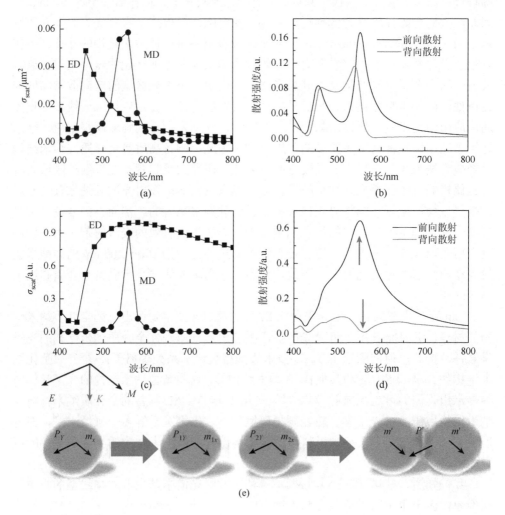

图 10-7 （a）单个硅纳米球的电磁耦合模式响应；（b）单个硅纳米球的前向散射和背向散射；（c）硅纳米球二聚体的电磁耦合模式响应；（d）硅纳米球二聚体的前向散射和背向散射（图中向下箭头所指为 Fano 凹陷）；（e）从单个硅纳米球到硅纳米球二聚体的电磁耦合模式演化示意图

子和电偶极子相互作用使得其具有方向性，即背向散射能够观察到 Fano 共振，而前向散射则表现为散射增强；②它的产生不依赖于对称破缺，即无论同质二聚体还是异质二聚体都能够产生 Fano 共振；③它的共振波长可以通过调整硅纳米球尺寸在可见光波长范围内调节，同时由于硅纳米球在可见光波段的低损耗，硅纳米球二聚体有着比金纳米颗粒大得多的散射截面。这些新发现丰富了全介质纳米结构 Fano 共振的物理认知，在全介质超结构、方向性纳米天线、全光开关、生物传感等方面有重要应用。

磁感应透明作为一种独特的量子干涉效应可以使得电磁波在介质中零反射和完美透射，它导致了零折射率超结构的出现及在光学隐身等方面的应用。作为一种经典类比，Yan 等[78]在硅纳米球二聚体 Fano 共振的基础上，首次提出并实验证实了硅纳米结构（LAL 组装的硅纳米球多聚体）的磁感应透明（magnetically induced transparency，MIT）效应，如图 10-8 所示。考虑到全介质纳米结构独有的磁偶极子与电偶极子共振的相互作用，通过对暗场反射谱分析并结合米氏（Mie）散射理论和偶极-偶极（dipole-dipole）模型，Yan 等建立了全介质纳米结构 MIT 的物理模型。紧密连接的硅纳米球之间产生不受聚集形态影响的宽谱电偶极子共振，同时，单个硅纳米球内部产生不受周围环境影响而只取决于颗粒尺寸的窄带磁偶极子共振，这种独特的电磁耦合模式分布及相互作用导致了硅纳米球多聚体的 MIT 效应。进一步的理论预测显示硅纳米球阵列会在特定波长实现零反射和完美透射，这意味着入射光可以穿过硅纳米球阵列而不产生附加相位，从而实现对阵列的"隐形"。显然，MIT 效应的发现不仅带来了新物理机制，而且极大地拓展了全介质纳米结构的潜在应用。

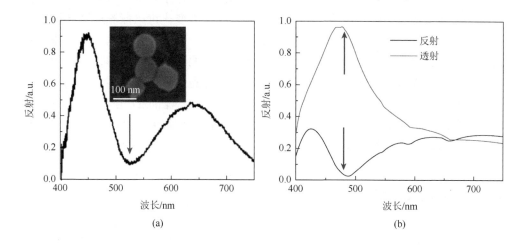

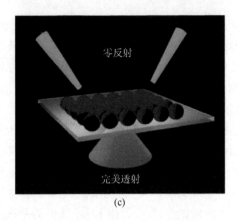

(c)

图 10-8　LAL 组装硅纳米球（a）多聚体反射谱、（b）MIT 效应和（c）完美透射示意图

贵金属等离子体纳米结构在无标记生物探测领域的应用受到关注，这类传感器的设计是基于表面等离子体共振对周围环境折射率改变的高灵敏度。但是，贵金属自身的损耗限制了更高品质因子的获得，同时，大的欧姆损耗、高成本及不可避免的生物毒性等阻碍了贵金属等离子体纳米结构在活体中的生物探测。Yan 等[84]首次提出并实验证实了硅纳米结构即硅纳米球二聚体可以实现环境折射率和附着生物大分子的超灵敏探测。这里需要指出的是，该器件工作在基于全新的传感原理上即基于散射峰强度随环境折射率的变化而非传统的基于散射峰位的偏移。这种新型全介质纳米结构传感器的物理机制为展宽电模式偏移（broadening electric mode shift，BEMS）和 Kerker 散射强度偏移（Kerker's scattering intensity shift）的协同作用。由于 BEMS 的作用能够削弱电偶极子和磁偶极子的耦合，并且极大地增强背向散射，所以，根据背向散射的强度变化可以探测周围环境的变化。图 10-9 显示了国际上第一个原理型检测生物大分子的全介质纳米结构传感器即硅纳米球二聚体传感器检测示意图。与基于峰位偏移的等离子体传感器相比，全介质纳米结构传感器的品质因子可以与当时的等离子体传感器相媲美，而且，全介质纳米结构传感器的灵敏度是最好的等离子体传感器的 27 倍。因此，我们相信，全介质纳米结构传感器会在单分子探测、表征单颗粒催化及活体动态过程原位表征方面有着重要的应用前景。

10.1.3　非线性光学

由于表面等离子体共振能够引起较大的非线性光学系数和超快的响应时间[87]，所以，贵金属等离子体纳米结构在光限幅和光通信等领域的应用引起了人们的广泛关注[88-91]。目前，研究人员采用 LAL 纳米制备技术合成了系列非线性光学

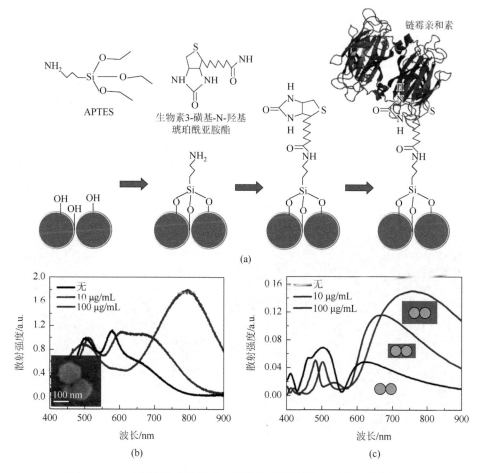

图 10-9 LAL 组装的硅纳米球二聚体传感器检测示意图（后附彩图）

(a) 传感器用于生物大分子检测的表面功能化的示意图；(b) 实验测量了使用不同浓度的链霉亲和素的背向散射光谱，插图是典型二聚体的 SEM 图像；(c) 模拟背向散射光谱，以揭示当增加与表面结合的链霉亲和素分子数量时的强度变化

纳米材料，包括纳米金属胶体[92-95]和纳米合金[49, 50]等。例如，Tan 等[96]发表了采用 800 nm 的飞秒脉冲激光 LAL 合成的碳纳米点具有很强的光限幅效应，阈值为 74 mJ/cm^2，与贵金属纳米复合材料的光限幅效应[97, 98]相当。Ma 等[99]采用 LAL 纳米制备技术发现了非晶 Se 纳米颗粒的光学二次谐波现象，发现其效率在 10^{-8} 数量级，并且证明了这种二次谐波具有波长可调谐性。

非晶材料是各向同性的，且拥有对称中心，由于其结构的短程全介质纳米颗粒的光学二次谐波有序、长程无序特性，理论上并不能产生光学二次谐波[100]。然而，Alexandrova 等于 2000 年首次在生长在熔融石英上的非晶 Si：H 薄膜中观察到光学二次谐波，他们认为这种光学二次谐波源于 Si 与 SiO$_2$ 衬底之间的应力层，并提出了一个应力模型。2002年,Lettieri等[101]也在非晶SiN:H微腔中观察到光学二次谐波,他们认为这种二次谐波源于在多层结构界面处诱导的表面二次谐波。截至目前,还没有关于单个非晶纳米颗粒的光学二次谐波报道。Ma等使用飞秒激光烧蚀去离子水中的固体Se靶合成了具有完美球形的非晶Se纳米颗粒,如图10-10所示。进一步地,他们研究了单个非晶Se纳米颗粒的光学二次谐波,发现尺寸大于300 nm的Se纳米颗粒可以在近红外飞秒激光激发下产生光学二次谐波,单个纳米颗粒的效率在10^{-8}数量级(图10-11)。他们认为该二次谐波源于非晶Se纳米颗粒表面结构的不连续性及电场强度在颗粒表面法向分量的不连续性。此外,他们还证明了非晶Se纳米颗粒的磁偶极子共振能够增强其光学二次谐波,揭示了非线性光学响应与线性光学响应的内在联系。

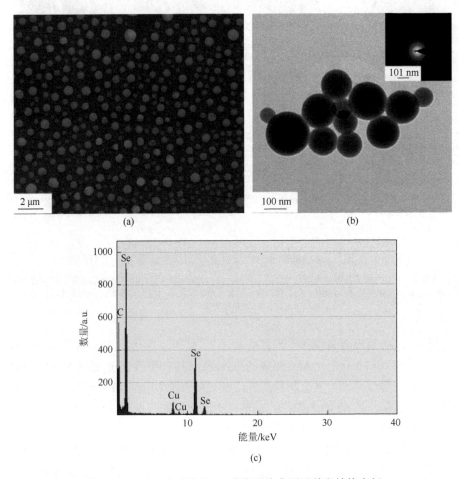

图10-10　LAL合成非晶Se纳米颗粒典型形貌和结构表征
(a)样品的SEM;(b)TEM图像,插图是相应的SAED图像;(c)样品的EDX分析

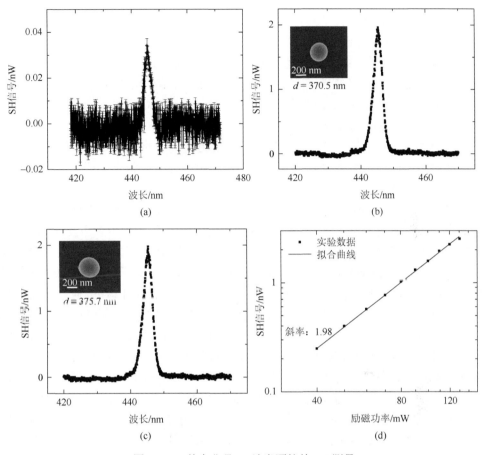

图 10-11 单个非晶 Se 纳米颗粒的 SH 测量

(a) 单个非晶 Se 纳米颗粒在 890 nm 激发波长下的二次谐波 (SH) 信号; (b) 在 110 mW 的激发功率下, 在 890 nm 的激发波长下, 直径 370.5 nm 的单个非晶 Se 纳米颗粒发射的 SH 信号, 插图是测量的非晶 Se 纳米颗粒的 SEM 图像; (c) 在 110 mW 的激发功率下, 在 890 nm 的激发波长下, 直径 375.7 nm 的单个非晶 Se 纳米颗粒发射的 SH 信号, 插图是测量的非晶 Se 纳米颗粒的 SEM 图像; (d) 测量的 SH 信号是励磁功率平方的函数, 坐标取对数

综上所述,考虑到 LAL 纳米制备技术的简单、清洁、绿色、高效等特点,这些利用 LAL 技术制备出来的碳纳米点、非晶 Se 纳米颗粒等纳米材料有望在超快、非线性等光学器件中得到应用。

10.1.4 光热转换

众所周知,高效的光热转换材料必须满足两个基本条件:一是要在整个太阳光谱范围内具有强烈的吸收,二是需要具有高的光热转化效率。在纳米光子学材料中,由于等离子体纳米材料激发的表面等离子体共振可以大幅增强光学吸收,加上其能够利用材料内部的热电子弛豫实现局域加热，因此等离子体纳米材料包括 Au、Al、TiO$_{1.67}$ 等已经用于太阳能光热转换[86, 102-105]。此外，近年来，高折射率全介质纳米材料如 Si、Ge 等也已经证实可以依靠 Mie 共振增强光吸收，实现高效加热，进而用于光热转换[104, 105]。所以，贵金属纳米结构表面等离子体共振和全介质纳米结构 Mie 共振这两种光学模式都可以大幅增强纳米光子学材料的光学吸收效率，是目前调控材料光吸收的主要方法。虽然这两类材料各自拥有优势，但并没有哪种材料能够结合他们的优势，实现更高效率的光热转换[106]。因此，在一种材料上同时实现两种光学模式的光吸收调控，无论在科学上还是在技术上，都是很大的挑战。

Ma 等[106]采用 LAL 纳米制备技术首次制备出在太阳光谱范围内展现光学二重性的全介质纳米材料碲（Te），其独特的"光学二重性"表现为在可见光范围内，随颗粒尺度而变化。碲纳米颗粒的介电常数会由负变化到正，从而实现从金属属性向全介质（all-dielectric）属性的转变。例如，尺寸小于等于 120 nm 的碲纳米颗粒，像金一样，表现出类表面等离子体共振；而当尺寸大于 120 nm 时，它又转变为像高折射率全介质材料硅一样，表现出 Mie 共振。这样，就可以通过调节纳米颗粒尺寸对其光吸收模式进行调控。全介质纳米材料碲"光学二重性"背后的物理机制就是这类材料具有独特的介电常数。如图 10-12 所示，金在整个区域的介电常数实部都为负值，硅的介电常数实部在整个区域都为正值；而碲的介电常数非常特殊，在 300～490 nm [图 10-12（a）阴影区]，它的介电常数实部为负值，而当大于 490 nm 时，它的介电常数转变为正值，这意味着它在 300～2000 nm 范围内经历了从类等离子体到高折射率全介纳米材料的转变。这里提到的全介质纳米颗粒的电磁共振特性是类等离子体，而非等离子体，这两者的差别在于，等

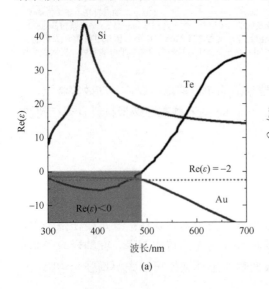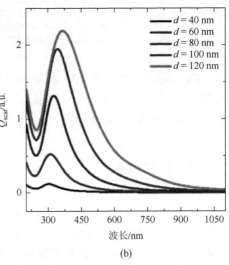

(a)   (b)

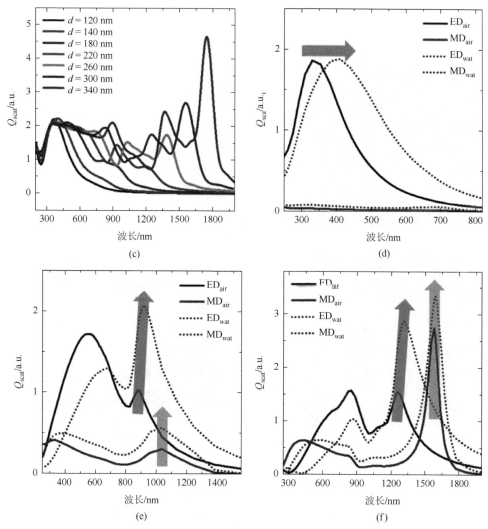

图 10-12 LAL 合成的碲纳米粒子的光学二重性（后附彩图）

（a）与 Au（等离子体材料）和 Si（全介质材料）相比，碲的介电常数的实部；(b) 直径小于等于 120 nm 的碲纳米颗粒的类等离子体行为；(c) 直径在 120~340 nm 的碲纳米颗粒的全介质材料行为；(d)~(f) 电偶极子（ED）和磁偶极子（MD）对于在空气（air）和水中（wat）直径分别为 100 nm、200 nm 和 300 nm 的碲纳米颗粒的散射效率的贡献

离子体纳米材料可以产生自由电子，而碲纳米颗粒只是由于其在 300 nm 附近存在着从价带的 p-键三重态（p-bonding triplet）到导带的 p-反键三重态（p-antibonding triplet）的直接跃迁，其本身虽然是半金属材料，并不能产生自由电子。所以，具有一定尺寸分布的碲纳米颗粒薄膜可以实现对全波段太阳光的完美吸收和高光热转换。如图 10-13 所示，该吸收层在整个太阳光谱范围内（蓝色区域）的吸收率（红色曲线）超过 85%，甚至在紫外区（300~400 nm）的吸收率超过 95%。

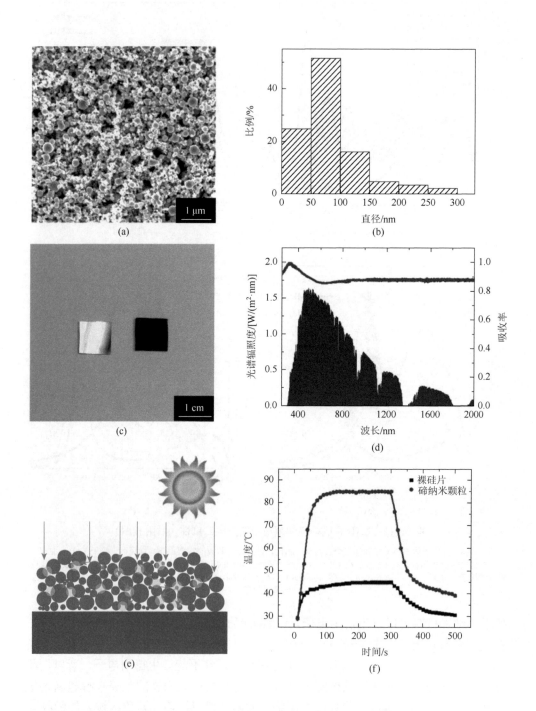

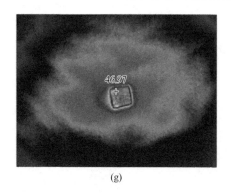 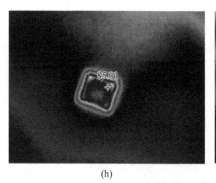

(g) (h)

图 10-13　沉积在硅衬底上的碲纳米颗粒层的光热效应（后附彩图）

(a) 碲纳米颗粒自组装的 SEM 图像；(b) 基于 (a) 中 SEM 图像的尺寸分布（来自 500 个颗粒的统计数据）；(c) 裸硅片（左）和沉积在硅衬底上的碲纳米颗粒层（右）的照片；(d) 碲纳米颗粒吸收剂的吸收光谱（红色曲线），蓝色区域是太阳光谱；(e) 太阳光照射下的碲纳米颗粒层的示意图；(f) 裸硅片（黑色曲线）和碲纳米颗粒吸收体（红色曲线）的随时间变化的温度变化；(g) 和 (h) 分别为硅晶片和碲纳米颗粒吸收体的稳态热图像

碲纳米颗粒的高效光热转化主要来自等离子体共振。对于等离子体纳米材料，其等离子体共振可以通过光子再发射这种辐射方式来弛豫，或者通过朗道阻尼（Landau damping）产生热电子-空穴对这种非辐射方式弛豫[107]。通过与等离子体纳米颗粒内部的晶格相互作用而发射声子，使热电子发生内部弛豫，可显著加热纳米颗粒本身及其周围环境。然而，对于半导体纳米颗粒，光热转换的机理很少被讨论[108]。通常，具有较大比表面积的半导体纳米颗粒在其表面具有丰富的悬挂键和缺陷，这就引入了表面态，导致带隙中出现电子能级，这与中间带（intermediate band）类似[109]。LAL 合成的碲纳米颗粒含有丰富的体内和表面缺陷，这就导致了碲纳米颗粒中间带的存在。当纳米颗粒被光激发时，激发能量被完全转移到载流子即电子和空穴上，产生具有特定动量状态和温度的非平衡载流子。随着系统向平衡态发展，这些载流子通过带内载流子-载流子散射或带间重组而弛豫。非平衡载流子导致库仑热化，形成热化载流子与声子耦合的热气体，并将其过剩能量传递到晶格，这导致了纳米颗粒的有效加热，产生高的光热转化效率[110]。

淡水资源短缺是当前世界面临的一大难题，海水淡化作为获得淡水的有效方法之一而备受关注。然而传统的海水淡化装置复杂、效率低、耗能多。考虑到太阳能是取之不尽的清洁能源，近年来，科学家们一直在寻找合适的材料用于太阳能海水淡化装置从而实现高效的海水淡化。Ma 等将 LAL 合成的碲纳米颗粒通过自组装技术组装出一定尺寸（10～300 nm）的全介质碲纳米颗粒薄膜，实现了整个太阳光谱范围内的完美吸收（在整个太阳光谱范围内的吸收率超过 85%，其中紫外区接近 100%）。在太阳光照射下，该吸收层的温度从 29℃ 上升到 85℃ 只需要 100 s；通过将碲纳米颗粒均匀分散到水中，在太阳光照射下，水的蒸发速率提升了 3 倍，碲纳米颗粒的表现超越了 2018 年报道的用于太阳能光热转换水蒸发的纳米光子学材料包括表面等离子体和全介质材料的表现,如图10-14所示[106]。这些研究表明,LAL 纳米制备技术已经成为探索新型光热转换纳米材料的有力工具。

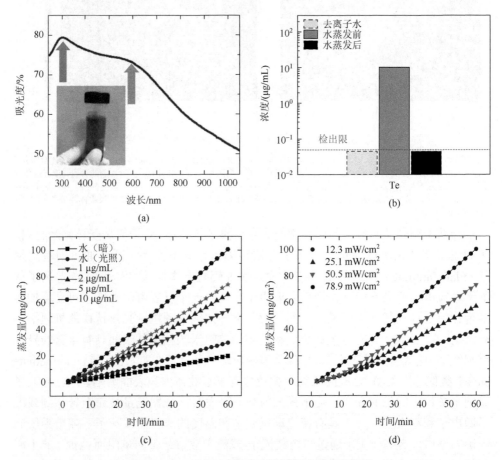

图 10-14　LAL 合成的碲纳米颗粒应用于水蒸发

（a）碲纳米颗粒胶体溶液（5 μg/mL）的紫外可见-近红外吸收光谱,插图为 ns-LAL 合成的碲纳米颗粒胶体溶液;
（b）蒸发前后水中碲的 ICP-AES 分析结果,去离子水作为参考标准,虚线表示浓度低于 0.05 μg/mL 的检出限;
（c）不同浓度的碲纳米颗粒溶液在 78.9 mW/cm² 模拟太阳光照射下的蒸发量;（d）在模拟太阳光的不同辐射照度下碲纳米颗粒溶液的蒸发量（10 μg/mL）

10.2　磁性功能纳米结构

众所周知,铁、钴、镍是三种重要的磁性材料,而磁性铁、钴和镍基纳米结构因其在铁磁流体、磁记录介质和磁性传感器中的应用而受到越来越多科学家的关注[111-113]。自 21 世纪以来,研究人员已经利用 LAL 纳米制备技术合成了大量的磁性纳米结构和混合磁性纳米结构,例如,铁和氧化铁的核-壳纳米结构[114]、双金属合金亚微米球[28]和洋葱状碳封装的碳化钴核-壳纳米颗粒[115]等。Liang 等[116-118]使用 MFLAL 组装了各种磁性纳米链，包括双金属合金颗粒链、Co_3C 纳米颗粒链和 Fe_3C 磁性纳米链束等。值得注意的是，Fe_3C 磁性纳米链束在室温下具有 261 emu/g 的饱和磁化强度，这是至 2013 年氮化铁和碳化物纳米结构所达到的最高值[118]。而常规 LAL（没有磁场辅助的情况下）合成的磁性纳米颗粒的饱和磁化强度仅为 124 emu/g[119]。Swiatkowska-Warkocka 等[28]发展了一种基于 LAL 纳米制备技术的磁性互不相溶合金的合成方法。他们通过脉冲激光辐照分散在乙醇中的金和氧化钴纳米颗粒，合成了球状形态的亚微米金钴合金颗粒。采用类似的 LAL 技术，其他研究人员合成了铁基纳米颗粒，并且发现合成的 Fe_3O_4/FeO 中存在有趣的交换偏向效应[120]。

Amendola 等[114]利用 LAL 纳米制备技术合成了系列磁性纳米颗粒，包括碳化铁、氧化铁、氧化铁包覆的铁及不同形貌和结构的金属铁，他们发现，LAL 合成磁性纳米颗粒中的一个重要参数是有机溶剂类型。例如，他们通过在硫醇聚乙二醇（PEG）和乙二胺四乙酸（EDTA）溶液中脉冲激光烧蚀 Fe-Ag 双金属靶材，组装了掺杂铁的银纳米颗粒，其磁性得到了增强并且具有等离子体特性，如图 10-15（a）所示，图 10-15（b）显示了 NdFeB 磁体对纳米颗粒的聚集作用。实验表征显示，这些 Fe-Ag 纳米复合材料具有松露状结构，是由在面心立方 Ag 支架中嵌入掺杂金属 Fe 所形成的无序 Ag-Fe 合金组成。重要的是，与未掺杂的样品相比，这些纳米颗粒表现出更显著的等离子体响应和光热效应。这是因为在外部磁场作用下，Fe-Ag 纳米颗粒有效地聚集在毫米级激光光斑尺寸大小的区域内，如图 10-15（c）和图 10-15（d）所示[121]。而且，具有等离子体和磁性的 $AuFeO_x$ 纳米团簇已经用于细胞引导和 SERS 成像[122]。由此可见，LAL 纳米制备技术在磁性多功能、多元素纳米颗粒制备方面大有作为。

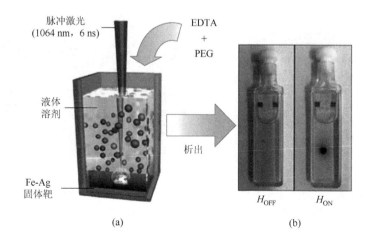

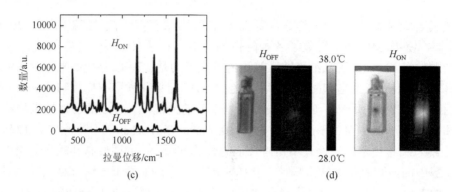

图 10-15 （a）LAL 纳米制备示意图和（b）Fe-Ag 纳米颗粒溶液的光学照片，黄绿色为 Ag 等离子体的典型颜色，黑色区域为通过在比色皿一侧放置小型 NdFeB 磁铁施加磁场所积累的磁性纳米颗粒；（c）磁聚焦之前（$H_{OFF}=0$，黑线）和磁聚焦之后（$H_{ON}>0$，蓝线），MG/Fe-Ag 纳米颗粒分散液的拉曼光谱；（d）在磁聚焦前后，使用波长 785 nm 连续激光辐照 Fe-Ag 纳米颗粒分散液，进行光热加热实验（后附彩图）

此外，我们通过选择性激光辐照分散在液体中的纳米颗粒，以及后期酸处理，可以合成多孔磁性金基亚微米球形颗粒[123]。图 10-16（a）描绘了激光辐照之前

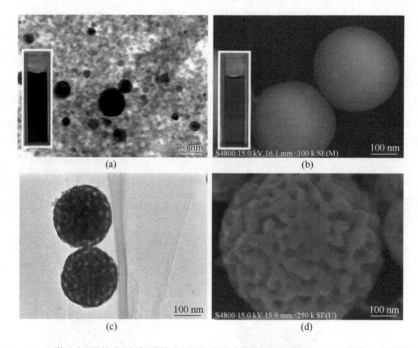

图 10-16 （a）激光辐照前金纳米颗粒和磁铁矿纳米颗粒混合物的 TEM 图像，插图的激光辐照前 $Au+Fe_3O_4$ 溶液为红棕色；（b）激光辐照合成的纳米颗粒的 SEM 图像，插图的激光辐照后溶液呈灰色；酸处理后 Au/AuFe 纳米颗粒的（c）TEM 图像和（d）SEM 图像

的金纳米颗粒（5 nm）和磁铁矿纳米颗粒（20 nm）的混合物。而在激光辐照后，我们仅能观察到较大的颗粒（450 nm）[图 10-16（b）]，同时，可以明显看到溶液的颜色从红棕色变为灰色[图 10-16（a）和图 10-16（b）的插图]。有趣的是，酸处理后的样品具有粗糙的表面和多孔结构，如图 10-16（c）和图 10-16（d）所示。形成多孔结构的原因主要是酸处理过程刻蚀了富铁壳，从而导致 Au/AuFe 纳米颗粒具有多孔表面。由于这些亚微米球形颗粒既有良好的磁性又具有多孔结构，所以可以用于生物分离和化学催化。

10.3 应用于环境科学的功能纳米结构

10.3.1 吸附

吸附是环境治理和污染物去除的重要手段。开发环境友好的吸附技术，通过简单地吸附去除水中的有机污染物对环境科学来说一直都是一项极具挑战性的工作，而且这种吸附技术对于制药和食品加工行业甚至人类健康都至关重要[124]。我们知道，过渡金属氧化物因其优异的吸附能力引起了人们的极大兴趣[125, 126]。最近，Li 等[127]率先采用 LAL 纳米制备技术合成无定形过渡金属氧化物纳米颗粒作为吸附剂。他们发现，过渡金属（Fe、Co 和 Ni）氧化物晶体颗粒可以通过简单且环保的 LAL 技术转化为非晶相（图 10-17），而这种无序结构可以极大地增强对染料的吸附性。例如，无定形 NiO 纳米颗粒对亚甲蓝（MB）的吸附容量最高，高达 10 584.6 mg/g[127]，这是截至目前已报道的最高吸附量。无定形 NiO 纳米颗粒能够在 1 min 内从溶液中去除 99%的亚甲蓝，而晶体 NiO 纳米颗粒则需要 150 min 才能做到这一点（图 10-18）。与相应的晶体纳米颗粒相比，非晶纳米颗粒由于其更高的表面积和表面电荷，使其吸附性能大大提高。这些研究证明了 LAL 纳米制备技术可以方便地用于新型高效染料吸附剂的合成。需要注意的是，在这种情况下，纳米颗粒通常被认为是吸附剂而不是吸附物。

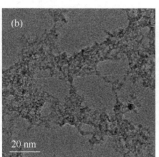

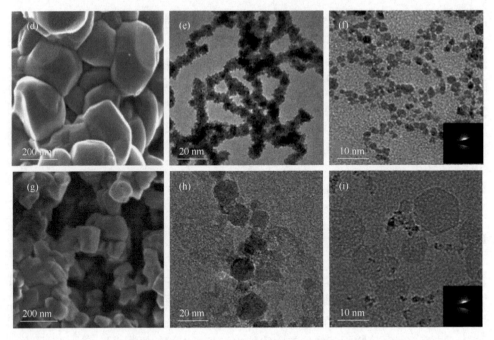

图 10-17 （a）、（d）和（g）分别为 Fe$_2$O$_3$、CoO 和 NiO 晶体纳米颗粒；（b）和（c）、（e）和（f）、（h）和（i）分别为 Fe$_2$O$_3$、CoO、NiO 非晶纳米颗粒的 TEM 图像

插图是相应 SAED 图像

  Wagener 等[128]也开展了采用 LAL 纳米制备技术合成多种功能纳米颗粒作为吸附剂。例如，他们合成了一种有无配体银纳米颗粒附着的硫酸钡微米颗粒的表面，这种复合微纳结构表现出了很强的吸附性［图 10-19（a）］。需要注意的是，这种吸附是一个快速、不可逆并且几乎定量的过程，表面覆盖率高达 17.5%。然而，将这种吸附剂浸入高浓度的配体溶液中（＞50 mmol/L 柠檬酸盐），可以得到较高的配体覆盖率，但是配体产生的静电力又几乎完全阻止吸附［图 10-19（b）］。图 10-19（b）中的 SEM 图像显示了覆盖无配体纳米颗粒的微米颗粒，以及纳米颗粒与配体结合后无吸附性的微米颗粒。重要的是，这种技术可以扩展到多种纳

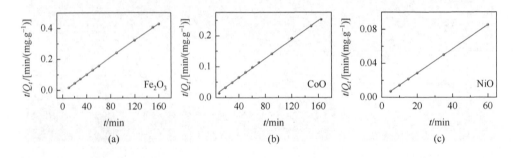

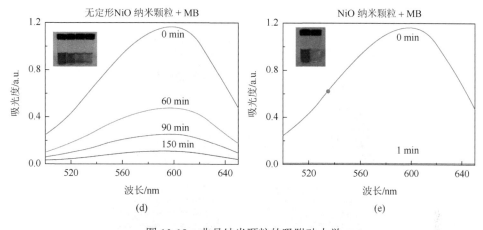

图 10-18 非晶纳米颗粒的吸附动力学

(a)~(c) 三个样品均表现出拟二阶吸附动力学行为；(d) 晶体和 (e) 非晶 NiO 纳米颗粒的吸光度与波长的关系，插图为溶液颜色变化

米颗粒（Ag、Au、Pt 和 Fe）和微米颗粒载体[$BaSO_4$、TiO_2 和 $Ca_3(PO_4)_2$]的复合微纳结构的组装[128]。这些研究表明 LAL 纳米制备技术在纳米吸附剂合成中有很大的应用潜力。

10.3.2 光催化降解

众所周知，虽然吸附是去除水中污染物如有机污染物的有效方法，但是它无法完全降解有机污染物，而只是将其从水中转移到了吸附剂表面。所以，从完全除去水中有机污染物的角度考虑的话，我们会发现光催化降解技术在废水处理

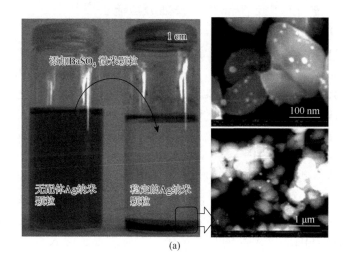

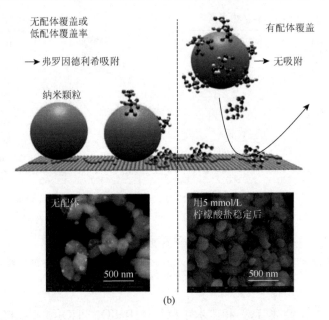

图 10-19 （a）无配体银纳米颗粒吸附到 $BaSO_4$ 微米颗粒上的吸附前及吸附后照片（左），用 Ag 纳米颗粒修饰 $BaSO_4$ 微米颗粒的前（右上）后（右下）SEM 图像；（b）不同表面覆盖量的纳米颗粒的吸附过程的示意图（上），SEM 图像显示覆盖无配体纳米颗粒的微米颗粒（左下）及有配体的纳米颗粒混合后未受影响的微米颗粒（右下）

领域变得越来越重要。光催化降解具有部分或完全矿化、无须废物处理、成本低等突出优点。应用于光催化降解的光催化剂材料应该满足以下要求：①具有光化学活性并且能够充分利用紫外光或可见光；②具有化学惰性且光稳定性；③廉价；④无毒[129, 130]。LAL 纳米制备技术合成的高化学活性纳米材料在降解有机染料方面表现出了优异的光催化活性和独特的表面反应活性[131-143]。

在五花八门的光催化剂材料中，纳米金属氧化物半导体似乎最具吸引力。Liu 等[139]采用 LAL 纳米制备技术合成了纯金红石（TiO_2）纳米颗粒并用于降解亚甲蓝溶液，展现出了高效的降解能力。与此同时，Zeng 等[140]应用 LAL 纳米制备技术合成了 ZnO、Au/ZnO、Pt/ZnO 和 Au/Pt/ZnO 等系列中空纳米颗粒，以提高其对甲基橙降解的光活性和光稳定性。例如，他们研究发现，Pt/ZnO 多孔纳米笼具有比 Pt/ZnO 纳米笼更有效的电子空穴分离效率和更大的比表面积，从而表现出更高的光催化活性[141]。Xiao 等[131]通过 LAL 方法制备了高稳定性的亚 5 nm 的水碱石 [$Sn_6O_4(OH)_4$]纳米晶作为先进的光催化材料（图 10-20），表现出了对甲基橙（MO）的高效快速降解及高的光稳定性（图 10-21）。我们知道，$Sn_6O_4(OH)_4$ 通常是通过化学沉淀法制备的，并且可以作为合成 Sn 基氧化物纳米材料的中间体。Xiao 等[131]对 $Sn_6O_4(OH)_4$ 纳米晶的物理和化学性质进行了深入研究，提出含有丰

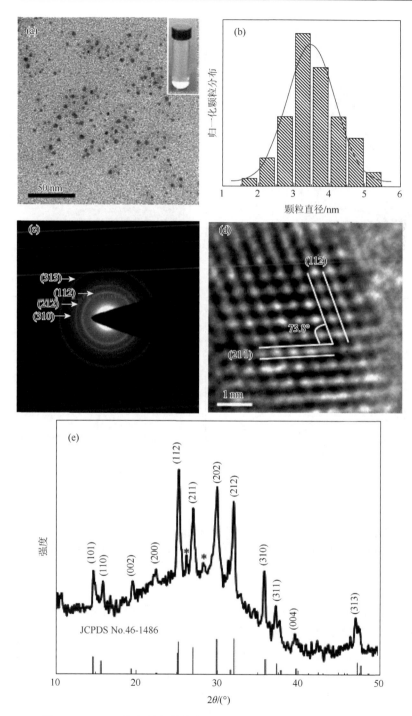

图 10-20 LAL 纳米制备技术合成的亚 5 nm $Sn_6O_4(OH)_4$ 纳米颗粒

(a) TEM 明场像，插图为溶液；(b) 粒径分布；(c) 电子衍射图像；(d) 高分辨 TEM 图像；(e) XRD 谱

富氧空位的清洁表面在 $Sn_6O_4(OH)_4$ 纳米晶光催化降解中发挥着关键作用,其中·OH 自由基是攻击有机分子的主要物质。

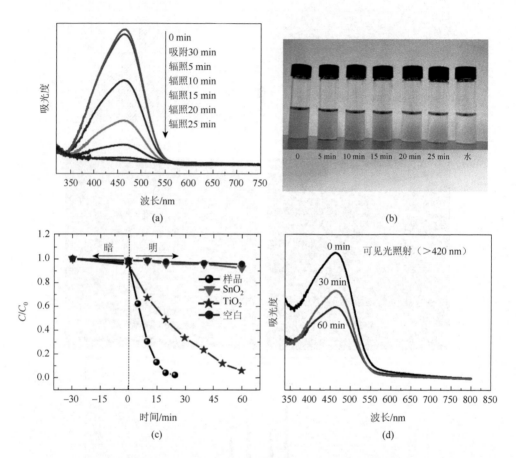

图 10-21　(a) $Sn_6O_4(OH)_4$ 纳米晶降解甲基橙的光谱图;(b) 对应的降解后溶液光学照片图;(c) $Sn_6O_4(OH)_4$ 纳米晶与 SnO_2 纳米晶、TiO_2 纳米晶及空白对照的降解效果对比图;(d) $Sn_6O_4(OH)_4$ 纳米晶可见光降解的效果(后附彩图)

除了上述简单的金属氧化物材料以外,研究人员利用 LAL 纳米制备技术还开发了多功能复杂纳米结构来进一步提高光催化剂的光降解性能,如分层球体[138]、立方状和海胆状三元氧化物[133]、核-壳纳米结构[137]和三维结构有序框架结构[136]等。Li 等[137]采用 LAL 纳米制备技术合成了 $Ta_xO@Ta_2O_5$ 核-壳纳米结构并用于光催化降解,发现其中低氧化物 Ta_xO 核在光催化活性中发挥着重要作用,因为它们能够捕获电子、延长电子空穴对的寿命并增强光吸收。为了在光催化反应后回收催化剂材料以便循环使用,研究人员应用 LAL 方法组装了栗子状 $Fe_3O_4@C@ZnSnO_3$

核-壳纳米结构，其中碳外壳的保护使具有催化活性的磁性 Fe_3O_4 核在回收过程中不易被破坏[136]。此外，Cai 等[142]将 LAL 纳米制备技术与水热合成法结合，在管状 TiO_2 阵列上组装出层状 $ZnTiO_3$ 纳米片的新型异质结构。这种异质结构在异质界面中创造了一个低电阻界面层，有利于电子转移。基于同样的合成路线，他们还组装出立方状 Zn_2SnO_4 和海胆状 $ZnSnO_3$ 等新颖纳米结构用于去除水中甲基橙和 2,5-二氯苯酚[133]。

虽然上面介绍的 LAL 纳米制备技术合成的光催化剂材料表现出优异的光催化性能，但是它们通常只能在紫外光下工作，而紫外光仅占入射太阳光的 4%，因此这类光催化剂材料的实际应用领域有限[144]。为了解决这个问题，研究人员投入了大量的精力来开发能够使用光子能量较低但占比更高的可见光的光催化剂材料[70, 72, 73, 81]。例如，LAL 组装的核-壳纳米结构中 TiO_2 和 α-Fe_2O_3 的能级可以匹配，以增强可见光范围的光催化作用[135]，因为 α-Fe_2O_3 是带隙为 2.2 eV 的 n 型半导体，其收集太阳光的能力是带隙为 3.2 eV 的 TiO_2 的 15 倍。当然，我们也可以缩小 TiO_2 的带隙，这样就可以赋予其相当大的太阳光收集能力。此外，这类纳米结构的光生电子空穴对分离能力因其良好匹配的界面和吸附的 O_2 量而得到提升。此外，引入等离子体纳米结构被认为是提升可见光光催化剂性能的一种重要手段[133]。例如，与商业光催化降解材料 Ag_3PO_4 相比，Ag/AgCl 异质结构光催化剂对甲基橙、罗丹明 B 和亚甲蓝等有机染料表现出更高的光降解性能，而 Ag/AgCl 异质结构的局域表面等离子体共振被认为是其超高可见光光降解效率的主要原因[132]。

下面，我们重点介绍如何通过 LAL 纳米制备技术在光催化剂材料中引入缺陷来调制其能带结构，进而改善其光催化性能[134]。Lin 等使用脉冲激光辐照去离子水中的商用 Ag_2WO_4 纳米粉末，实现了对 Ag_2WO_4 电子结构的重构，如图 10-22 所示，其中图 10-22（a）和图 10-22（b）分别展示了 Ag_2WO_4 的形貌和元素分布。我们从图 10-22（c）可以看到，LAL 合成的样品在 12 min 内有效降解了溶液中 90%的甲基橙。相比之下，未经 LAL 处理的原始样品几乎没有表现出任何光催化活性。通过深入研究后，Lin 等发现，LAL 处理后的 Ag_2WO_4 之所以具有高的光催化活性，是因为 LAL 过程中的高能量使晶体结构中的 WO_6 簇发生扭曲，导致其缺陷密度增加了 2.75 倍。同时，通过第一性原理计算表明，LAL 引入的缺陷使能带带隙缩小了 0.44 eV，从而实现可见光的光催化，如图 10-22（e）所示。重要的是，这种合成方法不需要添加其他化合物，如 Ag、Ag/AgCl、Ag_2S 或 Zn-Cr 等，来增加 Ag_2WO_4 对可见光的敏感性。所以，与商用光催化剂的化学合成法相比，LAL 纳米制备技术具有简单性和实用性。

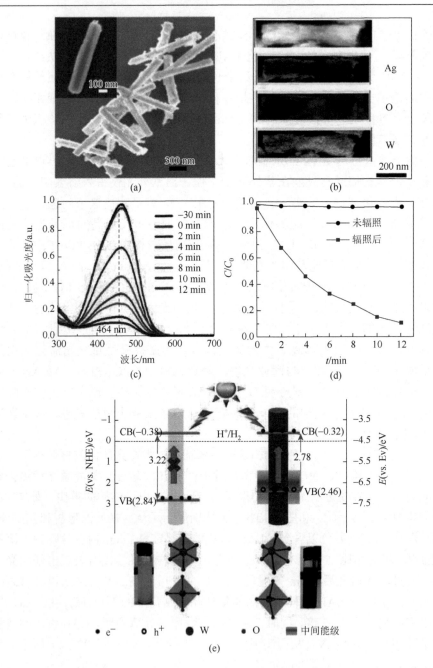

图 10-22 （a）LAL 处理后样品低倍和（插图）高倍放大的 SEM 图像；（b）EDX 元素分析图；（c）MO 溶液被降解的典型紫外可见吸收光谱变化（−30 min 表示样品配好后静止 30 min，再用光照）；（d）未 LAL 处理和 LAL 处理的 Ag_2WO_4 样品降解的 MO 溶液的相对浓度变化图；（e）基于团簇扭曲和中间能级导致带隙变窄的可见光光催化机制示意图（后附彩图）

Yang 等[145]报道了关于 LAL 制备的超细 Si 纳米颗粒具有异常氧化还原行为的重要科学发现。他们采用 LAL 纳米制备技术在乙醇中烧蚀固体 Si 靶,合成了由超细(4 nm)纳米晶组成的 Si 纳米颗粒胶体溶液(Si 纳米颗粒未暴露于氧气中)。但是,当他们将氯金酸溶液添加到 Si 纳米颗粒胶体溶液中时,发现 Au^{3+} 会发生还原而形成 Au 纳米晶体。相比之下,商业 Si 粉由于暴露在空气中时表面氧化而无法实现这种还原。此外,他们还发现表面清洁的 Si 纳米颗粒甚至可以直接还原 Ag^+,而不需要其他添加剂。然而,我们知道,这种反应是不应该发生的,因为 Au/Si 系统的整体电势为负(−0.13 eV)。根据混合势理论,块状 Si 在热力学上无法还原 Ag^+。所以,Yang 等[145]将这种异常的氧化还原行为归因于 Si 纳米颗粒的尺寸效应。随着尺寸的减小,Si 纳米颗粒的氧化还原电位变得高于标准值,最终变得足够高以还原 Au^{3+}。随后,Liu 等[146]报道了 Ge 纳米颗粒存在类似的现象。由于 LAL 合成的纳米颗粒具有新暴露的活性表面,因此它们通常在熟化过程中表现出自由生长,有时还伴随着相结构的变化。因此,大量研究已经表明,LAL 合成的纳米颗粒不同于传统化学途径合成的纳米颗粒,清洁且缺陷丰富的表面使得 LAL 制备的纳米材料非常适合光催化应用。

综上所述,我们介绍了四种典型的 LAL 方法合成光催化剂的途径。第一种是在去离子水中通过脉冲激光烧蚀单元素金属靶材直接合成二元金属氧化物纳米材料;第二种是采用 LAL 技术合成三元 POM 作为后续水热合成的前驱体来制备光催化剂材料;第三种方法是通过 LAL 方法组装复合结构,如核-壳纳米结构或分层纳米结构等;第四种方法就是通过 LAL 过程在材料中引入缺陷来调制其电子结构。这些 LAL 基光催化剂材料制备技术为发展新型可见光区的光催化纳米材料提供了重要的技术支撑。

10.3.3 传感

本节我们介绍 LAL 制备的纳米材料在传感器件中的应用[147, 148]。这里,根据传感器件所检测的物质,我们把 LAL 制备的传感材料检测方法分为三类:第一类是离子和液体检测;第二类是气体检测;第三类是分子检测。

众所周知,水污染治理事关人类的生命健康。当饮用水中二价铅、汞和镉离子的浓度达到百万分之一或更高时,会构成严重的公共健康风险[149]。而对于水中污染物的鉴别和总浓度的评估,常用方法是离子耦合等离子体光谱和其他光谱检测技术[150, 151]。然而,这些检测方法由于所用仪器设备复杂和昂贵,因此并不适合在室外环境中进行实时实地的原位分析检测。为了解决这些问题,研究人员开发了基于 Au 纳米颗粒/GC 电极的电化学检测法[152],通过优化电化学检测条件,这种新型检测法可以同时检测浓度为 3×10^{-7} mol/L 的 Cd^{2+}、Pb^{2+}、Cu^{2+} 和 Hg^{2+}。

但是,这一类传感器需要高化学活性和高灵敏度的纳米传感材料,这就为 LAL 纳米制备技术在传感技术领域的应用打开了一扇门。

Pan 等[153]通过在去离子水中应用脉冲激光辐照金红石(TiO_2)粉末合成了重 Ti^{3+} 掺杂的 TiO_2 纳米颗粒,然后,他们将所合成的 TiO_2 纳米颗粒用于 H_2O_2 检测,无须四甲基联苯胺或任何其他过氧化物酶底物及缓慢反应的过程,实现了快速、高灵敏检测,如图 10-23 所示。XPS 和 EPRS 分析证实了所合成纳米颗粒中 Ti^{3+} 的

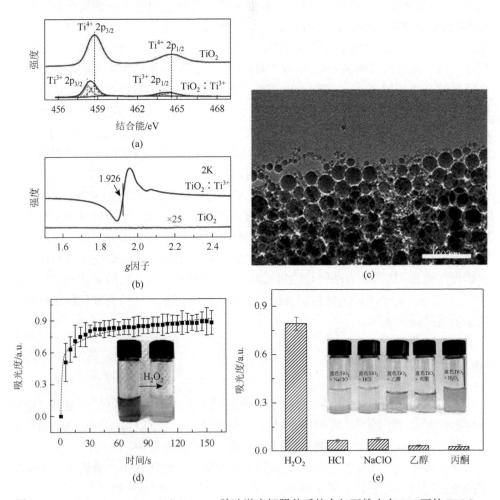

图 10-23 (a) Ti 2p XPS;(b) 355 nm 脉冲激光辐照前后的金红石粉末在 2 K 下的 EPRS;(c) $TiO_2:Ti^{3+}$ 纳米颗粒的 TEM 图像;(d) $TiO_2:Ti^{3+}$ 纳米颗粒和 H_2O_2 在 450 nm 处的典型吸光度动力学,插图为未添加(左)和添加(右)H_2O_2 溶液的 $TiO_2:Ti^{3+}$ 纳米颗粒的颜色变化;(e) 通过监测相对吸光度进行 H_2O_2 检测的选择性分析,插图的分析物浓度如下:1 mol/L 次氯酸钠(NaClO)、2 mol/L 盐酸(HCl)、1 mol/L 乙醇、1 mol/L 丙酮、0.5 mol/L 过氧化氢(H_2O_2)

(后附彩图)

存在 [图 10-23（a）和图 10-23（b）]，TEM 图像显示合成后的样品是尺寸为 30 nm 的球形纳米颗粒 [图 10-23（c）]。重要的是，这种无机纳米颗粒传感器的显色响应是瞬时的（<1 s）[图 10-23（d）]，其检出限达到 5×10^{-7} mol/L，甚至低于辣根过氧化物酶（一种常用来测定 H_2O_2 的天然酶）[153]。通过对照实验研究表明，TiO_2：Ti^{3+} 纳米颗粒在 H_2O_2 检测中具有高选择性，如图 10-23（e）所示。所以，由于材料的易于制备、无毒性和优异的生物相容性，这些 LAL 合成的 Ti^{3+} 掺杂 TiO_2 纳米颗粒可以作为用于 H_2O_2 直接测定的超快探针。

通过 LAL 纳米制备技术组装的量子尺寸 SnO_2 纳米颗粒与共轭还原氧化石墨烯复合纳米结构已经被用于非酶葡萄糖传感器。首先，通过 LAL 纳米制备技术合成具有高反应活性的亚稳态 SnO_x 纳米颗粒胶体溶液作为下一步的还原剂和复合材料的前驱体。然后，通过将 LAL 合成的 SnO_x 胶体溶液与氧化石墨烯水溶液混合，就可以将分散良好的量子尺寸 SnO_2 纳米颗粒与共轭还原的氧化石墨烯结合，组装出量子尺寸 SnO_2 纳米颗粒/石墨烯复合纳米结构[154]。最后，由所组装的复合纳米结构构筑的葡萄糖传感器的检测灵敏度为 1.93 $A/(mol\cdot L^{-1}\cdot cm^2)$，检出限为 13.35 μmol/L。由于正常人体血糖水平在 4~6 mmol/L，因此该葡萄糖传感器完全可以用于检测实际样品中的葡萄糖浓度。

众所周知，半导体气体传感器因其便宜、简单且可检测气体丰富而被广泛使用[155-159]。大量研究已经表明，气体传感过程与半导体表面反应密切相关[160]。由于 LAL 纳米制备技术合成的纳米材料具有清洁的表面和丰富的与气体分子相互作用的活性位点，因此它们作为气体传感器的传感材料可以提供高灵敏度。

Shaw 等首次利用由 LAL 合成的胶体溶液作为后续熟化过程的前驱体溶液，组装出多孔氧化钨纳米片并且用于气体传感器[161]，发现该传感器可以在相对较低的工作温度下检测出浓度低至 20 mg/L 的酒精气体。基于相同的 LAL 纳米制备技术，Xiao 等[162]合成了具有高纵横比的 β-MnO_2 纳米线（图 10-24），并且通过所合成的纳米材料构筑了可以检测爆炸性气体的气体传感器（图 10-25），例如，CO 检测 [图 10-25（a）]、乙醇气体检测 [图 10-25（b）] 和氢气检测 [图 10-25（c）]。更为重要的是，这些气体传感器可以以 10 s 的短响应时间和 20 mg/L 的低检出限检测出氢气，并且可通过升高检测温度来改善器件性能，如图 10-25（d）所示。Liang 等[163]采用 ECLAL 技术组装出具有刷状纳米结构的三元过渡金属钒酸盐（$Ag_2V_4O_{11}$）并将其用于检测乙醇气体，发现气体传感器在低工作温度下对 10~600 mg/L 的乙醇气体表现出较大的传感响应。同时，Niu 等[164]报道了 LAL 纳米制备技术组装的中空 ZnS 纳米颗粒与普通 ZnS 晶体相比，具有较大的表面积与体积之比，是一种非常有效的气敏纳米材料。Li 等[165]证明了由 LAL 制备的非晶过渡金属氧化物纳米材料制成的气体传感器，其活性与当前由晶体纳米材料制造的气体传感器的活性相当。上述这些研究表明，LAL

纳米制备技术是制造高比表面积、高气体反应活性半导体气敏纳米材料的好方法。

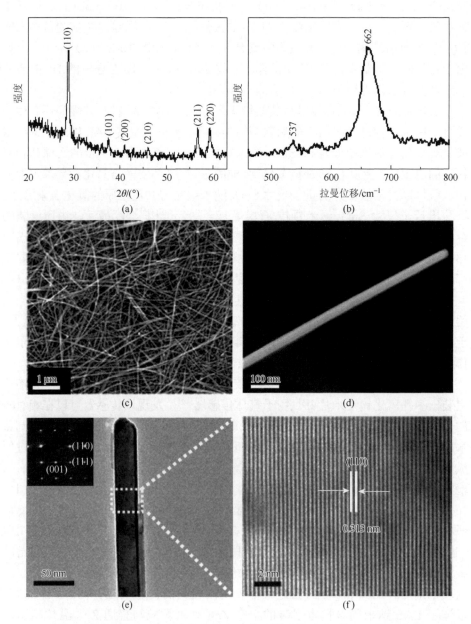

图 10-24 （a）高纵横比 β-MnO$_2$ 纳米线在空气中退火处理后的 XRD 谱；（b）拉曼光谱；β-MnO$_2$ 纳米线的典型（c）低倍和（d）高倍放大 SEM 图像；（e）单根 β-MnO$_2$ 纳米线的 TEM 图像和（插图）相应的 SAED 图像；（f）相应的 HRTEM 图像

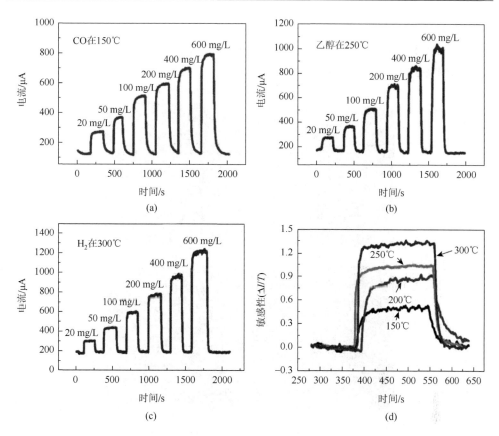

图 10-25 β-MnO$_2$ 纳米线在不同气体浓度下对（a）150℃的 CO 和（b）250℃的乙醇的传感器响应；（c）在空气中稀释的 20~600 mg/L 氢气的连续工作模式下，β-MnO$_2$ 纳米线气体传感器在 300℃下对氢气的传感器响应；（d）暴露于 50 mg/L 浓度氢气下的高纵横比 β-MnO$_2$ 纳米线的敏感性随工作温度的变化

LAL 制备的纳米传感材料也可用于制造 SERS 基底来检测分子。我们知道，LAL 合成的贵金属纳米颗粒具有清洁的表面和化学活性，作为 SERS 基底有利于和目标分子的接触，因此已经广泛用于分子传感的 SERS 基底[166, 167]。目前，大多数 SERS 基底是通过将贵金属纳米颗粒胶体直接滴到基底上来制造的。但是，这种组装技术很难控制 SERS 基底的重复性，并且由于吸附力相对较小，贵金属纳米颗粒很容易从基底上脱附。

为了提高 SERS 基底组装的可控性，增加分子与基底的吸附力，Cui 等[168]利用液相 PLD 在透明基底表面制备出高灵敏度的贵金属纳米颗粒 SERS 基底。这种方法的优点就是避免了表面官能化带来的污染，并且不需要在溶液中添加还原剂。将 LAL 纳米制备技术与电泳合成方法相结合，Yang 等[169]组装出了具有明确结构的大规模有序 Ag 纳米壳阵列作为 SERS 基底，如图 10-26 所示。令人感兴趣的是，组装的 Ag 纳米壳图案对罗丹明 6 G 的增强因子高达 10^6。显然，通过 LAL 纳米制备技术与电泳合成方法结合组装的 Ag 纳米壳具有干净、粗糙的表面，而且这种表面纳米图案化技术可以扩展到其他 SERS 基底。

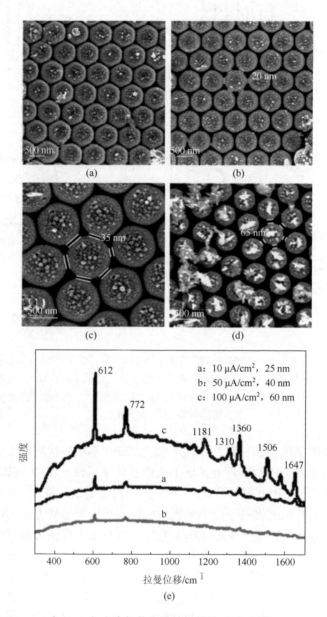

图 10-26 4 个 LAL 与电泳组装合成的样品的 SEM 图像和 SERS 光谱

（a）密堆积（无刻蚀）；（b）20 nm（刻蚀 1 min）；（c）35 nm（刻蚀 2 min）；（d）65 nm（刻蚀 3 min）；（e）吸附在不同尺寸 Ag 纳米壳上的 R6G 分子的 SERS 光谱

综上所述，LAL 纳米制备技术合成的纳米（传感或敏感）材料与纳米结构有如下特点：第一是高的比表面积，如一维纳米线、多孔纳米片、中空结构的多级纳米结构等，大的表面积为表面化学反应提供了重要的空间；第二是清洁的表面，LAL 纳米制备技术的简单和绿色为合成出清洁表面的纳米材料提供了重要保障，而清洁的表面是具备高化学活性的必要条件之一；第三是多活性位点，LAL 过程会在合成的纳米材料表面和体内制造缺陷，这些缺陷不仅能够调制材料的电子结构，更重要的是为表面反应提供更多的反应活性位点。这三方面的独特优势使得 LAL 纳米制备技术在传感器领域大放异彩。

10.4 应用于绿色能源中的功能纳米结构

10.4.1 超级电容器

超级电容器作为一种新型储能装置，由于其具有充电时间短、循环寿命长、功率密度高和维护成本低等优异性能，近年来引起了人们的极大兴趣[170, 171]。然而，超级电容器要想像它的主要竞争对手——电池一样成为主要的供电设备，那么它需要进一步发展以提高它提供高能量和功率的能力[172]，其中关键技术问题就是高性能电极材料的研制。最近，LAL 纳米制备技术展现了它在超级电容器新型电极材料研发中的重要作用。Liang 等采用 LAL 纳米制备技术研制出系列高性能新型超级电容器电极材料[173, 174]。例如，他们通过在 $NiCl_2$ 溶液中对 Mn 靶进行脉冲激光烧蚀从而制备出一种 Mn 掺杂的 α-$Ni(OH)_2$ 球形纳米结构，该纳米结构由许多小纳米片组装而成，如图 10-27 所示[175]。我们可以从样品的 XRD 谱看出，它与纯六方 α-$Ni(OH)_2$·0.75 H_2O 结构一致，并未探测到 Mn 相，同时，样品的 EDS 和 XPS 分析证实了 α-$Ni(OH)_2$ 纳米片中 Mn 的存在，这些结果说明 Mn 原子掺杂到了 α-$Ni(OH)_2$ 晶格中。重要的是，作为超级电容器电极材料（图 10-28），由 Mn 掺杂的 α-$Ni(OH)_2$ 纳米片组装的正极在 5 A/g 的充放电电流密度下表现出约 1000 F/g 的高比电容和良好的循环性能。此外，Liang 等[176]还通过脉冲激光烧蚀去离子水中 Co 靶组装出了具有由超细 Co_3O_4 纳米颗粒装饰的 MoS_2 纳米片表面的 MoS_2-Co_3O_4 复合纳米材料。有意义的是，相比纯 MoS_2 纳米片和纯 Co_3O_4 纳米颗粒，MoS_2-Co_3O_4 复合纳米材料作为超级电容器电极材料表现出更高的比容量［在电流密度为 0.5 A/g 时为 69 (mA·h)/g］和更长的循环寿命（在 1 mol/L KOH 溶液中循环 500 次后保留约 87%的比容量）。

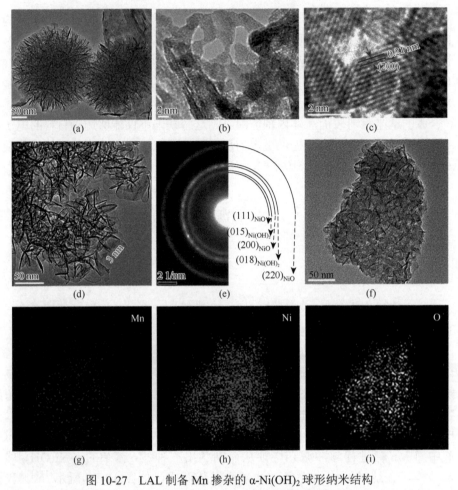

图 10-27 LAL 制备 Mn 掺杂的 α-Ni(OH)$_2$ 球形纳米结构

(a)、(b)、(d) 和 (f) 分别显示了样品在不同尺度下和不同区域的 TEM 图像；(c) 样品的 HRTEM 图像；(e) 样品相应的 SAED 图像；(g) ~ (i) 样品相应的 EDS 分析

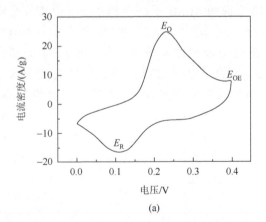

(a)

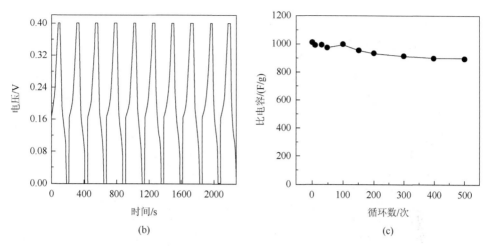

图 10-28 （a）由 Mn 掺杂的 Ni(OH)$_2$ 纳米结构构建的超级电容器的循环伏安图；（b）电流密度为 5 A/g 时的恒电流充放电曲线；（c）在 6 mol/L KOH 溶液中 4 mA/cm^2 下的循环性能

10.4.2 锂离子电池

除了超级电容器之外，具有更高能量、容量和更长循环寿命的可充电锂离子电池已经广泛用于新能源技术的多个领域如交通运输工具、储能电站等[177, 178]。近年来，LAL 纳米制备技术在锂离子电池电极材料研发中的应用也在不断发展。由聚苯胺（PANI）、无定形 TiO$_2$ 和氧化石墨烯网络组成的 TiO$_2$-氧化石墨烯夹层结构已经被用作锂离子电池的阳极材料[179]。采用 LAL 纳米制备技术在非晶 TiO$_2$-氧化石墨烯复合纳米片的两面生长聚苯胺纳米棒，我们可以组装出一个稳定的三明治纳米结构，也就是聚苯胺纳米棒/非晶 TiO$_2$ 纳米颗粒-氧化石墨烯复合纳米片/聚苯胺纳米棒。随后，将该复合纳米材料作为锂离子电池的正极材料组装成电池并进行系列表征，包括放电/充电容量、循环伏安法（CV）响应、循环稳定性和电化学阻抗谱（EIS）测量等，结果如图 10-29 所示。我们可以观察到，在 50 mA/g 下的初始放电容量为 1335 (mA·h)/g，然而，在 250 次循环后，该三明治纳米结构仍然表现出 444/436 (mA·h)/g 的稳定放电/充电容量 [图 10-29（a）]。在不同电流密度下循环 160 次后，复合纳米材料在 100 mA/g 下的放电容量为 547 (mA·h)/g [图 10-29（e）]。因此，这些结果表明，LAL 组装的三明治纳米结构电极在循环稳定性和高倍率方面表现出良好的性能。主要原因是因为三明治纳米结构提供了快速的电子传导路径以提高电导率，并且非晶 TiO$_2$ 结构为 Li$^+$ 和电子传输提供了许多通道，如图 10-29（f）所示。此外，LAL 纳米制备技术也用于锂离子电池的负极材料的开发。Nowak 等[180]将 LAL 合成的 SnO 纳米颗粒封装在碳基质中用作锂离子电池的负极材料，发现与其他负极材料相比，该负极材料的容量和稳定性不足[181, 182]。因此，LAL 制备电极材料的性能需要进一步提高，以满足商业设备的要求。

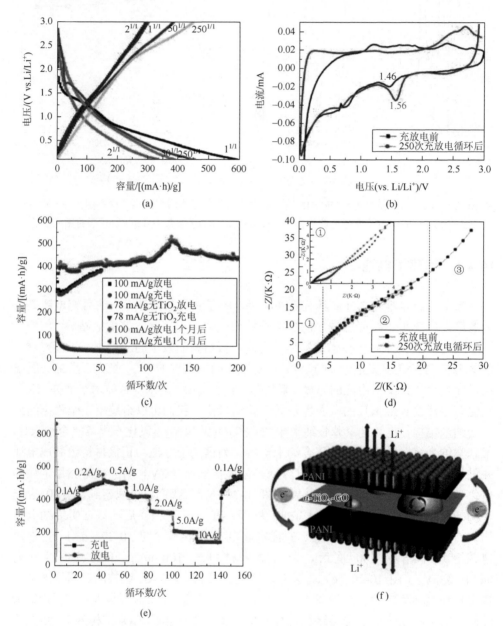

图 10-29 LAL 组装的聚苯胺纳米棒/非晶 TiO_2 纳米颗粒-氧化石墨烯复合纳米片/聚苯胺纳米棒三明治纳米结构作为锂离子电池正极材料的电化学性能（后附彩图）

（a）放电/充电容量；（b）CV 测试；（c）循环稳定性；（d）EIS 测试，插图为①区放大图，横纵表示实部，纵横表示虚部；（e）通过改变电流密度来提高三明治纳米结构的稳定性；（f）三明治纳米结构中 Li^+ 和 e^- 的传输示意图

10.4.3 太阳电池

LAL 纳米制备技术产生的纳米材料在太阳电池中具有广泛的应用。首先，我们以量子点（QD）敏化太阳电池为例介绍 LAL 技术的应用。我们知道，在 QD 敏化太阳电池中，配体会起到电子从 QD 转移到光电阳极过程的势垒作用。然而，配体的存在在 QD 的合成中常常是不可避免的。因此，Horoz 等[183]采用 LAL 纳米制备技术合成了平均尺寸为 5 nm 的无配体 CdSe QD，然后将它们用于制造 QD 敏化 ZnO 纳米线太阳电池，发现器件 0.48 V 的开路电压与使用化学溶液法合成的 CdSe QD 的太阳电池相当。在有机光伏电池研究中，人们发现含有 PbS 纳米颗粒的 P3HT：PCBM 混合物纳米复合薄膜与不含 PbS 纳米颗粒的相同混合物薄膜相比，在整个太阳光谱范围内表现出更低的透射率[184]。为此，Paci 等开展了系列创新工作，将 LAL 纳米制备技术合成的无配体 Au[185, 186]、Ag[187]和 Al[188, 189]纳米颗粒用于有机光伏发电中。他们发现，在 P3HT：PCBM 活性层中加入 Ag 纳米颗粒可以改善复合材料混合物的结构和形态特性，从而使这些活性材料在空气中长时间连续运行后依然保有良好的光伏性能和稳定性[187]。如果在该混合物中掺杂 Au 纳米颗粒，那么，活性层中这些不含表面活性剂的 Au 纳米颗粒可以通过局域表面等离子体共振和散射来显著增强器件性能[185]。同时，Au 纳米颗粒还可作为三线态激子的猝灭剂，阻碍光氧化过程并增强器件结构稳定性。此外，Kakavelakis 等[189]报道了添加 Al 纳米颗粒可以抑制光降解的发生。

上述的研究已经证明 LAL 纳米制备技术合成的表面清洁（不含配体或表面活性剂）的纳米颗粒在 QD 敏化太阳电池和有机光伏电池中发挥着重要作用。类似地，LAL 合成的不含表面活性剂且独立的 Si 纳米晶 Si-nc 也已经应用在这些光伏器件的活性材料中了，因为此类纳米晶可以通过调节彼此隔离的离散 QD 的表面特性来改善器件电输运和非线性光学特性。Švrček 等[190]的文章指出，在乙醇和水中使用纳秒激光辐照 Si 纳米颗粒，可以通过表面工程来调节 Si 纳米颗粒表面结构，如图 10-30 所示。例如，在纳秒激光加工过程中，Si 纳米颗粒表面悬挂键和二聚体被—OH 末端有效取代，从而演变成氧化物和—OH 末端［图 10-30（g）～图 10-30（j）］，这与空气中干燥的 Si 纳米晶［图 10-30（a）～图 10-30（c）］及在水或乙醇中的 Si 纳米晶［图 10-30（d）～图 10-30（f）］的情况不同。进一步地，他们将 LAL 处理过的 Si 纳米晶组装成两种不同的光伏器件结构，一种是 Si 纳米晶敏化太阳电池，另一种是 Si 纳米晶/富勒烯光敏界面的有机光伏电池。结果表明，不含表面活性剂的 Si 纳米晶足够稳定，并且可以促进与富勒烯的电子耦合。因此，这种规模化且便捷的 LAL 处理方法可以直接在不同的液体环境中合成

Si 纳米晶，并且可以通过调整纳米晶表面结构以改善太阳电池的性能。此外，该方法可以扩展到具有不寻常表面特性的其他纳米材料的合成。例如，Wang 等[191]采用 LAL 纳米制备技术合成了单晶金红石（TiO_2）空心纳米球，并将其用于 QD 敏化太阳电池，实现了 10%的转换效率。

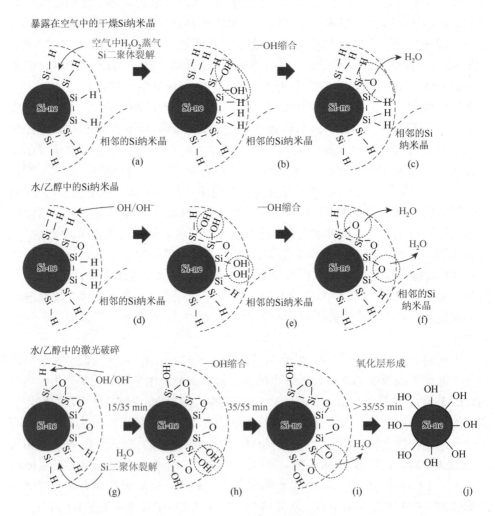

图 10-30　示意图描绘了当纳米晶（a）~（c）暴露于空气中、（d）~（f）分散在水或乙醇中，以及（g）~（j）在 LAL 过程中，Si 纳米晶表面的变化

除了用于 QD 敏化太阳电池和有机光伏电池外，研究人员还报道了将 LAL 制备和电泳沉积相结合，在胶体金属纳米颗粒上组装的新型 Cu(In, Ga)Se_2（CIGS）太阳电池[192]。这些表面电荷被调控的 LAL 合成的纳米颗粒适合通过电泳沉积制

造薄膜，且无须溶剂转移或黏合剂（图 10-31）。由于所制备材料的高纯度和精确的组分，组装的 CIGS 太阳电池的转换效率高达 7.37%，如图 10-31 所示。因此，LAL 纳米制备技术的应用有力地推动太阳电池的发展。

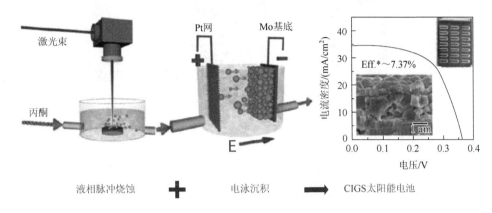

图 10-31　LAL 制备和电泳沉积结合制造高效 CIGS 太阳电池示意图

10.4.4　混合发光二极管

II-VI、III-V 和 IV-VI 族半导体纳米晶在光电子学领域受到了极大的关注，因为它们在可见光和红外光谱范围内表现出可调谐的电致发光[193, 194]。然而，因为 Si 是间接带隙半导体，本身不发光，所以目前人们对包括 Si 在内的 IV 族半导体纳米晶的关注较少。事实上，尺寸小于 5 nm 的 Si 纳米晶由于量子限域效应和表面缺陷亦可展现出极高的光致发光，并且它们具有 Si 材料固有的低毒性、高丰度和低成本[195]。

Xin 等[196]使用 LAL 纳米制备技术合成了胶体 Si QD 并通过溶液工艺制造了 Si QD 的无机/有机杂化材料发光二极管（light emitting diode，LED），当受到紫外光照射时，组装的 LED 显示白蓝色荧光［图 10-32（a）］。图 10-32（b）总结了 Si QD 的激发-发射光谱的三维图。由于 Si QD 可发出强烈的可见光，因此，研究人员将其组装为混合式 LED 器件，图 10-32（c）和图 10-32（d）分别显示了 Si QD 混合式 LED 的照片和示意图及能级图。非常重要的是，这种混合式 LED 可以产生高的电流密度（280 mA/cm^2）和光功率密度（700 nW/cm^2），它们比已有报道的含有相同外加电压的蓝光 Si QD 混合式 LED 的电流密度（约 1 mA/cm^2）和光功率密度（约 2 nW/cm^2）分别高 279 和 349 倍[197]。请注意，在这里 78% 的有效发射来自 Si QD。所以，该器件的优越性能主要归功于 LAL 合成的 Si 量子点的强烈发光、整个 LED 中载流子的有效迁移及 Si QD 层中发生的载流子复合。

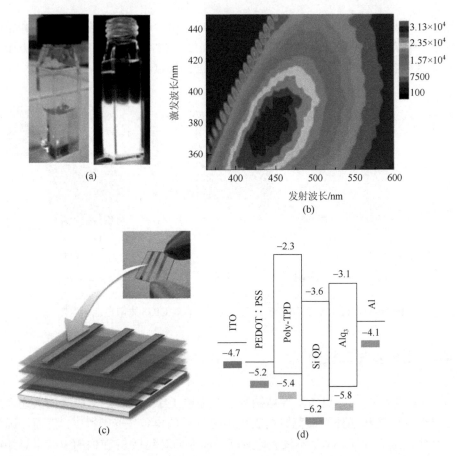

图 10-32 （a）在环境光下分散在异丙醇中的 Si QD 的照片（左）和波长 365 nm 的 UV LED 激发下胶体溶液的照片（右）；（b）分散在异丙醇中的 Si QD 的激发-发射光谱三维图；（c）组装的夹层结构 Si QD 混合式 LED 的照片和示意图；（d）Si QD 混合式 LED 的能级图（单位：eV）（后附彩图）

10.4.5 光催化分解水产氢

通过化学能收集和储存太阳能是解决能源问题的一种非常理想的方法。因此，利用太阳能光催化分解水产生的 H_2 和 O_2 作为可再生能源，是目前发展绿色能源的一条重要途径，已经引起了全球科学家的广泛关注[198]。由于 LAL 纳米制备技术合成的纳米材料普遍具有清洁表面、高化学活性、富含缺陷等特点，因此这项技术已经被广泛用于包括光催化剂材料在内的各种催化纳米材料的制备。Liao 等[199]采用 fs-LAL 技术合成了先进的光催化分解水产氢催化剂材料，他们的研究表明，通过 LAL 技术合成的 CoO 纳米晶，无须任何助催化剂或牺牲剂，在太阳光辐照下，可以将水分解成化学计量的 H_2 和 O_2，并且实现了约 5%的从太阳能到氢气的转化效率，这是光催化分解水产氢研究的一项重要进展。在太阳光分解水产氢的过程中，纳米晶催化剂和微米晶催化剂的差异在于它们导带边缘和价带边缘与水的氧化还原电位的相对位置不同。CoO 微米晶的导带边缘位于析氢电位下方，而 CoO 纳米晶的导带边缘位于析氢电位上方，并且 CoO 纳米晶的价带边缘位于析氧电位下方。所以，CoO 纳米晶与水的氧化还原电位的能带位置解释了为什么纳米晶表现出远比微米晶优越的光催化活性。显然，这一发现凸显了纳米晶作为太阳光分解水产氢催化剂的优势。

Lin 等[200]将 LAL 合成的纳米金刚石（ND）与 p 型 Cu_2O 纳米晶结合，组装的 ND-Cu_2O 复合纳米结构作为光催化分解水产氢催化剂，如图 10-33 所示。图 10-33（a）显示了合成后的 ND-Cu_2O 和 Cu_2O 样品的 XRD 谱。我们可以看出，ND-Cu_2O 复合纳米结构具有准立方体形状［图 10-33（b）］，并且有许多小 ND 颗粒嵌入 Cu_2O 纳米晶内［图 10-33（c）］。为了获得最佳光催化性能，研究人员可以设计 Cu_2O 的最佳 ND 负载量［图 10-33（d）］。重要的是，ND-Cu_2O 复合纳米结构在大气质量（atmospheric mass，AM）1.5 照射下表现出 1597 μmol/(h·g) 的高析氢反应速率［图 10-33（e）］。ND-Cu_2O 复合纳米结构在不同波长光照射

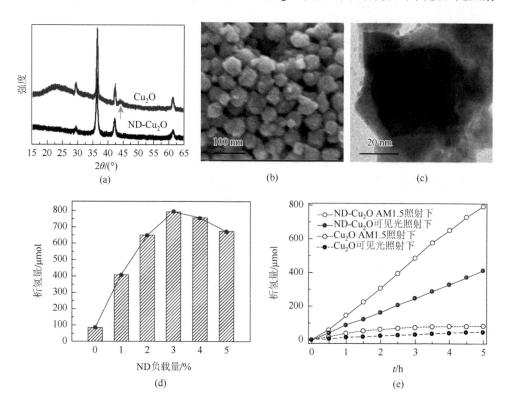

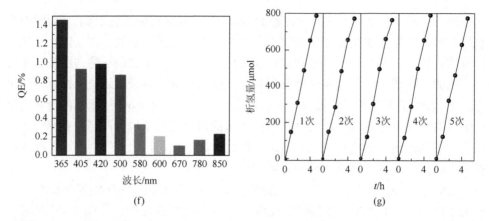

图 10-33 （a）ND-Cu$_2$O 复合纳米结构和 Cu$_2$O 纳米颗粒的 XRD 谱；（b）和（c）分别是 ND-Cu$_2$O 复合纳米结构的 SEM 和 TEM 图像；（d）不同 ND 负载量的 ND-Cu$_2$O 复合纳米结构的析氢量；（e）AM 1.5 和可见光照射下催化剂分解水析氢的典型进程；（f）ND-Cu$_2$O 复合纳米结构在不同波长光照射下的量子效率比较；（g）ND-Cu$_2$O 复合纳米结构在 AM 1.5 照射下的循环稳定性

下的量子效率比较如图 10-33（f）所示。同时，ND-Cu$_2$O 复合纳米结构表现出良好的稳定性，在 5 次循环后活性没有明显下降 [图 10-33（g）]。ND 的给电子能力和异质结构合适的能带结构促进了由 ND 到 Cu$_2$O 的电子注入，从而实现广谱光催化析氢。

10.4.6 电化学催化剂

LAL 纳米制备技术已经在高活性电化学纳米催化剂材料制造领域发挥着重要作用[201-205]。LAL 合成的催化剂材料的高活性主要来自所制备纳米材料的清洁表面即表面不存在任何有机污染物或保护性配体、小尺寸、形成优先晶面等。Blakemore 等[205]报道了采用 LAL 纳米制备技术合成出非常小的（<5 nm）、不含表面活性剂且尺寸可控的 Co$_3$O$_4$ 纳米颗粒，并发现该纳米颗粒对于从碱性水溶液中析出氧气表现出高活性。他们首先使用波长 355 nm 的脉冲激光聚焦辐照分散在去离子水中的商用 Co 纳米粉末，合成了超细 Co$_3$O$_4$ 纳米颗粒的水悬浮液；然后使用强磁铁将剩余的 Co 粉末从合成的纳米颗粒溶液中分离出来进行纯化；最后进行电化学催化表征。研究结果表明，LAL 合成的催化剂的过电势为 314 mV，与电沉积氧化钴的最佳过电势相媲美[205]。更重要的是，该催化剂的周转频率为 0.21mol O$_2$/[(mol Co$_{surface}$)·s]，这是截至目前 Co$_3$O$_4$ 纳米颗粒析氧催化剂报道的最高值。LAL 合成 Co$_3$O$_4$ 纳米颗粒的高活性归因于表面没有表面活性剂，确保了表

面反应位点的充分利用。基于同样的技术路线，Hunter 等[201]还利用 LAL 技术制备了其他高活性电化学催化剂如混合金属纳米片水氧化催化剂等。

实际上，在 LAL 纳米制备技术中，我们可以很容易地通过改变金属固体靶（如 Ni 或 Fe）、环境液体中金属离子类型（如 Fe、Ni、Ti 或 La）及离子浓度（如 0.01^{-3} mol/L）、脉冲激光能量（如 90 mJ 或 210 mJ）等，合成多种混合金属催化剂[201]。例如，具有层状双氢氧化物结构的富 Ni 纳米颗粒即 $Ni_{1-x}Fe_x(OH)_2(NO_3)_y(OH)_{x-y}\cdot nH_2O$，研究表明 Ti^{4+} 和 La^{3+} 的添加进一步增强了催化活性，该催化剂在 260 mV 的过电势下达到 10 mA/cm^2[201]，这是截至目前报道的平面电极上混合金属氧化物催化剂的最低过电势。

除了电催化中的水氧化之外，LAL 纳米制备技术合成的新型电催化剂还对葡萄糖、甲醇和甲酸氧化表现出高活性[202-204]。众所周知，Au 纳米颗粒是最常见的高活性电催化剂之一。Hebie 等[204]报道了 LAL 合成的 Au 纳米颗粒表现出适合生物燃料电池应用的独特性能，如图 10-34（a）所示，图 10-34（b）描绘了典型的 Au 纳米胶体及其相应的紫外可见吸收光谱。研究人员将 LAL 合成的无配体的 Au 纳米颗粒作为葡萄糖电氧化中的电极，揭示了与十六烷基三甲基溴化铵（CTAB）涂层的 Au 纳米颗粒相比，LAL 合成的具有清洁表面 Au 纳米颗粒的优点 [图 10-34（c）]。此外，LAL 合成 Au 纳米颗粒的效率几乎比化学溶液法合成的对应物 [例如，CTAB 稳定的 Au 纳米颗粒和柠檬酸盐法合成的 Au 纳米颗粒（Cit-Au NP）] 高一个数量级 [图 10-34（d）]。LAL 合成 Au 纳米颗粒的电流密度高达 2.65 A/(cm^2·mg)[204]，超过了目前已经报道的所有金属和金属合金基电催化剂的电流密度。

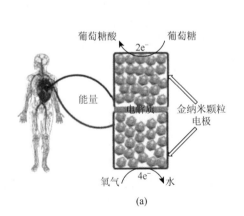
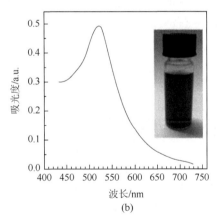

(a)　　　　　　　　　　　　　　(b)

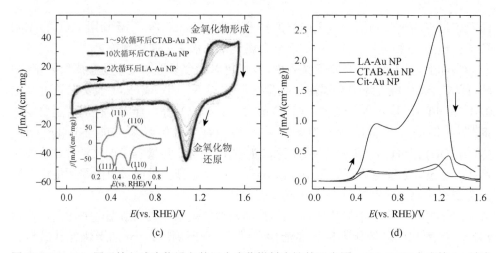

图 10-34 （a）用于植入式生物设备的混合生物燃料电池的示意图；（b）LAL 合成的 Au 纳米颗粒的典型消光光谱，插图为 LAL 合成 Au 纳米颗粒的去离子水溶液的玻璃比色皿；（c）LAL 合成 Au 纳米颗粒（LA-Au 纳米颗粒，红色）和化学溶液法合成 Au 纳米颗粒（CTAB-Au 纳米颗粒）形成的 Au 电极的不同循环次数的循环伏安图，插图为局部放大图；（d）CTAB-Au 纳米颗粒（红色）、Cit-Au 纳米颗粒（蓝色）和 LA-Au 纳米颗粒（黑色）电极的伏安图（后附彩图）

此外，这些 LAL 合成的高活性电催化剂材料还可以与其他纳米材料进行复合以更进一步地增强其催化活性。Wu 等[202]已经证明，与商业 Pt/C 电催化剂相比，LAL 组装的 Pt/C 还原氧化石墨烯复合电催化剂对于甲醇电氧化具有更高的活性和更好的循环稳定性。这里需要提到的是，Zhang 等[206]系统综述了具有独特性质的无配体 LAL 纳米制备技术合成的贵金属纳米颗粒及它们在电化学催化中的应用，有兴趣的读者可阅读参考。对于电化学催化过程所涉及到的几乎所有金属或合金纳米材料催化剂的制造，LAL 纳米制备技术都是最佳选择之一。

10.5 生物医学功能纳米结构

10.5.1 生物分子载体

众所周知，贵金属纳米颗粒由于其低的毒性、高的稳定性、好的生物相容性在生物医学领域有着广泛的应用。尤其是 Au 纳米颗粒在生命系统中具有很高的接受度，并且很容易与寡核苷酸等生物分子缀合。所以，Au 纳米颗粒作为用于生物医学领域的典型贵金属纳米颗粒受到了极大的关注和系统的研究，并且已经用于生物大分子识别、重大疾病诊疗、药物输送载体等。与常用的化学溶液法合成的 Au 纳米颗粒相比，LAL 纳米制备技术合成的 Au 纳米颗粒在生物医学应用中已经显示出了独特的优越性。根据 Sylvestre 等[207]的研究，采用在纯水中应用脉冲激光烧蚀 Au 靶合成的 Au 纳米颗粒的表面是带正电的，原因是生成的氧化态 Au^+ 和 Au^{3+} 是由水溶液中存在的氧气引起的部分氧化产生的。这些带正电的 Au 纳米颗粒可以作为电子受体，很容易与—COOH 和—SH 等基团进行配位。这就是 LAL 合成的 Au 纳米颗粒能够在一步合成法中与小分子配位的基本机制[207]。显然，这种 LAL 合成方法不依赖于吸附在纳米颗粒表面上的化学前驱体，也不产生化学反应副产物，并且肽、寡核苷酸和抗体等生物大分子可以通过共价硫醇键连接到 Au 纳米颗粒表面。所以，我们可以看到，无配体 LAL 合成 Au 纳米颗粒的显著优点就是生物分子可以轻松附着到 Au 纳米颗粒的自由表面上。因此，人们已经探索了 LAL 诱导的与多种生物大分子的生物共轭作用[209-214]。Petersen 等[215]报道通过原位和非原位方法共轭的 LAL 合成的 Au 纳米晶的最大表面覆盖率比传统配体交换方法制备的 Au 纳米颗粒的高出 5 倍。这里需要注意的是，如此高的表面覆盖率使得这种生物缀合技术成为一种具有潜在应用前景的有效方法，它可以使纳米颗粒与 DNA 或适体缀合进行药物输送。

Petersen 等[216]已经证明可以通过 LAL 纳米制备技术实现 Au 纳米颗粒与细胞穿透肽的原位生物共轭，如图 10-35（a）所示。他们通过系列实验研究了这种穿透肽共轭对 Au 纳米颗粒的细胞内化的影响［图 10-35（b）］，发现在 2 h 内，共孵育细胞 100%摄取了穿透共轭 Au 纳米颗粒，成功实现了细胞穿透纳米标记物的设计和快速细胞内生物成像。类似地，Barchanski 等[217]报道采用 ps-LAL 合成了用于细胞染色的 Au 抗体纳米缀合物，并且发现，与 fs-LAL 技术相比，使用 ps-LAL 合成的 Au 纳米缀合物的产额高一个数量级，同时，通过印迹和细胞免疫标记实验证实，这种 LAL 合成的纳米缀合物可以获得优异的染色效果。所以，考虑到高缀合效率的实现，无配体 LAL 技术合成的 Au 纳米颗粒比化学溶液法合成的对应物更适合生物缀合。

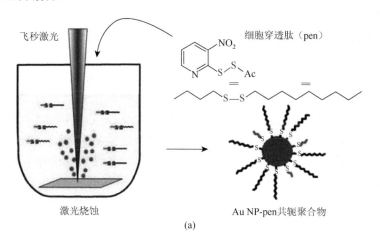

(a)

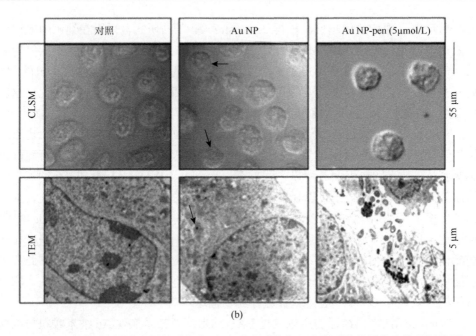

图 10-35 （a）在 LAL 纳米制备过程中，Au 纳米颗粒与细胞穿透肽的原位生物共轭作用；（b）穿透蛋白缀合对 Au 纳米颗粒细胞内化的影响：永生化牛内皮细胞的激光扫描共聚焦显微镜（CLSM）图像（上）和 TEM 图像（下）

10.5.2 阳性造影剂

核磁共振成像（MRI）技术由于具有高空间分辨率、良好的软组织对比度、良好的感知能力、丰富的解剖信息和方位信息，已经成为现代临床医学的常规诊断工具[218-220]。但是，MRI 技术灵敏度偏低，因此，为了对 MRI 的低灵敏度进行补偿，我们通常需要使用阳性造影剂（又称 T_1 造影剂），通过加速水质子的纵向弛豫 r_1 来增加目标器官和正常器官之间的对比度，从而产生更明亮的 MRI 影像[221-223]。众所周知，由于具有 7 个不成对的 4f 电子，Gd^{3+} 被认为是目前用作阳性造影剂的最佳金属材料[224]。Gd_2O_3 表面的 Gd^{3+} 可以将它的 7 个不成对电子全部用于水合反应，而 Gd^{3+} 螯合物只能提供一个水合物位置，因为它的其他 6 个不成对电子是通过螯合配体配位的。考虑到这一原因，研究人员认为 Gd_2O_3 纳米颗粒的 r_1 可能比 Gd^{3+} 螯合物的更大[169, 170]。这样，用 MRI 技术合成的 Gd_2O_3 纳米颗粒引起了人们的关注。Luo 等[225]发展了一种基于 LAL 纳米制备技术的简单、绿色策略来合成无配体的 Gd_2O_3 纳米颗粒，并且他们应用所发展的 LAL 纳米制备技术合成了直径为 3.8 nm 的超细 Gd_2O_3 纳米颗粒作为 MRI 阳性造影剂，实现了高达 9.76 s^{-1}/(mmol·L^{-1}) 的 r_1[226]。显然，与较大纳米颗粒相比，超细纳米颗粒表面积与体积的增加及水合数降低等是较小纳米颗粒显示较大的 r_1 的主要原因。此外，为了设计多功能生物医学纳米探针，Luo 等[42]还通过将标准固态反应方法和 LAL 纳米制备技术相结合，研制了稀土掺杂的 Gd_2O_3 纳米颗粒，它可以作为 MRI 和荧光成像中的双模态造影剂。

考虑到 Gd 有一定的毒性，Mn 基纳米颗粒被认为有望替代 Gd 基纳米颗粒作为造影剂，因为它的内在毒性比 Gd 基纳米颗粒低很多，因此，Mn 基纳米颗粒造影剂在神经科学研究中受到越来越多的关注[227-229]。Xiao 等[230]采用在去离子水中应用脉冲激光烧蚀 Mn 靶合成了直径为 9 nm 的 Mn_3O_4 纳米颗粒，如图 10-36（a）所示，插图中的高分辨 TEM 图像（右下）表明合成的样品是纯四方相 Mn_3O_4。研究人员将 LAL 合成的 Mn_3O_4 纳米颗粒作为 MRI 阳性造影剂，证明它们具有 8.26 L/(mmol·s) 的超高弛豫率，是商用 Gd 基阳性造影剂 [4.11 L/(mmol·s)] 的两倍，也是截至

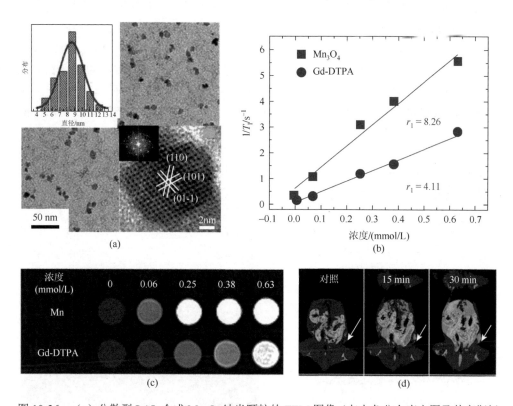

图 10-36 （a）分散型 LAL 合成 Mn_3O_4 纳米颗粒的 TEM 图像（左上角分布直方图及其高斯拟合曲线证明样品的平均尺寸约为 9 nm）和高分辨 TEM 图像（右下的插图）；（b）Mn_3O_4 纳米颗粒和商用 Gd-DTPA 的弛豫率（r_1）的比较；（c）不同浓度 Mn_3O_4 纳米颗粒（上排）和 Gd-DTPA（下排）的 T_1 加权 MRI；（d）鼻咽癌（NPC）CNE-2 移植瘤的代表性动态对比增强的 T_1 加权 MRI（白色箭头）

目前 Mn 基 MRI 阳性造影剂报道的最高弛豫率[图 10-36（b）]。同时，这些纳米颗粒的超高弛豫率也被肿瘤的 T_1 加权 MRI 和动态对比增强的 T_1 加权 MRI 所证实，如图 10-36（c）和图 10-36（d）所示。最重要的是，根据包括人体鼻咽癌细胞的细胞活力和体内免疫毒性实验在内的毒性测试表明，这些 Mn_3O_4 纳米颗粒在体外和体内均表现出令人满意的生物相容性。所以，这些研究充分证明了 LAL 合成的 Mn_3O_4 纳米颗粒是一种高效安全的 MRI 阳性造影剂。

10.5.3 生物识别

研究人员发现，当 LAL 合成的 Au 纳米颗粒被靶向剂和热响应聚合物修饰后，这些复合纳米结构可以用于通过温差识别分子和细胞[231]。聚合物 N-异丙基丙烯酰胺-共-丙烯酰胺是一种热响应共聚物，它的属性可以随着共聚物温度改变而切换，其最低临界共溶温度（LCST）为 37℃。而 LAL 合成的 Au 纳米颗粒上的细胞特异性配体的功能可以通过共接枝热响应聚合物的作用来控制。研究人员发现，在低于 37℃（温度 T<LCST）的情况下，这些复合纳米结构会排列为链延长的刷状聚合物，而将配体暴露在高于 37℃（温度 T>LCST）的外部环境，则会导致聚合物塌陷成难溶的球体[231]。图 10-37（a）显示了用靶向剂和热响应聚合物修饰的

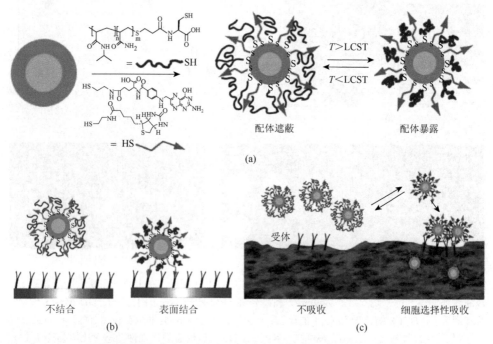

图 10-37　（a）用靶向剂和热响应聚合物修饰 LAL 合成的 Au 纳米颗粒的示意图；（b）和（c）分别为温度控制的生物识别作用

Au 纳米颗粒，以及它们在不同温度下的形貌演化。这样，我们就可以将热响应复合纳米结构的线圈-球体的形态转变用于检测癌症或炎症组织，因为这些病变组织的局部温度高于正常组织[231]。

10.5.4 纳米酶

酶作为生物体内最为重要的催化剂之一与自然界中的生命体系息息相关，它们调控了大部分活细胞中催化反应和机体的新陈代谢，具有较高的催化活性和底物特异性，可以在较为温和的条件下高效、专一地进行催化反应[232]。大部分酶是蛋白质，它们容易受到环境中的物理和化学因素（例如，温度和酸碱度）影响而失去酶活性。此外，酶的制备、纯化和保存过程烦杂，需要耗费大量的人力、物力和财力[233,234]。因此，研究人员从仿生学的角度开发出一系列非生物材料来模拟天然酶的催化功能，包括金属配合物、环糊精、卟啉和聚合物等，这类有机材料被称为模拟酶（mimic enzyme）[234-237]。有机模拟酶具有较高的稳定性和较低的制备成本，能够在一定程度上克服天然酶的缺点，但同时也会产生新的问题，有机模拟酶的催化效率与天然酶相比有较大的差距，而且与天然酶一样，有机模拟酶的可塑性较差，不能满足人们在生产生活中对于酶性能和应用场景的不同需要[238,239]。2007 年，Gao 等[240]发现了无机 Fe_3O_4 纳米颗粒具有类似辣根过氧化物酶（horseradish peroxidase，HRP）的催化活性，后来，研究人员又发现了纳米氧化铈、纳米金及氧化石墨烯等一系列具有类酶催化活性的无机纳米材料，并把这类材料称为无机模拟酶（inorganic mimic enzyme）材料或纳米酶（nanozyme）材料[238,241-249]。这些开创性的工作打破了长久以来人们对无机纳米材料生物惰性的固有认知，开启了人工酶研究的新天地。相对于天然酶，纳米酶具有诸多独特优点，如稳定性高、比表面积大、成本低、催化活性可调、易于表面修饰等。同时，它又兼有纳米材料特有的光、电、热、磁等物性，这就使得纳米酶成为一类多功能纳米材料。

正如我们介绍的 LAL 纳米制备技术在光化学催化剂和电化学催化剂材料制备中的应用一样，LAL 合成的高活性纳米材料作为纳米酶也得到了广泛应用[250-256]。一般来说，作为纳米酶的无机纳米材料应该具有如下特点：具有高化学活性即类似天然酶催化属性如氧化、还原等；具有清洁表面，易于进行化学或生物学修饰；富含表面缺陷，可以提供丰富的表面活性位点；无毒或低毒并具有好的生物相容性。所以，LAL 纳米制备技术十分匹配纳米酶材料合成的需要。Chen 等[253]采用 LAL 纳米制备技术合成了球状类富勒烯结构的 MoS_2（fullerence-like MoS_2，F-MoS_2）纳米颗粒，并证明它具有内在的类双重酶活性，包括超氧化物歧化酶（superoxide dismutase，SOD）和过氧化氢酶（catalase，CAT）的活性，揭示了这种纳米酶的抗氧化功能。图 10-38 展示了采用 LAL 纳米制备技术在去离子水中激光烧蚀固体 MoS_2 靶合成的 $F-MoS_2$ 纳米颗粒,可以清晰看到洋葱状结构 [图 10-38 (f)]。

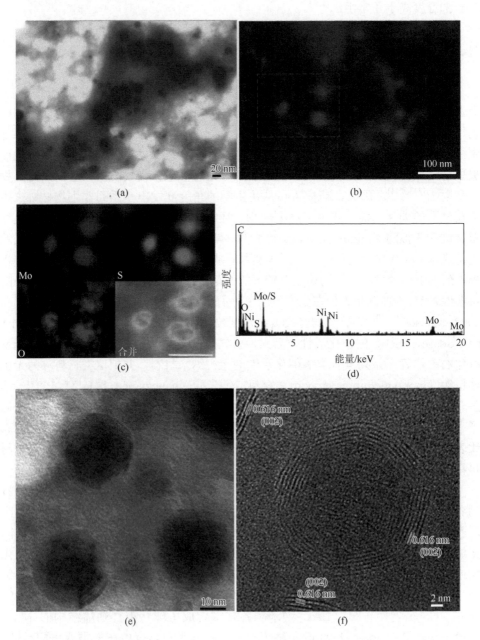

图 10-38　LAL 合成 $F-MoS_2$ 纳米颗粒的形貌和结构
(a) TEM 明场像；(b) SEM 图像；(c) ERS 图像；(d) EDS 分析；(e) 和 (f) HRTEM 图像

进一步地,他们证明了 F-MoS$_2$ 纳米颗粒是具有多重类酶活性(SOD 和 CAT)的纳米酶,如图 10-39 所示。众所周知,在生物细胞中,SOD 处于活性氧稳态调

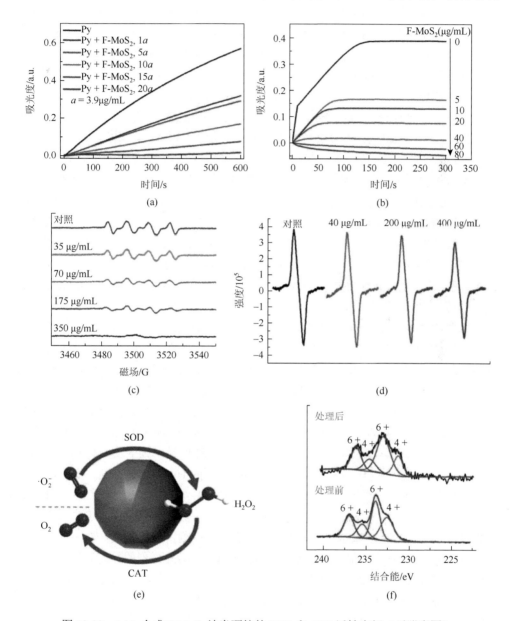

图 10-39 LAL 合成 F-MoS$_2$ 纳米颗粒的 SOD 和 CAT 活性表征(后附彩图)

(a)不同浓度的 F-MoS$_2$ 纳米颗粒与邻苯三酚(Py)共孵育体系在 318 nm 处的吸光度随时间的变化曲线;(b)与黄嘌呤氧化酶、黄嘌呤、DTPA 共孵育体系在 550 nm 处吸光度随时间的变化曲线;(c)不同浓度的 F-MoS$_2$ 纳米颗粒与黄嘌呤氧化酶、黄嘌呤、DTPA、BMPO 共孵育体系的 ESR 谱;(d)与 CTPO、H$_2$O$_2$ 共孵育体系的 ESR 谱;(e)基于 F-MoS$_2$ 纳米颗粒的级联催化过程的示意图;(f)处理前后的 F-MoS$_2$ 纳米颗粒高分辨 XPS 的 Mo 3d 峰

节的最前沿,它可以催化 $\cdot O_2^-$（活性氧）的歧化反应产生 O_2 和 H_2O_2；同时,CAT 是生物细胞抵抗氧化应激的防线的重要组成部分之一,它可以催化 H_2O_2 的分解产生 O_2 和 H_2O。所以,F-MoS_2 纳米颗粒是一种级联类酶催化剂,该级联催化过程的原理如图 10-39（e）描述,具体包括两个过程:第一,F-MoS_2 纳米颗粒催化 $\cdot O_2^-$ 的歧化反应产生 O_2 和 H_2O_2；第二,限制在 F-MoS_2 纳米颗粒附近的 H_2O_2 分子形成局部高浓度,F-MoS_2 纳米颗粒催化 H_2O_2 歧化反应,最终产 O_2 和 H_2O。因此,通过该级联催化体系,活性氧转化为低毒的 H_2O_2 和 O_2。

F-MoS_2 纳米颗粒的类酶催化活性源于表面化学反应。我们可以从样品的 XPS 来分析 $\cdot O_2^-$ 对 F-MoS_2 纳米颗粒表面的影响。如图 10-39（f）所示,在 F-MoS_2 纳米颗粒经过 $\cdot O_2^-$ 处理前后,Mo 3d 峰的结合能没有明显的变化。我们通过对 XPS 进行拟合可以看到,F-MoS_2 纳米颗粒的表面上存在一部分 Mo^{6+}。Mo 的氧化态（6+）可能源自表面形成的氧化物或缺陷,这也可以从 F-MoS_2 纳米颗粒的 ERS 图像得到验证［图 10-38（c）］。由于受带负电荷的 $\cdot O_2^-$ 的静电排斥作用,歧化反应在中性和碱性环境下进行得相对缓慢。然而,F-MoS_2 纳米颗粒表面带的正电荷的 Mo^{4+} 将吸引 $\cdot O_2^-$ 聚集在一起,从而促进 $\cdot O_2^-$ 之间的电子转移[257]。因此,我们认为 Mo^{6+}/Mo^{4+} 氧化还原对是催化活性的来源,Mo^{6+} 和 Mo^{4+} 两个状态的切换促进着电子的转移[245]。

重要的是,LAL 合成的 F-MoS_2 纳米颗粒具有好的生物相容性和强的抗氧化能力,如图 10-40 所示。这里我们用人脐静脉内皮细胞（HUVEC）来建立细胞模型,研究对氧化应激和炎症的影响,细胞活性采用 MTT 法来测定,再用不同浓度的 F-MoS_2 纳米颗粒与 HUVEC 共孵育 24 h 后分析细胞的活性。我们可以看到,即便高浓度（300 μg/mL）的 F-MoS_2 纳米颗粒在氧化后,HUVEC 仍保持超过 90%的活性［图 10-40（a）］,这些结果表明 F-MoS_2 纳米颗粒具有良好的生物相容性。另外,作为级联催化剂,F-MoS_2 纳米颗粒显示出强的抗氧化活性。这里我们采用 H_2O_2 作为氧化应激源,在用 200 μmol/L 的 H_2O_2 处理 1 h 后,HUVEC 的活性下降了约 50%。然而,用 200 μg/mL 的 F-MoS_2 纳米颗粒预处理细胞 24 h,结果显示 HUVEC 的活性增大到 90%［图 10-40（b）］。接着,我们通过 DCFH-DA 探针直接检测细胞内活性氧水平,使用流式细胞仪来检测荧光的强度。如图 10-40（c）和图 10-40（d）所示,用 H_2O_2 处理的 HUVEC 发出的强烈的绿色荧光,而加入 F-MoS_2 纳米颗粒孵育后,HUVEC 的荧光信号则大大减弱,只比对照组略高,这表明 F-MoS_2 纳米颗粒可以明显降低 HUVEC 的活性氧水平。所以,这些结果表明 F-MoS_2 纳米颗粒具有强的细胞抗氧化能力。

我们知道,骨关节炎（osteoarthritis,OA）是一种滑膜关节疾病,会导致关节疼痛、僵硬和肿胀等[258]。通常,骨关节炎被认为与关节过度的机械受力和炎症应激有关[259]。透明质酸（HA）是关节滑液的主要成分,为关节滑液提供黏弹性

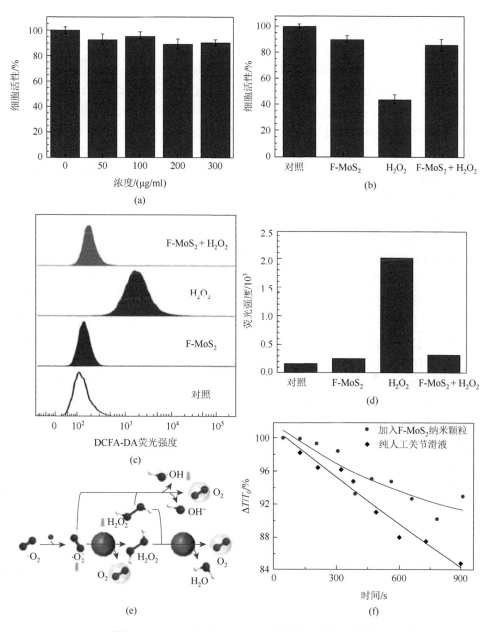

图 10-40 LAL 合成 F-MoS$_2$ 纳米颗粒的抗氧化作用

(a) MTT 法 F-MoS$_2$ 纳米颗粒的 HUVEC 毒性表征；(b) MTT 法衡量 H$_2$O$_2$ 和 F-MoS$_2$ 纳米颗粒对 HUVEC 活性的影响；(c) 和 (d) DCFH-DA 探针确定 F-MoS$_2$ 纳米颗粒对 HUVEC 中的活性氧的作用；(e) F-MoS$_2$ 纳米颗粒的类酶级联催化作用：·O$_2^-$ 经过歧化作用产生 H$_2$O$_2$，以及紧接着的 H$_2$O$_2$ 分解成 O$_2$ 和水；(f) 纯人工关节滑液和加入 F-MoS$_2$ 纳米颗粒的人工关节滑液的黏度随时间的变化

(viscoelasticity)。但是，透明质酸在患病关节中含量会明显降低[260]。因此，补充

含透明质酸的人工关节滑液常常作为非手术治疗手段用于治疗骨关节炎，重建关节的健康环境[261]。一般认为，活性氧尤其是 $\cdot O_2^-$，对于骨关节炎的恶化起到重要作用[262]，可能导致补充人工关节滑液手段的失败。

具体而言，过量产生的 $\cdot O_2^-$ 通过与 H_2O_2 反应产生强氧化性的 $\cdot OH$，降解关节中的滑液，而注入的透明质酸易受该反应的影响而解聚[263]。为了调节体内 $\cdot O_2^-$ 的平衡，生物体配备有超氧化物歧化酶，将 $\cdot O_2^-$ 转化为低毒性的 H_2O_2 和 O_2[264]。然而，骨关节炎患者的关节腔内超氧化物歧化酶的含量会显著降低，这表明在患病关节内缺乏对 $\cdot O_2^-$ 的调节作用[265]。因此，添加外源的超氧化物歧化酶和过氧化氢酶，可以减少对人工关节滑液的降解程度，保护人工关节滑液[263]，是一种有效的治疗手段。

因此，我们认为，具有类过氧化物歧化酶活性和类过氧化氢酶活性的 LAL 合成的 $F-MoS_2$ 纳米颗粒，可以作为调节过量活性氧含量和防止透明质酸解聚的抗氧化剂，用于氧化应激诱导的骨关节炎治疗。这里我们选择黄嘌呤氧化酶、黄嘌呤提供 $\cdot O_2^-$，作用于以透明质酸为主要成分的人工关节滑液，测试其黏度随时间变化的关系［图10-40（f）］。我们可以清楚地看到，纯人工关节滑液在 $\cdot O_2^-$ 作用下，黏度下降明显；加入 $F-MoS_2$ 纳米颗粒后，人工关节滑液黏度下降减缓，这说明 $F-MoS_2$ 纳米颗粒具有延缓人工关节滑液降解的作用，主要原因就是 $F-MoS_2$ 纳米颗粒作为纳米酶能够部分催化淬灭人工关节滑液中的 $\cdot O_2^-$ 和 H_2O_2。上述研究结果充分表明注射具有抗氧化能力的 $F-MoS_2$ 纳米颗粒是治疗骨关节炎的有效手段之一。

综上所述，LAL 纳米制备技术在无机纳米材料合成中有着独特的优势：①LAL 过程易于形成富含缺陷的表面，为表面酶催化反应提供了丰富、高活性反应位点；②无配体 LAL 合成可以得到用于生物医学上"干净"的材料，保证了产物的无毒或低毒，并具有好的生物相容性；③LAL 合成的无机纳米材料通常具有清洁表面，这样易于进行生物学修饰。

参 考 文 献

[1] Schwenke A, Dalüge H, Kiyan R, et al. Non-agglomerated gold-PMMA nanocomposites by in situ-stabilized laser ablation in liquid monomer for optical applications[J]. Applied Physics A, 2013, 111: 451-457.

[2] Nachev P, van'T Zand D D, Coger V, et al. Synthesis of hybrid microgels by coupling of laser ablation and polymerization in aqueous medium[J]. Journal of Laser Applications, 2012, 24 (4): 042012.

[3] Mastrotto F, Caliceti P, Amendola V, et al. Polymer control of ligand display on gold nanoparticles for multimodal switchable cell targeting[J]. Chemical Communications, 2011, 47 (35): 9846-9848.

[4] Makridis S S, Gkanas E I, Panagakos G, et al. Polymer-stable magnesium nanocomposites prepared by laser ablation for efficient hydrogen storage[J]. International Journal of Hydrogen Energy, 2013, 38 (26): 11530-11535.

[5] Kalyva M, Kumar S, Brescia R, et al. Electrical response from nanocomposite PDMS-Ag NPs generated by in situ laser ablation in solution[J]. Nanotechnology, 2012, 24 (3): 035707.

[6] Hahn A, Brandes G, Wagener P, et al. Metal ion release kinetics from nanoparticle silicone composites[J]. Journal of Controlled Release, 2011, 154(2): 164-170.

[7] Wagener P, Brandes G, Schwenke A, et al. Impact of in situ polymer coating on particle dispersion into solid laser-generated nanocomposites[J]. Physical Chemistry Chemical Physics, 2011, 13(11): 5120-5126.

[8] Zhang D S, Barcikowski S. Rapid nanoparticle-polymer composites prototyping by laser ablation in liquids[M]//Kobayashi S, Müllen K. Encyclopedia of polymeric nanomaterials. Heidelberg: Springer, 2015: 2131-3141.

[9] Zhang D S, Gökce B. Perspective of laser-prototyping nanoparticle-polymer composites[J]. Applied Surface Science, 2017, 392: 991-1003.

[10] Erwin S C, Zu L J, Haftel M I, et al. Doping semiconductor nanocrystals[J]. Nature, 2005, 436: 91-94.

[11] Jain P K, Manthiram K, Engel J H, et al. Doped nanocrystals as plasmonic probes of redox chemistry[J]. Angewandte Chemie International Edition, 2013, 52(51): 13671-13675.

[12] Norris D J, Efros A L, Erwin S C. Doped nanocrystals[J]. Science, 2008, 319(5871): 1776-1779.

[13] Dalpian G M, Chelikowsky J R. Self-purification in semiconductor nanocrystals[J]. Physical Review Letters, 2006, 96(22): 226802.

[14] Liu J, Liang C H, Zhang H M, et al. Silicon-doped hematite nanosheets with superlattice structure[J]. Chemical Communications, 2011, 47(28): 8040-8042.

[15] Liu J, Liang C H, Zhang H M, et al. General strategy for doping impurities (Ge, Si, Mn, Sn, Ti) in hematite nanocrystals[J]. The Journal of Physical Chemistry C, 2012, 116(8): 4986-4992.

[16] Liu J, Cai Y Y, Tian Z F, et al. Highly oriented Ge-doped hematite nanosheet arrays for photoelectrochemical water oxidation[J]. Nano Energy, 2014, 9: 282-290.

[17] Liu J, Liang C H, Xu G P, et al. Ge-doped hematite nanosheets with tunable doping level, structure and improved photoelectrochemical performance[J]. Nano Energy, 2013, 2(3): 328-336.

[18] Ruan G S, Wu S L, Wang P P, et al. Simultaneous doping and growth of Sn-doped hematite nanocrystalline films with improved photoelectrochemical performance[J]. RSC Advances, 2014, 4(108): 63408-63413.

[19] Zhang H M, Liang C H, Liu J, et al. Defect-mediated formation of Ag cluster-doped TiO_2 nanoparticles for efficient photodegradation of pentachlorophenol[J]. Langmuir, 2012, 28(8): 3938-3944.

[20] Wang H Q, Kawaguchi K, Pyatenko A, et al. General bottom-up construction of spherical particles by pulsed laser irradiation of colloidal nanoparticles: A case study on CuO[J]. Chemistry: A European Journal, 2012, 18(1): 163-169.

[21] Li X Y, Pyatenko A, Shimizu Y, et al. Fabrication of crystalline silicon spheres by selective laser heating in liquid medium[J]. Langmuir, 2011, 27(8): 5076-5080.

[22] Zhang D S, Lau M, Lu S W, et al. Germanium sub-microspheres synthesized by picosecond pulsed laser melting in liquids: Educt size effects[J]. Scientific Reports, 2017, 7: 40355.

[23] Fujiwara H, Niyuki R, Ishikawa Y, et al. Low-threshold and quasi-single-mode random laser within a submicrometer-sized ZnO spherical particle film[J]. Applied Physics Letters, 2013, 102(6): 061110.

[24] Kawasoe K, Ishikawa Y, Koshizaki N, et al. Preparation of spherical particles by laser melting in liquid using TiN as a raw material[J]. Applied Physics B, 2015, 119: 475-483.

[25] Li X Y, Shimizu Y, Pyatenko A, et al. Tetragonal zirconia spheres fabricated by carbon-assisted selective laser heating in a liquid medium[J]. Nanotechnology, 2012, 23(11): 115602.

[26] Li X Y, Shimizu Y, Pyatenko A, et al. Carbon-assisted fabrication of submicrometre spheres for

low optical-absorbance materials by selective laser heating in liquid[J]. Journal of Materials Chemistry, 2011, 21 (38): 14406-14409.

[27] Song X Y, Qiu Z W, Yang X P, et al. Submicron-lubricant based on crystallized Fe_3O_4 spheres for enhanced tribology performance[J]. Chemistry of Materials, 2014, 26 (17): 5113-5119.

[28] Swiatkowska-Warkocka Z, Koga K, Kawaguchi K, et al. Pulsed laser irradiation of colloidal nanoparticles: A new synthesis route for the production of non-equilibrium bimetallic alloy submicrometer spheres[J]. RSC Advances, 2013, 3 (1): 79-83.

[29] Tsuji T, Higashi Y, Tsuji M, et al. Preparation of submicron-sized spherical particles of gold using laser-induced melting in liquids and low-toxic stabilizing reagent[J]. Applied Surface Science, 2015, 348: 10-15.

[30] Tsuji T, Yahata T, Yasutomo M, et al. Preparation and investigation of the formation mechanism of submicron-sized spherical particles of gold using laser ablation and laser irradiation in liquids[J]. Physical Chemistry Chemical Physics, 2013, 15 (9): 3099-3107.

[31] Wang H Q, Jia L C, Li L, et al. Photomediated assembly of single crystalline silver spherical particles with enhanced electrochemical performance[J]. Journal of Materials Chemistry A, 2013, 1 (3): 692-698.

[32] Ishikawa Y, Feng Q, Koshizaki N. Growth fusion of submicron spherical boron carbide particles by repetitive pulsed laser irradiation in liquid media[J]. Applied Physics A, 2010, 99: 797-803.

[33] Pyatenko A, Wang H Q, Koshizaki N, et al. Mechanism of pulse laser interaction with colloidal nanoparticles[J]. Laser & Photonics Reviews, 2013, 7 (4): 596-604.

[34] Yang S K, Kiraly B, Wang W Y, et al. Fabrication and characterization of beaded SiC quantum rings with anomalous red spectral shift[J]. Advanced Materials, 2012, 24 (41): 5598-5603.

[35] Liu P, Wang C X, Chen J, et al. Localized nanodiamond crystallization and field emission performance improvement of amorphous carbon upon laser irradiation in liquid[J]. The Journal of Physical Chemistry C, 2009, 113 (28): 12154-12161.

[36] Wang F, Liu X G. Multicolor tuning of lanthanide-doped nanoparticles by single wavelength excitation[J]. Accounts of Chemical Research, 2014, 47 (4): 1378-1385.

[37] Liu Y S, Tu D T, Zhu H M, et al. Lanthanide-doped luminescent nanoprobes: Controlled synthesis, optical spectroscopy, and bioapplications[J]. Chemical Society Reviews, 2013, 42 (16): 6924-6958.

[38] Tu D T, Zheng W, Liu Y S, et al. Luminescent biodetection based on lanthanide-doped inorganic nanoprobes[J]. Coordination Chemistry Reviews, 2014, 273-274: 13-29.

[39] Zhou L J, Gu Z J, Liu X X, et al. Size-tunable synthesis of lanthanide-doped Gd_2O_3 nanoparticles and their applications for optical and magnetic resonance imaging[J]. Journal of Materials Chemistry, 2012, 22(3): 966-974.

[40] Petoral Jr R M, Soderlind F, Klasson A, et al. Synthesis and characterization of Tb^{3+}-doped Gd_2O_3 nanocrystals: A bifunctional material with combined fluorescent labeling and MRI contrast agent properties[J]. The Journal of Physical Chemistry C, 2009, 113 (17): 6913-6920.

[41] Xing H Y, Bu W B, Zhang S J, et al. Multifunctional nanoprobes for upconversion fluorescence, MR and CT trimodal imaging[J]. Biomaterials, 2012, 33 (4): 1079-1089.

[42] Luo N Q, Yang C, Tian X M, et al. A general top-down approach to synthesize rare earth doped-Gd_2O_3 nanocrystals as dualmodal contrast agents[J]. Journal of Materials Chemistry B, 2014, 2 (35): 5891-5897.

[43] Smith A M, Mancini M C, Nie S M. Second window for *in vivo* imaging[J]. Nature Nanotechnology, 2009, 4: 710-711.

[44] Wang F, Deng R R, Wang J, et al. Tuning upconversion through energy migration in core-shell nanoparticles[J].

Nature Materials, 2011, 10: 968-973.

[45] Weissleder R. A clearer vision for *in vivo* imaging[J]. Nature Biotechnology, 2001, 19: 316-317.

[46] Zeng H B, Li Z G, Cai W P, et al. Microstructure control of Zn/ZnO core/shell nanoparticles and their temperature-dependent blue emissions[J]. The Journal of Physical Chemistry B, 2007, 111 (51): 14311-14317.

[47] Zeng H B, Liu P S, Cai W P, et al. Aging-induced self-assembly of Zn/ZnO treelike nanostructures from nanoparticles and enhanced visible emission[J]. Crystal Growth & Design, 2007, 7 (6): 1092-1097.

[48] Zeng H B, Yang S K, Xu X X, et al. Dramatic excitation dependence of strong and stable blue luminescence of ZnO hollow nanoparticles[J]. Applied Physics Letters, 2009, 95 (19): 191904.

[49] Zeng H B, Yang S K, Cai W P. Reshaping formation and luminescence evolution of ZnO quantum dots by laser-induced fragmentation in liquid[J]. The Journal of Physical Chemistry C, 2011, 115 (12): 5038-5043.

[50] Zeng H B, Li Z G, Cai W P, et al. Strong localization effect in temperature dependence of violet-blue emission from ZnO nanoshells[J]. Journal of Applied Physics, 2007, 102 (10): 104307.

[51] Zeng H B, Cai W P, Hu J L, et al. Violet photoluminescence from shell layer of Zn/ZnO core-shell nanoparticles induced by laser ablation[J]. Applied Physics Letters, 2006, 88 (17): 171910.

[52] Zeng H B, Duan G T, Li Y, et al. Blue luminescence of ZnO nanoparticles based on non-equilibrium processes: Defect origins and emission controls[J]. Advanced Functional Materials, 2010, 20 (4): 561-572.

[53] Yang S K, Cai W P, Zeng H B, et al. Polycrystalline Si nanoparticles and their strong aging enhancement of blue photoluminescence[J]. Journal of Applied Physics, 2008, 104 (2): 023516.

[54] Wolkin M V, Jorne J, Fauchet P M, et al. Electronic states and luminescence in porous silicon quantum dots: The role of oxygen[J]. Physical Review Letters, 1999, 82 (1): 197.

[55] Lee M K, Peng K R, Blue emission of porous silicon[J]. Applied Physics Letters, 1993, 62 (24): 3159-3160.

[56] Švrček V, Sasaki T, Shimizu Y, et al. Blue luminescent silicon nanocrystals prepared by ns pulsed laser ablation in water[J]. Applied Physics Letters, 2006, 89 (21): 213113.

[57] Zou J, Baldwin R K, Pettigrew K A, et al. Solution synthesis of ultrastable luminescent siloxane-coated silicon nanoparticles[J]. Nano Letters, 2004, 4 (7): 1181-1186.

[58] Ding Z F, Quinn B M, Haram S K, et al. Electrochemistry and electrogenerated chemiluminescence from silicon nanocrystal quantum dots[J]. Science, 2002, 296 (5571): 1293-1297.

[59] Sato K, Hirakuri K. Influence of paramagnetic defects on multicolored luminescence from nanocrystalline silicon[J]. Journal of Applied Physics, 2006, 100 (11): 114303.

[60] Yang S K, Li W Z, Cao B Q, et al. Origin of blue emission from silicon nanoparticles: Direct transition and interface recombination[J]. The Journal of Physical Chemistry C, 2011, 115 (43): 21056-21062.

[61] Liu J, Deng H W, Huang Z Y, et al. Phonon-assisted energy back transfer-induced multicolor upconversion emission of Gd_2O_3 : Yb^{3+}/Er^{3+} nanoparticles under near-infrared excitation[J]. Physical Chemistry Chemical Physics, 2015, 17 (23): 15412-15418.

[62] Li X Y, Wang H Q, Shimizu Y, et al. Preparation of carbon quantum dots with tunable photoluminescence by rapid laser passivation in ordinary organic solvents[J]. Chemical Communications, 2010, 47 (3): 932-934.

[63] Tan D Z, Yamada Y, Zhou S F, et al. Photoinduced luminescent carbon nanostructures with ultra-broadly tailored size ranges[J]. Nanoscale, 2013, 5 (24): 12092-12097.

[64] Xiao J, Liu P, Li L H, et al. Fluorescence origin of nanodiamonds[J]. The Journal of Physical Chemistry C, 2015, 119 (4): 2239-2248.

[65] Hu S L, Niu K Y, Sun J, et al. One-step synthesis of fluorescent carbon nanoparticles by laser irradiation[J].

Journal of Materials Chemistry, 2009, 19 (4): 484-488.

[66] Tan D Z, Zhou S F, Xu B B, et al. Simple synthesis of ultra-small nanodiamonds with tunable size and photoluminescence[J]. Carbon, 2013, 62: 374-381.

[67] Shen J H, Zhu Y H, Yang X L, et al. Graphene quantum dots: Emergent nanolights for bioimaging, sensors, catalysis and photovoltaic devices[J]. Chemical Communications, 2012, 48 (31): 3686-3699.

[68] Wang L, Zhu S J, Wang H Y, et al. Common origin of green luminescence in carbon nanodots and graphene quantum dots[J]. ACS nano, 2014, 8 (3): 2541-2547.

[69] Liu F, Jang M H, Ha H D, et al. Facile synthetic method for pristine graphene quantum dots and graphene oxide quantum dots: Origin of blue and green luminescence[J]. Advanced Materials (Deerfield Beach, Fla.), 2013, 25 (27): 3657-3662.

[70] Mueller M L, Yan X, McGuire J A, et al. Triplet states and electronic relaxation in photoexcited graphene quantum dots[J]. Nano Letters, 2010, 10 (7): 2679-2682.

[71] Yan J H, Lin Z Y, Liu P, et al. A design of Si-based nanoplasmonic structure as an antenna and reception amplifier for visible light communication[J]. Journal of Applied Physics, 2014, 116 (15): 154307.

[72] Fu Y H, Kuznetsov A I, Miroshnichenko A E, et al. Directional visible light scattering by silicon nanoparticles[J]. Nature Communications, 2013, 4: 1527.

[73] Staude I, Miroshnichenko A E, Decker M, et al. Tailoring directional scattering through magnetic and electric resonances in subwavelength silicon nanodisks[J]. ACS Nano, 2013, 7 (9): 7824-7832.

[74] Evlyukhin A B, Novikov S M, Zywietz U, et al. Demonstration of magnetic dipole resonances of dielectric nanospheres in the visible region[J]. Nano Letters, 2012, 12 (7): 3749-3755.

[75] García-Etxarri A, Gómez-Medina R, Froufe-Pérez L S, et al. Strong magnetic response of submicron silicon particles in the infrared[J]. Optics Express, 2011, 19 (6): 4815-4826.

[76] Shi L, Harris J T, Fenollosa R, et al. Monodisperse silicon nanocavities and photonic crystals with magnetic response in the optical region[J]. Nature Communications, 2013, 4: 1904.

[77] Yan J H, Liu P, Lin Z Y, et al. Directional Fano resonance in a silicon nanosphere dimer[J]. ACS Nano, 2015, 9 (3): 2968-2980.

[78] Yan J H, Liu P, Lin Z Y, et al. Magnetically induced forward scattering at visible wavelengths in silicon nanosphere oligomers[J]. Nature Communications, 2015, 6: 7042.

[79] Liu P, Chen H J, Wang H, et al. Fabrication of Si/Au core/shell nanoplasmonic structures with ultrasensitive surface-enhanced Raman scattering for monolayer molecule detection[J]. The Journal of Physical Chemistry C, 2015, 119 (2): 1234-1246.

[80] Huang X Q, Lai Y, Hang Z H, et al. Dirac cones induced by accidental degeneracy in photonic crystals and zero-refractive-index materials[J]. Nature Materials, 2011, 10: 582-586.

[81] Maas R, Parsons J, Engheta N, et al. Experimental realization of an epsilon-near-zero metamaterial at visible wavelengths[J]. Nature Photonics, 2013, 7: 907-912.

[82] Moitra P, Yang Y M, Anderson Z, et al. Realization of an all-dielectric zero-index optical metamaterial[J]. Nature Photonics, 2013, 7: 791-795.

[83] Liu P, Yan J H, Ma C R, et al. Midrefractive dielectric modulator for broadband unidirectional scattering and effective radiative tailoring in the visible region[J]. ACS Applied Materials & Interfaces, 2016, 8 (34): 22468-22476.

[84] Yan J H, Liu P, Lin Z Y, et al. New type high-index dielectric nanosensors based on the scattering intensity shift[J].

Nanoscale, 2016, 8 (11): 5996-6007.

[85] Yan J H, Lin Z Y, Ma C R, et al. Plasmon resonances in semiconductor materials for detecting photocatalysis at the single-particle level[J]. Nanoscale, 2016, 8 (32): 15001-15007.

[86] Yan J H, Liu P, Ma C R, et al. Plasmonic near-touching titanium oxide nanoparticles to realize solar energy harvesting and effective local heating[J]. Nanoscale, 2016, 8 (16): 8826-8838.

[87] Guillet Y, Rashidi-Huyeh M, Palpant B. Influence of laser pulse characteristics on the hot electron contribution to the third-order nonlinear optical response of gold nanoparticles[J]. Physical Review B, 2009, 79 (4): 045410.

[88] Ho-Wu R, Yau S H, Goodson T. Linear and nonlinear optical properties of monolayer-protected gold nanocluster films[J]. ACS Nano, 2016, 10 (1): 562-572.

[89] Philip R, Chantharasupawong P, Qian H F, et al. Evolution of nonlinear optical properties: From gold atomic clusters to plasmonic nanocrystals[J]. Nano Letters, 2012, 12 (9): 4661-4667.

[90] Marinica D C, Kazansky A K, Nordlander P, et al. Quantum plasmonics: Nonlinear effects in the field enhancement of a plasmonic nanoparticle dimer[J]. Nano Letters, 2012, 12 (3): 1333-1339.

[91] Harutyunyan H, Volpe G, Quidant R, et al. Enhancing the nonlinear optical response using multifrequency gold-nanowire antennas[J]. Physical Review Letters, 2012, 108 (21): 217403.

[92] Krishna Podagatlapalli G, Hamad S, Tewari S P, et al. Silver nano-entities through ultrafast double ablation in aqueous media for surface enhanced Raman scattering and photonics applications[J]. Journal of Applied Physics, 2013, 113 (7): 073106.

[93] Hamad S, Podagatlapalli G K, Tewari S P, et al. Influence of picosecond multiple/single line ablation on copper nanoparticles fabricated for surface enhanced Raman spectroscopy and photonics applications[J]. Journal of Physics D: Applied Physics, 2013, 46 (48): 485501.

[94] Hamad S, Podagatlapalli G K, Sreedhar S, et al. Femtosecond and picosecond ablation of aluminum for synthesis of nanoparticles and nanostructures and their optical characterization[C]//Synthesis and Photonics of Nanoscale Materials IX. SPIE, 2012, 8245: 82450L.

[95] Podagatlapalli G K, Hamad S, Sreedhar S, et al. Fabrication and characterization of aluminum nanostructures and nanoparticles obtained using femtosecond ablation technique[J]. Chemical Physics Letters, 2012, 530: 93-97.

[96] Tan D Z, Yamada Y, Zhou S F, et al. Carbon nanodots with strong nonlinear optical response[J]. Carbon, 2014, 69: 638-640.

[97] Porel S, Venkatram N, Rao D N, et al. Optical power limiting in the femtosecond regime by silver nanoparticle: Embedded polymer film[J]. Journal of Applied Physics, 2007, 102 (3): 033107.

[98] Polavarapu L, Venkatram N, Ji W, et al. Optical-limiting properties of oleylamine-capped gold nanoparticles for both femtosecond and nanosecond laser pulses[J]. ACS Applied Materials & Interfaces, 2009, 1 (10): 2298-2303.

[99] Ma C R, Yan J H, Wei Y M, et al. Second harmonic generation from an individual amorphous selenium nanosphere[J]. Nanotechnology, 2016, 27 (42): 425206.

[100] Boyd R W. Nonlinear optics[M]. 2nd ed. New York: Elsevier, 2003.

[101] Lettieri S, Finizio S D, Maddalena P, et al. Second-harmonic generation in amorphous silicon nitride microcavities[J]. Applied Physics Letters, 2002, 81 (25): 4706-4708.

[102] Guo A K, Fu Y, Wang G, et al. Diameter effect of gold nanoparticles on photothermal conversion for solar steam generation[J]. RSC Advances, 2017, 7: 4815-4824.

[103] Zhou L, Tan Y L, Wang J Y, et al. 3D self-assembly of aluminium nanoparticles for plasmon-enhanced solar desalination[J]. Nature Photonics, 2016, 10: 393-398.

[104] Brongersma M L, Cui Y, Fan S H. Light management for photovoltaics using high-index nanostructures[J]. Nature Materials, 2014, 13: 451-460.

[105] Zograf G P, Petrov M I, Zuev D A, et al. Resonant nonplasmonic nanoparticles for efficient temperature-feedback optical heating[J]. Nano Letters, 2017, 17 (5): 2945-2952.

[106] Ma C R, Yan J H, Huang Y C, et al. The optical duality of tellurium nanoparticles for broadband solar energy harvesting and efficient photothermal conversion[J]. Science Advances, 2018, 4 (8): eaas9894.

[107] Brongersma M L, Halas N J, Nordlander P. Plasmon-induced hot carrier science and technology[J]. Nature Nanotechnology, 2015, 10: 25-34.

[108] Ghosh S, Avellini T, Petrelli A, et al. Colloidal $CuFeS_2$ nanocrystals: Intermediate Fe d-band leads to high photothermal conversion efficiency[J]. Chemistry of Materials, 2016, 28 (13): 4848-4858.

[109] Gaspari R, Della Valle G, Ghosh S, et al. Quasi-static resonances in the visible spectrum from all-dielectric intermediate band semiconductor nanocrystals[J]. Nano Letters, 2017, 17 (12): 7691-7695.

[110] Othonos A. Probing ultrafast carrier and phonon dynamics in semiconductors[J]. Journal of Applied Physics, 1998, 83 (4): 1789-1830.

[111] Xu Y H, Lv Z Z, Xia Y, et al. Highly porous magnetite/graphene nanocomposites for a solid-state electrochemiluminescence sensor on paper-based chips[J]. Analytical and Bioanalytical Chemistry, 2013, 405: 3549-3558.

[112] Reiss G, Hütten A. Applications beyond data storage[J]. Nature Materials, 2005, 4: 725-726.

[113] Holm C, Weis J J. The structure of ferrofluids: A status report[J]. Current Opinion in Colloid & Interface Science, 2005, 10 (3-4): 133-140.

[114] Amendola V, Riello P, Meneghetti M. Magnetic nanoparticles of iron carbide, iron oxide, iron@iron oxide, and metal iron synthesized by laser ablation in organic solvents[J]. The Journal of Physical Chemistry C, 2011, 115 (12): 5140-5146.

[115] Zhang H M, Liang C H, Liu J, et al. The formation of onion-like carbon-encapsulated cobalt carbide core/shell nanoparticles by the laser ablation of metallic cobalt in acetone[J]. Carbon, 2013, 55: 108-115.

[116] Liang Y, Liu P, Yang G W. Fabrication of one-dimensional chain of iron-based bimetallic alloying nanoparticles with unique magnetizations[J]. Crystal Growth & Design, 2014, 14 (11): 5847-5855.

[117] Liang Y, Liu P, Xiao J, et al. A general strategy for one-step fabrication of one-dimensional magnetic nanoparticle chains based on laser ablation in liquid[J]. Laser Physics Letters, 2014, 11 (5): 056001.

[118] Liang Y, Liu P, Xiao J, et al. A microfibre assembly of an iron-carbon composite with giant magnetisation[J]. Scientific Reports, 2013, 3: 3051.

[119] Franzel L, Bertino M F, Huba Z J, et al. Synthesis of magnetic nanoparticles by pulsed laser ablation[J]. Applied Surface Science, 2012, 261: 332-336.

[120] Swiatkowska-Warkocka Z, Kawaguchi K, Wang H Q, et al. Controlling exchange bias in Fe_3O_4/FeO composite particles prepared by pulsed laser irradiation[J]. Nanoscale Research Letters, 2011, 6: 1-7.

[121] Amendola V, Scaramuzza S, Agnoli S, et al. Laser generation of iron-doped silver nanotruffles with magnetic and plasmonic properties[J]. Nano Research, 2015, 8: 4007-4023.

[122] Bertorelle F, Ceccarello M, Pinto M, et al. Efficient $AuFeO_x$ nanoclusters of laser-ablated nanoparticles in water for cells guiding and surface-enhanced resonance Raman scattering imaging[J]. The Journal of Physical Chemistry C, 2014, 118 (26): 14534-14541.

[123] Swiatkowska-Warkocka Z, Kawaguchi K, Shimizu Y, et al. Synthesis of Au-based porous magnetic spheres by

selective laser heating in liquid[J]. Langmuir, 2012, 28 (11): 4903-4907.

[124] Suzuki M. Role of adsorption in water environment processes[J]. Water Science and Technology, 1997, 35 (7): 1-11.

[125] Kung M C, Kung H H. IR studies of NH_3, pyridine, CO, and NO adsorbed on transition metal oxides[J]. Catalysis Reviews Science and Engineering, 1985, 27 (3): 425-460.

[126] Kiselev V F, Krylov O V. Adsorption and catalysis on oxides of transition metals[M]//Kiselev V F, Krylov O V. Adsorption and catalysis on transition metals and their oxides. Springer Series in Surface Sciences, vol 9. Heidelberg: Springer, 1989.

[127] Li L H, Xiao J, Liu P, et al. Super adsorption capability from amorphousization of metal oxide nanoparticles for dye removal[J]. Scientific Reports, 2015, 5: 9028.

[128] Wagener P, Schwenke A, Barcikowski S. How citrate ligands affect nanoparticle adsorption to microparticle supports[J]. Langmuir, 2012, 28 (14): 6132-6140.

[129] Nakata K, Fujishima A. TiO_2 photocatalysis: Design and applications[J]. Journal of Photochemistry and Photobiology C: Photochemistry Reviews, 2012, 13 (3): 169-189.

[130] Schneider J, Matsuoka M, Takeuchi M, et al. Understanding TiO_2 photocatalysis: Mechanisms and materials[J]. Chemical Reviews, 2014, 114 (19): 9919-9986.

[131] Xiao J, Wu Q L, Liu P, et al. Highly stable sub-5 nm $Sn_6O_4(OH)_4$ nanocrystals with ultrahigh activity as advanced photocatalytic materials for photodegradation of methyl orange[J]. Nanotechnology, 2014, 25 (13): 135702.

[132] Lin Z Y, Xiao J, Yan J H, et al. Ag/AgCl plasmonic cubes with ultrahigh activity as advanced visible-light photocatalysts for photodegrading dyes[J]. Journal of Materials Chemistry A, 2015, 3 (14): 7649-7658.

[133] Tian Z F, Liang C H, Liu J, et al. Zinc stannate nanocubes and nanourchins with high photocatalytic activity for methyl orange and 2, 5-DCP degradation[J]. Journal of Materials Chemistry, 2012, 22 (33): 17210-17214.

[134] Lin Z Y, Li J L, Zheng Z Q, et al. Electronic reconstruction of α-Ag_2WO_4 nanorods for visible-light photocatalysis[J]. ACS Nano, 2015, 9 (7): 7256-7265.

[135] Lin Z Y, Liu P, Yan J H, et al. Matching energy levels between TiO_2 and α-Fe_2O_3 in a core-shell nanoparticle for visible-light photocatalysis[J]. Journal of Materials Chemistry A, 2015, 3 (28): 14853-14863.

[136] Liang D W, Wu S L, Wang P P, et al. Recyclable chestnut-like Fe_3O_4@C@ $ZnSnO_3$ core-shell particles for the photocatalytic degradation of 2, 5-dichlorophenol[J]. RSC Advances, 2014, 4 (50): 26201-26206.

[137] Li Q, Liang C H, Tian Z F, et al. Core-shell Ta_xO@Ta_2O_5 structured nanoparticles: Laser ablation synthesis in liquid, structure and photocatalytic property[J]. CrystEngComm, 2012, 14 (9): 3236-3240.

[138] Zhang H M, Liang C H, Tian Z F, et al. Hydrothermal treatment of colloids induced via liquid-phase laser ablation: A new approach for hierarchical titanate nanostructures with enhanced photodegradation performance[J]. CrystEngComm, 2011, 13 (14): 4676-4682.

[139] Liu P S, Cai W P, Fang M, et al. Room temperature synthesized rutile TiO_2 nanoparticles induced by laser ablation in liquid and their photocatalytic activity[J]. Nanotechnology, 2009, 20 (28): 285707.

[140] Zeng H B, Cai W P, Liu P S, et al. ZnO-based hollow nanoparticles by selective etching: Elimination and reconstruction of metal-semiconductor interface, improvement of blue emission and photocatalysis[J]. ACS Nano, 2008, 2 (8): 1661-1670.

[141] Zeng H B, Liu P S, Cai W P, et al. Controllable Pt/ZnO porous nanocages with improved photocatalytic activity[J]. The Journal of Physical Chemistry C, 2008, 112 (49): 19620-19624.

[142] Cai Y Y, Ye Y X, Tian Z F, et al. In situ growth of lamellar $ZnTiO_3$ nanosheets on TiO_2 tubular array with

[143] Wu S L, Wang P P, Cai Y Y, et al. Reduced graphene oxide anchored magnetic $ZnFe_2O_4$ nanoparticles with enhanced visible-light photocatalytic activity[J]. RSC Advances, 2015, 5 (12): 9069-9074.

[144] Zou Z G, Ye J H, Sayama K, et al. Direct splitting of water under visible light irradiation with an oxide semiconductor photocatalyst[J]. Nature, 2001, 414: 625-627.

[145] Yang S K, Cai W P, Liu G Q, et al. Optical study of redox behavior of silicon nanoparticles induced by laser ablation in liquid[J]. The Journal of Physical Chemistry C, 2009, 113 (16): 6480-6484.

[146] Liu J, Liang C H, Tian Z F, et al. Spontaneous growth and chemical reduction ability of Ge nanoparticles[J]. Scientific Reports, 2013, 3: 1741.

[147] Zhang H, Wu S L, Liu J, et al. Laser irradiation-induced Au-ZnO nanospheres with enhanced sensitivity and stability for ethanol sensing[J]. Physical Chemistry Chemical Physics, 2016, 18 (32): 22503-22508.

[148] Cai Y Y, Ye Y X, Wu S L, et al. Simultaneous Cu doping and growth of TiO_2 nanocrystalline array film as a glucose biosensor[J]. RSC Advances, 2016, 6 (81): 78219-78224.

[149] Darwish I A, Blake D A. Development and validation of a one-step immunoassay for determination of cadmium in human serum[J]. Analytical Chemistry, 2002, 74 (1): 52-58.

[150] Liu H W, Jiang S J, Liu S H. Determination of cadmium, mercury and lead in seawater by electrothermal vaporization isotope dilution inductively coupled plasma mass spectrometry[J]. Spectrochimica Acta Part B: Atomic Spectroscopy, 1999, 54 (9): 1367-1375.

[151] Wan Z, Xu Z R, Wang J H. Flow injection on-line solid phase extraction for ultra-trace lead screening with hydride generation atomic fluorescence spectrometry[J]. Analyst, 2006, 131 (1): 141-147.

[152] Xu X X, Duan G T, Li Y, et al. Fabrication of gold nanoparticles by laser ablation in liquid and their application for simultaneous electrochemical detection of Cd^{2+}, Pb^{2+}, Cu^{2+}, Hg^{2+}[J]. ACS Applied Materials & Interfaces, 2014, 6 (1): 65-71.

[153] Pan S S, Lu W, Zhao Y H, et al. Self-doped rutile titania with high performance for direct and ultrafast assay of H_2O_2[J]. ACS Applied Materials & Interfaces, 2013, 5 (24): 12784-12788.

[154] Ye Y X, Wang P P, Dai E, et al. A novel reduction approach to fabricate quantum-sized SnO_2-conjugated reduced graphene oxide nanocomposites as non-enzymatic glucose sensors[J]. Physical Chemistry Chemical Physics, 2014, 16 (19): 8801-8807.

[155] Barsan N, Koziej D, Weimar U. Metal oxide-based gas sensor research: How to? [J]. Sensors and Actuators B: Chemical, 2007, 121 (1): 18-35.

[156] Korotcenkov G. Metal oxides for solid-state gas sensors: What determines our choice? [J]. Materials Science and Engineering: B, 2007, 139 (1): 1-23.

[157] Wang B, Zhu L F, Yang Y H, et al. Fabrication of a SnO_2 nanowire gas sensor and sensor performance for hydrogen[J]. The Journal of Physical Chemistry C, 2008, 112 (17): 6643-6647.

[158] Liu X H, Zhang J, Guo X Z, et al. Enhanced sensor response of Ni-doped SnO_2 hollow spheres[J]. Sensors and Actuators B: Chemical, 2011, 152 (2): 162-167.

[159] Li Z P, Zhao Q Q, Fan W L, et al. Porous SnO_2 nanospheres as sensitive gas sensors for volatile organic compounds detection[J]. Nanoscale, 2011, 3 (4): 1646-1652.

[160] Wang C X, Yin L W, Zhang L Y, et al. Metal oxide gas sensors: Sensitivity and influencing factors[J]. Sensors, 2010, 10 (3): 2088-2106.

[161] Xiao J, Liu P, Liang Y, et al. Porous tungsten oxide nanoflakes for highly alcohol sensitive performance[J].

Nanoscale, 2012, 4 (22): 7078-7083.

[162] Xiao J, Liu P, Liang Y, et al. High aspect ratio β-MnO_2 nanowires and sensor performance for explosive gases[J]. Journal of Applied Physics, 2013, 114 (7): 073513.

[163] Liang Y, Zhu L F, Liu P, et al. $Ag_2V_4O_{11}$ nanostructures for highly ethanol sensitive performance[J]. CrystEngComm, 2013, 15 (31): 6131-6135.

[164] Niu K Y, Yang J, Kulinich S A, et al. Hollow nanoparticles of metal oxides and sulfides: Fast preparation via laser ablation in liquid[J]. Langmuir, 2010, 26 (22): 16652-16657.

[165] Li L H, Xiao J, Yang G W. Amorphization of cobalt monoxide nanocrystals and related explosive gas sensing applications[J]. Nanotechnology, 2015, 26 (41): 415501.

[166] Fazio E, Trusso S, Ponterio R C. Surface-enhanced Raman scattering study of organic pigments using silver and gold nanoparticles prepared by pulsed laser ablation[J]. Applied Surface Science, 2013, 272: 36-41.

[167] Nguyen T B, Thu Vu T K, Nguyen Q D, et al. Preparation of metal nanoparticles for surface enhanced Raman scattering by laser ablation method[J]. Advances in Natural Sciences: Nanoscience and Nanotechnology, 2012, 3 (2): 025016.

[168] Cui H, Liu P, Yang G W. Noble metal nanoparticle patterning deposition using pulsed-laser deposition in liquid for surface-enhanced Raman scattering[J]. Applied Physics Letters, 2006, 89 (15): 153124.

[169] Yang S K, Cai W P, Kong L C, et al. Surface nanometer-scale patterning in realizing large-scale ordered arrays of metallic nanoshells with well-defined structures and controllable properties[J]. Advanced Functional Materials, 2010, 20 (15): 2527-2533.

[170] Zhang L L, Zhao X S. Carbon-based materials as supercapacitor electrodes[J]. Chemical Society Reviews, 2009, 38 (9): 2520-2531.

[171] Wang G P, Zhang L, Zhang J J. A review of electrode materials for electrochemical supercapacitors[J]. Chemical Society Reviews, 2012, 41 (2): 797-828.

[172] Zhang Y, Feng H, Wu X B, et al. Progress of electrochemical capacitor electrode materials: A review[J]. International Journal of Hydrogen Energy, 2009, 34 (11): 4889-4899.

[173] Liang D W, Wu S L, Liu J, et al. Co-doped Ni hydroxide and oxide nanosheet networks: Laser-assisted synthesis, effective doping, and ultrahigh pseudocapacitor performance[J]. Journal of Materials Chemistry A, 2016, 4 (27): 10609-10617.

[174] Zhang H M, Liu J, Tian Z F, et al. A general strategy toward transition metal carbide/carbon core/shell nanospheres and their application for supercapacitor electrode[J]. Carbon, 2016, 100: 590-599.

[175] Zhang H M, Liu J, Ye Y X, et al. Synthesis of Mn-doped α-$Ni(OH)_2$ nanosheets assisted by liquid-phase laser ablation and their electrochemical properties[J]. Physical Chemistry Chemical Physics, 2013, 15 (15): 5684-5690.

[176] Liang D W, Tian Z F, Liu J, et al. MoS_2 nanosheets decorated with ultrafine Co_3O_4 nanoparticles for high-performance electrochemical capacitors[J]. Electrochimica Acta, 2015, 182: 376-382.

[177] Wu H, Cui Y. Designing nanostructured Si anodes for high energy lithium ion batteries[J]. Nano Today, 2012, 7 (5): 414-429.

[178] Goodenough J B, Kim Y. Challenges for rechargeable Li batteries[J]. Chemistry of Materials, 2010, 22 (3): 587-603.

[179] Ye Y X, Wang P P, Sun H M, et al. Structural and electrochemical evaluation of a TiO_2-graphene oxide based sandwich structure for lithium-ion battery anodes[J]. RSC Advances, 2015, 5 (56): 45038-45043.

[180] Nowak A P, Lisowska-Oleksiak A, Siuzdak K, et al. Tin oxide nanoparticles from laser ablation encapsulated in a

carbonaceous matrix: A negative electrode in lithium-ion battery applications[J]. RSC Advances, 2015, 5 (102): 84321-84327.

[181] Chan C K, Peng H L, Liu G, et al. High-performance lithium battery anodes using silicon nanowires[J]. Nature Nanotechnology, 2008, 3: 31-35.

[182] Magasinski A, Dixon P, Hertzberg B, et al. High-performance lithium-ion anodes using a hierarchical bottom-up approach[J]. Nature Materials, 2010, 9: 353-358.

[183] Horoz S, Lu L Y, Dai Q L, et al. CdSe quantum dots synthesized by laser ablation in water and their photovoltaic applications[J]. Applied Physics Letters, 2012, 101 (22): 223902.

[184] Semaltianos N G, Maximova K A, Aristov A I, et al. Nanocomposites composed of P_3HT : PCBM and nanoparticles synthesized by laser ablation of a bulk PbS target in liquid[J]. Colloid and Polymer Science, 2014, 292: 3347-3354.

[185] Spyropoulos G D, Stylianakis M M, Stratakis E, et al. Organic bulk heterojunction photovoltaic devices with surfactant-free Au nanoparticles embedded in the active layer[J]. Applied Physics Letters, 2012, 100(21): 213904.

[186] Paci B, Generosi A, Albertini V R, et al. Enhancement of photo/thermal stability of organic bulk heterojunction photovoltaic devices via gold nanoparticles doping of the active layer[J]. Nanoscale, 2012, 4 (23): 7452-7459.

[187] Paci B, Spyropoulos G D, Generosi A, et al. Enhanced structural stability and performance durability of bulk heterojunction photovoltaic devices incorporating metallic nanoparticles[J]. Advanced Functional Materials, 2011, 21 (18): 3573-3582.

[188] Sygletou M, Kakavelakis G, Paci B, et al. Enhanced stability of aluminum nanoparticle-doped organic solar cells[J]. ACS Applied Materials & Interfaces, 2015, 7 (32): 17756-17764.

[189] Kakavelakis G, Stratakis E, Kymakis E. Aluminum nanoparticles for efficient and stable organic photovoltaics[J]. RSC Advances, 2013, 3 (37): 16288-16291.

[190] Švrček V, Mariotti D, Nagai T, et al. Photovoltaic applications of silicon nanocrystal based nanostructures induced by nanosecond laser fragmentation in liquid media[J]. The Journal of Physical Chemistry C, 2011, 115 (12): 5084-5093.

[191] Wang H Q, Miyauchi M, Ishikawa Y, et al. Single-crystalline rutile TiO_2 hollow spheres: Room-temperature synthesis, tailored visible-light-extinction, and effective scattering layer for quantum dot-sensitized solar cells[J]. Journal of the American Chemical Society, 2011, 133 (47): 19102-19109.

[192] Guo W, Liu B. Liquid-phase pulsed laser ablation and electrophoretic deposition for chalcopyrite thin-film solar cell application[J]. ACS Applied Materials & Interfaces, 2012, 4 (12): 7036-7042.

[193] Mashford B S, Stevenson M, Popovic Z, et al. High-efficiency quantum-dot light-emitting devices with enhanced charge injection[J]. Nature Photonics, 2013, 7: 407-412.

[194] Anikeeva P O, Halpert J E, Bawendi M G, et al. Quantum dot light-emitting devices with electroluminescence tunable over the entire visible spectrum[J]. Nano Letters, 2009, 9 (7): 2532-2536.

[195] Dohnalová K, Gregorkiewicz T, Kůsová K. Silicon quantum dots: Surface matters[J]. Journal of Physics: Condensed Matter, 2014, 26 (17): 173201.

[196] Xin Y Z, Nishio K, Saitow K. White-blue electroluminescence from a Si quantum dot hybrid light-emitting diode[J]. Applied Physics Letters, 2015, 106 (20): 201102.

[197] Tu C C, Tang L, Huang J D, et al. Visible electroluminescence from hybrid colloidal silicon quantum dot-organic light-emitting diodes[J]. Applied Physics Letters, 2011, 98 (21): 213102.

[198] Hunter B M, Gray H B, Müller A M. Earth-abundant heterogeneous water oxidation catalysts[J]. Chemical

Reviews, 2016, 116 (22): 14120-14136.

[199] Liao L, Zhang Q H, Su Z H, et al. Efficient solar water-splitting using a nanocrystalline CoO photocatalyst[J]. Nature Nanotechnology, 2014, 9: 69-73.

[200] Lin Z Y, Xiao J, Li L H, et al. Nanodiamond-embedded p-type copper (I) oxide nanocrystals for broad-spectrum photocatalytic hydrogen evolution[J]. Advanced Energy Materials, 2015, 6 (4): 1501865.

[201] Hunter B M, Blakemore J D, Deimund M, et al. Highly active mixed-metal nanosheet water oxidation catalysts made by pulsed-laser ablation in liquids[J]. Journal of the American Chemical Society, 2014, 136 (38): 13118-13121.

[202] Wu S L, Liu J, Tian Z F, et al. Highly dispersed ultrafine Pt nanoparticles on reduced graphene oxide nanosheets: In situ sacrificial template synthesis and superior electrocatalytic performance for methanol oxidation[J]. ACS Applied Materials & Interfaces, 2015, 7 (41): 22935-22940.

[203] Oko D N, Zhang J M, Garbarino S, et al. Formic acid electro-oxidation at PtAu alloyed nanoparticles synthesized by pulsed laser ablation in liquids[J]. Journal of Power Sources, 2014, 248: 273-282.

[204] Hebie S, Holade Y, Maximova K, et al. Advanced electrocatalysts on the basis of bare Au nanomaterials for biofuel cell applications[J]. ACS Catalysis, 2015, 5 (11): 6489-6496.

[205] Blakemore J D, Gray H B, Winkler J R, et al. Co_3O_4 nanoparticle water-oxidation catalysts made by pulsed-laser ablation in liquids[J]. ACS Catalysis, 2013, 3 (11): 2497-2500.

[206] Zhang J M, Chaker M, Ma D L. Pulsed laser ablation based synthesis of colloidal metal nanoparticles for catalytic applications[J]. Journal of Colloid and Interface Science, 2017, 489: 138-149.

[207] Sylvestre J P, Poulin S, Kabashin A V, et al. Surface chemistry of gold nanoparticles produced by laser ablation in aqueous media[J]. The Journal of Physical Chemistry B, 2004, 108 (43): 16864-16869.

[208] Amendola V, Meneghetti M. Controlled size manipulation of free gold nanoparticles by laser irradiation and their facile bioconjugation[J]. Journal of Materials Chemistry, 2007, 17 (44): 4705-4710.

[209] Petersen S, Jakobi J, Barcikowski S. In situ bioconjugation: Novel laser based approach to pure nanoparticle-conjugates[J]. Applied Surface Science, 2009, 255 (10): 5435-5438.

[210] Sajti C L, Petersen S, Menéndez-Manjón A, et al. In-situ bioconjugation in stationary media and in liquid flow by femtosecond laser ablation[J]. Applied Physics A, 2010, 101: 259-264.

[211] Barchanski A, Taylor U, Klein S, et al. Golden perspective: Application of laser-generated gold nanoparticle conjugates in reproductive biology[J]. Reproduction in Domestic Animals, 2011, 46 (s3): 42-52.

[212] Intartaglia R, Barchanski A, Bagga K, et al. Bioconjugated silicon quantum dots from one-step green synthesis[J]. Nanoscale, 2012, 4 (4): 1271-1274.

[213] Barchanski A, Hashimoto N, Petersen S, et al. Impact of spacer and strand length on oligonucleotide conjugation to the surface of ligand-free laser-generated gold nanoparticles[J]. Bioconjugate Chemistry, 2012, 23 (5): 908-915.

[214] Joshi D, Soni R K. Laser-induced synthesis of silver nanoparticles and their conjugation with protein[J]. Applied Physics A, 2014, 116: 635-641.

[215] Petersen S, Barcikowski S. Conjugation efficiency of laser-based bioconjugation of gold nanoparticles with nucleic acids[J]. The Journal of Physical Chemistry C, 2009, 113 (46): 19830-19835.

[216] Petersen S, Barchanski A, Taylor U, et al. Penetratin-conjugated gold nanoparticles: Design of cell-penetrating nanomarkers by femtosecond laser ablation[J]. The Journal of Physical Chemistry C, 2011, 115 (12): 5152-5159.

[217] Barchanski A, Funk D, Wittich O, et al. Picosecond laser fabrication of functional gold: Antibody nanoconjugates for biomedical applications[J]. The Journal of Physical Chemistry C, 2015, 119 (17): 9524-9533.

[218] Sereno M I, Dale A M, Reppas J B, et al. Borders of multiple visual areas in humans revealed by functional magnetic resonance imaging[J]. Science, 1995, 268 (5212): 889-893.

[219] Weissleder R, Pittet M J. Imaging in the era of molecular oncology[J]. Nature, 2008, 452: 580-589.

[220] Belliveau J W, Kennedy D N, McKinstry R C, et al. Functional mapping of the human visual cortex by magnetic resonance imaging[J]. Science, 1991, 254 (5032): 716-719.

[221] Park Y II, Kim J H, Lee K T, et al. Nonblinking and nonbleaching upconverting nanoparticles as an optical imaging nanoprobe and T_1 magnetic resonance imaging contrast agent[J]. Advanced Materials, 2009, 21 (44): 4467-4471.

[222] Ananta J S, Godin B, Sethi R, et al. Geometrical confinement of gadolinium-based contrast agents in nanoporous particles enhances T_1 contrast[J]. Nature Nanotechnology, 2010, 5: 815-821.

[223] Cheng Z L, Thorek D L J, Tsourkas A. Gadolinium-conjugated dendrimer nanoclusters as a tumor-targeted T_1 magnetic resonance imaging contrast agent[J]. Angewandte Chemie International Edition, 2010, 49 (2): 346-350.

[224] Caravan P. Strategies for increasing the sensitivity of gadolinium based MRI contrast agents[J]. Chemical Society Reviews, 2006, 35 (6): 512-523.

[225] Luo N Q, Tian X M, Yang C, et al. Ligand-free gadolinium oxide for in vivo T_1-weighted magnetic resonance imaging[J]. Physical Chemistry Chemical Physics, 2013, 15 (29): 12235-12240.

[226] Luo N Q, Tian X M, Xiao J, et al. High longitudinal relaxivity of ultra-small gadolinium oxide prepared by microsecond laser ablation in diethylene glycol[J]. Journal of Applied Physics, 2013, 113 (16): 164306.

[227] Idée J M, Port M, Dencausse A, et al. Involvement of gadolinium chelates in the mechanism of nephrogenic systemic fibrosis: An update[J]. Radiologic Clinics, 2009, 47 (5): 855-869.

[228] Langer R D, Lorke D E, Van Gorkom K F N, et al. In an animal model nephrogenic systemic fibrosis cannot be induced by intraperitoneal injection of high-dose gadolinium based contrast agents[J]. European Journal of Radiology, 2012, 81 (10): 2562-2567.

[229] Thomsen H S, Morcos S K, Almén T, et al. Nephrogenic systemic fibrosis and gadolinium-based contrast media: Updated ESUR contrast medium safety committee guidelines[J]. European Radiology, 2013, 23: 307-318.

[230] Xiao J, Tian X M, Yang C, et al. Ultrahigh relaxivity and safe probes of manganese oxide nanoparticles for in vivo imaging[J]. Scientific Reports, 2013, 3: 3424.

[231] Salmaso S, Caliceti P, Amendola V, et al. Cell up-take control of gold nanoparticles functionalized with a thermoresponsive polymer[J]. Journal of Materials Chemistry, 2009, 19 (11): 1608-1615.

[232] Hu X N, Liu J B, Hou S, et al. Research progress of nanoparticles as enzyme mimetics[J]. Science China Physics, Mechanics and Astronomy, 2011, 54: 1749.

[233] Kuah E, Toh S, Yee J, et al. Enzyme mimics: Advances and applications[J]. Chemistry: A European Journal, 2016, 22 (25): 8404-8430.

[234] Breslow R. Biomimetic chemistry and artificial enzymes: Catalysis by design[J]. Accounts of Chemical Research, 1995, 28 (3): 146-153.

[235] Fruk L, Niemeyer C M. Covalent hemin: DNA adducts for generating a novel class of artificial heme enzymes[J]. Angewandte Chemie International Edition, 2005, 44 (17): 2603-2606.

[236] Ju H X, Zhang X J, Wang J, et al. Nanostructured mimic enzymes for biocatalysis and biosensing[J]. NanoBiosensing, 2011: 85-109.

[237] Dong Z Y, Luo Q, Liu J Q. Artificial enzymes based on supramolecular scaffolds[J]. Chemical Society Reviews, 2012, 41 (23): 7890-7908.

[238] Wei H, Wang E. Nanomaterials with enzyme-like characteristics (nanozymes): Next-generation artificial enzymes[J]. Chemical Society Reviews, 2013, 42 (14): 6060-6093.

[239] Ragg R, Tahir M N, Tremel W. Solids go bio: Inorganic nanoparticles as enzyme mimics[J]. European Journal of Inorganic Chemistry, 2016, 2016 (13-14): 1906-1915.

[240] Gao L Z, Zhuang J, Nie L, et al. Intrinsic peroxidase-like activity of ferromagnetic nanoparticles[J]. Nature Nanotechnology, 2007, 2: 577-583.

[241] Asati A, Santra S, Kaittanis C, et al. Oxidase-like activity of polymer-coated cerium oxide nanoparticles[J]. Angewandte Chemie International Edition, 2009, 48 (13): 2308-2312.

[242] Jv Y, Li B X, Cao R. Positively-charged gold nanoparticles as peroxidiase mimic and their application in hydrogen peroxide and glucose detection[J]. Chemical Communications, 2010, 46 (42): 8017-8019.

[243] Song Y J, Qu K G, Zhao C, et al. Graphene oxide: Intrinsic peroxidase catalytic activity and its application to glucose detection[J]. Advanced Materials, 2010, 22 (19): 2206-2210.

[244] Chen Q, Liu M L, Zhao J N, et al. Water-dispersible silicon dots as a peroxidase mimetic for the highly-sensitive colorimetric detection of glucose[J]. Chemical Communications, 2014, 50 (51): 6771-6774.

[245] Su H, Liu D D, Zhao M, et al. Dual-enzyme characteristics of polyvinylpyrrolidone-capped iridium nanoparticles and their cellular protective effect against H_2O_2-induced oxidative damage[J]. ACS Applied Materials & Interfaces, 2015, 7 (15): 8233-8242.

[246] Sobańska K, Pietrzyk P, Sojka Z. Generation of reactive oxygen species via electroprotic interaction of H_2O_2 with ZrO_2 gel: Ionic sponge effect and pH-switchable peroxidase-and catalase-like activity[J]. ACS Catalysis, 2017, 7 (4): 2935-2947.

[247] Mu J S, Zhao X, Li J, et al. Novel hierarchical NiO nanoflowers exhibiting intrinsic superoxide dismutase-like activity[J]. Journal of Materials Chemistry B, 2016, 4 (31): 5217-5221.

[248] Zuo X L, Peng C, Huang Q, et al. Design of a carbon nanotube/magnetic nanoparticle-based peroxidase-like nanocomplex and its application for highly efficient catalytic oxidation of phenols[J]. Nano Research, 2009, 2: 617-623.

[249] Sun X L, Guo S J, Chung C S, et al. A sensitive H_2O_2 assay based on dumbbell-like PtPd-Fe_3O_4 nanoparticles[J]. Advanced Materials, 2013, 25 (1): 132-136.

[250] Chen T M, Xiao J, Yang G W. Cubic boron nitride with an intrinsic peroxidase-like activity[J]. RSC Advances, 2016, 6 (74): 70124-70132.

[251] Chen T M, Tian X M, Huang L, et al. Nanodiamonds as pH-switchable oxidation and reduction catalysts with enzyme-like activities for immunoassay and antioxidant applications[J]. Nanoscale, 2017, 9 (40): 15673-15684.

[252] Wu X J, Tian X M, Chen T M, et al. Inorganic fullerene-like molybdenum selenide with good biocompatibility synthesized by laser ablation in liquids[J]. Nanotechnology, 2018, 29 (29): 295604.

[253] Chen T M, Zou H, Wu X J, et al. Fullerene-like MoS_2 nanoparticles as cascade catalysts improving lubricant and antioxidant abilities of artificial synovial fluid[J]. ACS Biomaterials Science & Engineering, 2019, 5 (6): 3079-3088.

[254] Chen T M, Yang F, Liu P, et al. General top-down strategy for generating single-digit nanodiamonds for bioimaging[J]. Nanotechnology, 2020, 31 (48): 485601.

[255] Chen T M, Yang F, Wu X J, et al. A fluorescent and colorimetric probe of carbyne nanocrystals coated Au nanoparticles for selective and sensitive detection of ferrous ions[J]. Carbon, 2020, 167: 196-201.

[256] Yang F, Chen T M, Wu X J, et al. A hybrid gold-carbyne nanocrystals platform for light-induced crossover of

redox enzyme-like activities[J]. Chemical Engineering Journal, 2021, 408: 127244.

[257] Chen T M, Zou H, Wu X J, et al. Nanozymatic antioxidant system based on MoS_2 nanosheets[J]. ACS Applied Materials & Interfaces, 2018, 10 (15): 12453-12462.

[258] Dieppe P A, Lohmander L S. Pathogenesis and management of pain in osteoarthritis[J]. The Lancet, 2005, 365 (9463): 965-973.

[259] Berenbaum F. Osteoarthritis as an inflammatory disease (osteoarthritis is not osteoarthrosis!) [J]. Osteoarthritis and Cartilage, 2013, 21 (1): 16-21.

[260] Corvelli M, Che B, Saeui C, et al. Biodynamic performance of hyaluronic acid versus synovial fluid of the knee in osteoarthritis[J]. Methods, 2015, 84: 90-98.

[261] Balazs E A, Watson D, Duff I F, et al. Hyaluronic acid in synovial fluid. I. Molecular parameters of hyaluronic acid in normal and arthritic human fluids[J]. Arthritis & Rheumatism, 1967, 10 (4): 357-376.

[262] Henrotin Y, Kurz B, Aigner T. Oxygen and reactive oxygen species in cartilage degradation: Friends or foes? [J]. Osteoarthritis and Cartilage, 2005, 13 (8): 643-654.

[263] McCord J M. Free radicals and inflammation: Protection of synovial fluid by superoxide dismutase[J]. Science, 1974, 185 (4150): 529-531.

[264] Ślesak I, Ślesak H, Zimak-Piekarczyk P, et al. Enzymatic antioxidant systems in early anaerobes: Theoretical considerations[J]. Astrobiology, 2016, 16 (5): 348-358.

[265] Regan E, Flannelly J, Bowler R, et al. Extracellular superoxide dismutase and oxidant damage in osteoarthritis[J]. Arthritis & Rheumatism, 2005, 52 (11): 3479-3491.

第 11 章 结论和展望

通过上述各章的论述，我们介绍了 LAL 纳米制备技术的产生和发展，主要包括 LAL 中纳米晶形成的物理和化学过程、LAL 合成纳米金刚石及新碳相纳米材料、新颖亚稳相纳米材料的 LAL 探索等，并在此基础上总结了目前国际上广泛应用的 LAL 纳米材料制备技术和纳米结构组装方法及所制备的功能纳米结构与纳米材料，并指出了 LAL 基纳米材料与纳米结构在光子、微电子、能源、环境、生物医学等诸多领域的应用。显然，我们可以看到，LAL 作为一种纳米制备技术有许多独特的地方。第一是 LAL 过程创造了局域极端的热力学环境，涉及到液、固、气三相的相互作用，高温化学反应和高压物质相变同时存在，这些奇异的物理和化学过程在其他纳米制备技术中是很难出现的。而正是这些独特的物理和化学过程，例如，远离热力学平衡的高温高压高密度等离子体羽的产生和快速的非平衡淬灭过程等，使得 LAL 方法常常用来制备其他传统纳米材料合成方法如化学溶液法等无法制备的纳米材料与纳米结构如亚稳纳米相等。第二是 LAL 纳米制备技术的多维度调控使得它成为一种几乎"万能"的纳米材料制备技术。我们可以通过环境液体的选择、固体靶组分的设计、脉冲激光参数的调节及辅助外场的施加等，面向基础研究和工业应用，对所制备纳米材料的结构、尺寸、形貌及功能进行可控合成。研究人员应用 LAL 纳米制备技术已经制备出了具有力、热、光、电、磁及生物医学等物性的一大批功能纳米结构与纳米材料，充分展示了 LAL 技术的普适性。第三是 LAL 制备的纳米材料具有化学上"干净"的表面和高活性。LAL 纳米材料合成是在液体环境中进行的，这样液体就起到一个隔离作用，使得纳米材料在合成过程中不被"氧化"和"还原"，从而保持新鲜清洁的表面。LAL 方法中的快速非平衡淬灭过程会让大量的缺陷留在合成的纳米材料体内和表面，这就为表面化学反应提供了丰富的反应位点，进而使得这些纳米材料表现出化学上的高活性。第四是 LAL 制备的纳米材料在应用上常常表现出"超常"的优越性能。一方面是 LAL 所制备的那些具有奇异物性的新型纳米材料，在器件应用上会表现出传统材料无法比拟的优越性，原因就是 LAL 过程中形成的独特结构引起了新奇的物理和化学性质，而这些性质是通过传统材料制备技术无法实现的；另一方面就是 LAL 基纳米材料本身固有的一些属性如清洁表面和高活性等，使得它们在涉及化学反应的相关应用领域如催化等，表现出传统材料制备技术制备的相同纳米材料不具备的优越性能。

但是，与传统纳米制备技术如化学溶液法等相比，LAL 纳米制备技术也有一些自身的短板。首先是材料制备的可控性不够。LAL 是一个远离热力学平衡的超快速制备过程，它犹如一把双刃剑，一方面创造了局域极端的热力学环境为合成新结构、新材料提供了机会；另一方面也使得制备过程的精准调控变得异常困难，从而限制了 LAL 纳米制备技术在一些领域的应用。其次是 LAL 制备纳米材料的产额不够高。LAL 过程中脉冲激光与固体靶的相互作用决定了所制备纳米材料的产额。由于常用的固体脉冲激光器的束斑较小（毫米量级），激光与固体靶的作用面积就小，自然合成产物的产额就低。所以，LAL 纳米制备技术很难应用于常规纳米材料的大规模工业化制备。

本章我们具体讨论 LAL 纳米制备技术的主要优势和有待改进的不足，并且展望未来的发展。

11.1　LAL 纳米制备技术的主要优势

LAL 纳米制备技术是目前国际上最快的、最"干净"的合成纳米胶体溶液的方法，并且它适用于几乎所有材料（无机、有机、有机-无机杂化等）纳米胶体溶液的合成（除了一些挥发性和不稳定的材料）[1]。因此，自 1987 年，Patil 等[2]在液体环境中通过脉冲激光烧蚀固体靶用于新材料研发的开创性工作发表以来，LAL 纳米制备技术就以令人难以置信的速度发展。回顾过去，我们可以总结出 LAL 纳米制备技术在纳米材料合成中的一些主要优点。我们认为，LAL 纳米制备技术具有两个最独特的优势，使其在众多纳米材料合成方法中脱颖而出：合成产物的清洁表面，以及亚稳纳米相的合成。

11.1.1　清洁表面

大多数研究在谈论 LAL 纳米制备技术的优点时都会提到合成产物的清洁表面，特别是对于金和银等贵金属材料纳米颗粒。然而，事实上，金纳米晶的表面并不是 100%干净的。Sylvestre 等[3]研究了 LAL 合成的金纳米颗粒的表面，他们发现 LAL 合成的金纳米颗粒被部分氧化，Au-O 化合物以 Au—O—形式存在于纳米颗粒表面，导致了纳米颗粒带有负净电荷。这就是为什么 LAL 合成的金纳米颗粒可以形成稳定胶体的原因。尽管如此，LAL 合成的纳米颗粒表面仍然可以视为无配体。

LAL 合成的纳米晶无配体表面的优势主要体现在以下 4 个方面：①高催化活性；②高吸附能力；③超高弛豫率；④生物共轭效率高。下面我们对这 4 个方面进行更详细的讨论。

（1）高催化活性。Hebie[4]提供了强有力的实验证据来支持 LAL 合成的金纳米颗粒的高催化活性，他们的实验表明，LAL 合成的金纳米颗粒比化学溶液法合成的纳米颗粒表现出更高的电催化效率。化学溶液法可使用 CTAB 作为稳定剂通过化学还原氯金酸合成的金纳米颗粒（CTAB-Au NP）及使用柠檬酸盐法合成的 Au 纳米颗粒（Cit-Au NP），其中 CTAB-Au NP 和 Cit-Au NP 具有一种有趣的自清洁机制。在自清洁机制的作用下，它们的电流密度从第一个循环到第十个循环会逐渐增加。相比之下，LAL 合成的金纳米颗粒从第一个循环就表现出很高的催化活性。LAL 合成的 Au 纳米颗粒的电流密度 2.65 A/(cm^2·mg)，比相应化学合成的样品高将近一个数量级［图 11-1（a）］，该值分别比 Au/Vulcan 和双金属 $Au_{70}Pt_{30}$/Vulcan 催化剂大 3.3 倍和 2.6 倍，是迄今为止报道的最好的催化剂之一。

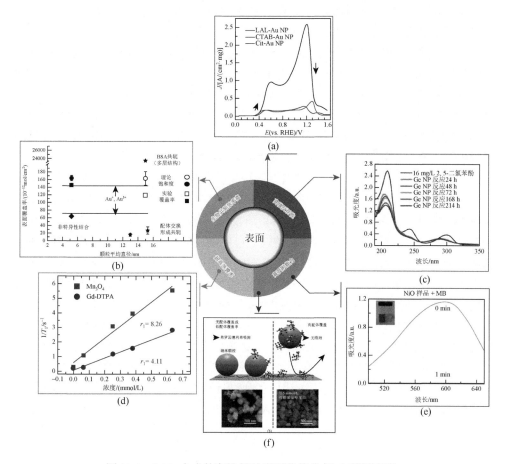

图 11-1　LAL 合成的高性能纳米颗粒催化剂（后附彩图）

（a）具有高催化活性的 Au 纳米颗粒；（b）具有高共轭效率的无配体 Au 纳米颗粒；（c）具有高催化活性的 Ge 纳米颗粒；（d）具有高弛豫率的 Mn_3O_4 纳米颗粒；（e）具有高吸附能力的 NiO 纳米颗粒；（f）具有高吸附能力的无配体纳米颗粒

除了贵金属纳米颗粒之外，LAL 合成的无表面活性剂的氧化物纳米颗粒催化剂如 Co_3O_4 纳米颗粒也同样表现出极高的水裂解析氧反应催化活性[10]。在催化过程中，这些 LAL 合成的 Co_3O_4 纳米颗粒的每个钴表面位点的转化频率为 $0.21\ mol\ O_2/[(mol\ Co_{surface})·s]$，这是已报道的纳米颗粒的析氧催化剂的最高转化频率之一。尽管高活性纳米颗粒会产生聚集，但是它们的催化活性并未受到抑制，其根本原因就是这类 LAL 合成催化剂的裸露、不含表面活性剂的表面。

上面论述了 LAL 合成的贵金属和氧化物纳米颗粒的高催化活性，实际上，我们还可以看到包括 LAL 合成的 Si 和 Ge 纳米颗粒在内的单原子半导体也表现出异常的还原能力。例如，Si 纳米颗粒可以将 Ag^+ 和 Au^{3+} 分别还原为 Ag 和 Au，这是普通商用 Si 无法实现的，因为其表面存在氧化[11]。此外，在 LAL 过程中，Ge 纳米颗粒经历了从无定形到四方再到立方的结构相变[12]，这些高活性的 Ge 纳米颗粒可以用于去除有机染料和减少重金属离子［图 11-1（c）］。

（2）高吸附能力。LAL 合成的无定形过渡金属氧化物可以表现出很强的吸附能力。例如，NiO 非晶纳米结构对亚甲蓝的最大吸附容量高达 $10\ 584.6\ mg/g$，这是截至目前报道的最大值[6]，这些纳米结构可以在 $1\ min$ 内超快去除亚甲蓝［图 11-1（e）］。另外，LAL 合成的无配体银纳米颗粒通过相反的过程，表现出强吸附能力［图 11-1（f）］。对比实验表明，过量的配体（表面覆盖率约 50%）会完全消除纳米颗粒的吸附能力，有力地证明了无配体表面确实可以提供高吸附能力[7]。

（3）超高弛豫率。LAL 合成的具有洁净表面的 Mn_3O_4 纳米颗粒可以作为性能优异的 MRI 的阳性造影剂[8]。这种纳米颗粒具有高达 $8.26\ L/(mmol·s)$ 的超高弛豫率，是商用 Gd 基阳性造影剂［$4.11\ L/(mmol·s)$］的两倍，也是截至目前报道的 Mn 基纳米颗粒的最高值。由于水和造影剂距离的关系可以描述为 $r_1 \propto d^{-6}$，其中，d 是 Mn 离子和水质子的距离，因此距离 d 是影响弛豫率的主要因素。具体而言，外部 Mn 离子与水质子的距离越短，r_1 越高。由于 LAL 合成的 Mn_3O_4 纳米颗粒的表面不会被任何化学配体或任何还原剂的残留物阻隔，因此 d 更短，r_1 更高［图 11-1（d）］。

（4）生物共轭效率高。与通过配体交换法等传统生物共轭技术获得的纳米颗粒相比，LAL 制备的无配体金纳米颗粒的最大表面覆盖度可高出 5 倍[8]。这种高共轭效率表明 LAL 过程中的生物缀合是一种很有前景的高效的技术［图 11-1（b）］。所以，我们可以看到，上述这些不同应用领域所报道的超强性能及奇异现象说明 LAL 合成纳米材料的表面比其他方法合成纳米材料的表面（化学）活性更高，这正是 LAL 技术独一无二之处。

11.1.2　亚稳纳米相

LAL 是一个非常快且远离热力学平衡的过程，因此，在过程的初始、中间和最终阶段形成的所有亚稳相和稳定相都可以存在于最终产物中,特别是亚稳态的中间相。换句话说,LAL 过程的淬灭时间非常短,以至于在转化中间阶段形成的亚稳相都有可能冻结在最终产物中[13]。例如,在 LAL 合成纳米金刚石研究中,我们已经证明了 LAL 过程提供的温度和压强可以覆盖碳相图中的绝大部分区域,这就意味着大量的可能存在的新颖碳纳米结构(理论上预测的新亚稳相碳纳米材料)可以通过碳-碳相变的中间态捕获。正如在我们的实验中,我们通过 LAL 方法合成了具有立方和六方结构的纳米金刚石、体心立方结构碳(C_8)、C_8-like、面心立方结构的新金刚石、carbyne 及碳葱等一系列亚稳相碳纳米材料[14-16]。

所以,大量研究证明,LAL 方法已经成为探索新型亚稳相纳米材料的有力工具。例如,LAL 方法可以用于制备亚稳相金属(双层六方 Fe、A15 相的 W)[17, 18]、亚稳相半导体(面心立方结构 Si、立方 GaN、四方结构 Ge)[19-21]及亚稳相绝缘体(立方相 ZrO_2、c-BN、立方相 C_3N_4)[22-24]。与传统的合成方法不同,LAL 是一个远离热力学平衡的过程,在环境液体的束缚作用下,反应体系可达到数千开尔文的温度和十几吉帕的压强,而大多数亚稳相形成需要相对高的温度和压强氛围。当激光诱导的等离子体羽快速淬灭时,那些在合成的中间阶段形成的亚稳相会被冻结在最终产物中。值得注意的是,环境液体也是 LAL 合成中的重要组成部分,例如,来自水溶液的羟基或氢原子,对亚稳相的形成有很大的促进作用。

虽然 LAL 纳米制备技术可以实现一些传统化学途经合成难以获得的亚稳相纳米结构,但是在合成过程中对具体亚稳相的精准控制仍然有些困难。我们知道,LAL 过程创造的高温高压高密度热力学状态显然有利于热力学平衡相图中高温高压区域的亚稳相的形成,然而,激光诱导的等离子体羽的寿命太短了,这就造成了合成产物中多相共存现象的存在。此外,LAL 过程中的温度和压强变化很大,难以调控。因此,增强 LAL 合成亚稳相的可控性是该领域的一项具有挑战性的任务。

11.2 LAL 纳米制备技术的不足和解决方案

11.2.1 产额

在 LAL 纳米制备技术领域,几乎每一篇综述文章都会提到 LAL 合成产物的产额问题[1, 27, 28]。事实上,对于大多数 LAL 合成的纳米材料来说,产额仅限于每小时几毫克或几十毫克的规模。因此,这种较低的产额严重阻碍了 LAL 技术在实际和工业领域的应用与发展。为了解决这个关键技术的瓶颈问题,Barcikowski 等通过研究 LAL 过程中激光参数、靶材制备和环境液体选择等，提出了提高 LAL 合成产额的重要技术路线。

靶材形状。一般情况下，我们会将靶材的几何形状制备成圆柱体或立方体，这样的靶材可以旋转以防止长时间激光照射同一点产生不均匀烧蚀。例如，Messina 等[29]提出，将靶材几何形状改变为细丝状能够显著提高激光烧蚀效率和合成纳米颗粒的产额，这是由于与立方体或圆柱体相比，细丝状具有更高的表面积与体积之比，从而会向周围液体散发更多热量。此外，细丝状靶材的空泡动力学与块状靶材的空泡动力学不同。与细丝状靶材相比，激光烧蚀块状靶材时会形成更大的空泡，从而导致消耗更多的能量。例如，使用细丝状靶材时，直径为 750 μm 的细丝状靶材的最高激光烧蚀效率比块状靶材提高了 15 倍[29]。

产额在很大程度上还取决于激光功率密度。然而，产额和激光功率密度的关系并不是线性的。当激光功率密度低于某个阈值时，纳米颗粒产额几乎随激光功率密度线性增加。然而，如果高于阈值，就会观察到与其平方值呈线性关系，表明反应机制发生了变化[30]。

此外，研究人员建议使用流动的环境液体来提高 LAL 合成纳米颗粒的产额。例如，研究人员通过搅拌旁路流通池来调整液体的流速，液体流速的提高有助于加速产物的冷却并促进纳米颗粒和空泡的去除，从而保护表面以防激光进一步烧蚀。与静止的环境液体相比，流动的液体提高了纳米颗粒产物合成的可重复性，并将纳米颗粒的产额提高了 4 倍[31]。由于纳米颗粒在环境液体中具有很强的吸收和散射效应，因此环境液体高度（液面距靶材表面的距离）是决定纳米颗粒产额的主要参数之一。例如，2.5 mm 的液体层是提升 LAL 合成 Al_2O_3 纳米颗粒产额的最佳高度[32]。Jiang 等[33]的研究表明水层厚度是影响 LAL 合成 Ge 纳米颗粒产额的重要参数，他们发现，当水层厚度为 1.2 mm 时，纳米颗粒的产额最大。因此，最佳的环境液体高度取决于 LAL 合成中的靶材。

除了上述参数外，Sajti 等[32]还发现，在水中激光烧蚀 α-Al_2O_3 靶时，合成的纳米颗粒尺寸、晶体结构和其他因素也会影响最终产物产额。因此，他们通过对激光、环境液体等参数的精细调整和优化，实现了 1.3 g/h 的最高产额。而此前，LAL 合成产额仅限于每小时几或几十毫克的规模。近年还出现了另外两种提高 LAL 合成产额的方法：在流动的环境液体中对线状靶材进行激光烧蚀[34]、改用高速旋转的靶材[35]。

近年来，研究人员通过使用高功率和高频率的脉冲激光及高速运行的多边形扫描仪来研究纳米材料的 LAL 合成，产额相关的研究取得了显著的进展。为了提高产额，Streubel 等[36]重点研究如何解决空泡屏蔽问题。ps-LAL 中空泡的寿命为 100 μs，其中在空泡破裂期间会生成纳米颗粒并分散到液体中。然而，空泡也可以吸收、散射或反射激光，换句话说，需要保护靶材以免受到后续脉冲激光的影响。他们提出了一种创新方法，使用高功率和高重复频率的脉冲激光（500 W、10 MHz、3 ps）及扫描速度高达 500 m/s 的多边形扫描仪，在空间上绕过空泡。令人高兴的是，该方法使 LAL 合成的纳米颗粒的产额提高至 4 g/h。他们还研究了高速脉冲 LAL 合成的胶体的粒径、形态和氧化态。相关的表征如 XRD、XPS 和 TEM 表明这种半工业规模的方法是一种稳定的纳米颗粒合成过程，其产生的微晶尺寸和氧化态在 LAL 进行 1 h 内保持恒定[37]。因此，这项研究大大提高了 LAL 制备纳米颗粒的产额。当 LAL 合成贵金属纳米颗粒的产额达到每小时克级时，就已经具有工业化的潜力。目前，LAL 合成的 Au、Ag、Pt 等贵金属纳米颗粒胶体皆已商业化，例如，来自德国的 Particular GmbH 的贵金属纳米胶体产品。重要的是，这些研究为我们提供了一条明确的途径，使用高功率和高重复频率的脉冲激光及高速扫描仪，为未来进一步提高 LAL 合成纳米颗粒的产量提供了方向。

11.2.2　尺寸和分散度控制

2000 年，Mafuné 等[38]的研究表明，可以通过在 LAL 过程中引入 SDS 来调节 Ag 纳米颗粒的尺寸，他们发现，随着 SDS 浓度的增加和激光功率密度的降低，纳米颗粒向着更小的尺寸转变。这是控制 LAL 合成的纳米颗粒尺寸的早期尝试。然而，尽管该方法对尺寸控制有所改善，但纳米颗粒尺寸的分散度仍然很大，并且纳米颗粒的表面会被 SDS 覆盖。

近年来，更多的技术被开发来实现 LAL 合成纳米材料的可控性。例如，Wang 等[39]使用非聚焦纳秒激光来合成具有光滑表面的亚微米球体，他们通过仔细选择激光参数和靶材的固有特性（例如，熔点和沸点），可以控制这些产物实现较窄的尺寸分布范围。重要的是，该方法适用于各种半导体、金属和氧化物，例如，Ag、Au、Fe、Ni、Fe_2O_3、Co_3O_4、NiO 和 WO_3。然而，该技术的缺点是选择性脉冲激光加热只能合成直径为 100 nm 甚至更大的亚微米球体。因此，这些微米级球体将失去小尺寸纳米材料的独特优势。

LAL 合成单分散纳米晶的方法还有使用长脉宽（μs）激光辐照含有大颗粒多分散的纳米晶悬浮液[40]。该方法适合生产约 5 nm 单分散 PbS 或 PbSe CQD，尺寸分散度＜10%。然而，只有少量材料适合这种方法。此外，还有一些有机添加剂可充当表面活性剂和硫源，例如，1-十二烷硫醇。

因此，LAL 合成无配体的单分散纳米颗粒胶体仍然是一个挑战。无配体和单分散似乎是相互矛盾的要求，要获得无配体且单分散的纳米颗粒非常困难。LAL 合成的纳米颗粒的表面几乎是裸露的。由于小尺寸纳米颗粒具有高表面能，因此这些产物的聚集似乎是不可避免的，所以很难获得单分散的纳米颗粒，反之亦然。尽管如此，研究人员还是可以通过液相激光碎裂、电解质的原位淬灭、液体流中的延迟共轭、LAL 等技术，在没有表面活性剂的条件下实现对纳米材料尺寸的控制[41]。另外，尽管 LAL 合成的纳米颗粒容易聚集，但聚集后的纳米颗粒如 Co_3O_4 纳米颗粒依然具有高催化活性，这表明它们的高活性表面仍然得到了保留[10]，换句话说，LAL 合成纳米颗粒的聚集不会严重影响它们的表面反应活性。

11.3 LAL 纳米制备技术的未来发展

11.3.1 物理化学机制探索

首先，与已经广泛应用的 ns-LAL 技术相比，人们对 fs-LAL 过程的基本物理和化学研究相对较少。目前，大多数研究人员对 fs-LAL 机制的理解都是基于真空或气体环境的情况，而实际上，LAL 机制与激光烧蚀在真空或气体环境中的情况有很大不同。如前文所知，fs-LAL 过程非常复杂，它包括了相爆炸、碎裂、库仑爆炸和等离子体烧蚀等多物理过程。如果将环境液体反应也考虑在内的话，那么其机制的确定就会成为一项复杂的科学任务。幸运的是，目前研究人员已经开发出多种先进的表征方法用于描述 LAL 纳米颗粒的合成。例如，用于检测脉冲激光诱导的等离子体羽的光学发射光谱和快速成像，用于探索空泡的阴影图，用于研究纳米颗粒在溶液中的位置、释放的双脉冲 LAL 和激光散射，用于研究纳米颗粒形成过程的小角度 X 射线散射等。尽管纳秒激光的 LAL 过程是 LAL 技术中研究最多的，但是 ns-LAL 中空泡产生、膨胀和破裂的热力学和动力学仍不明晰，尤其是空泡内部的温度和压强。总的来说，LAL 过程的实验与理论存在明显差距[42]。

其次，人们对 μs-LAL 过程的研究相对更少。与 fs-LAL 和 ns-LAL 不同，目前还没有利用 μs-LAL 方法合成贵金属纳米颗粒的报道。事实上，我们进行的一些实验研究表明，当微秒激光烧蚀固体靶时，不会形成纳米贵金属胶体。这表明 μs-LAL 作用机制与常见的 ns-LAL 和 fs-LAL 有很大不同。我们知道，μs-LAL 有较长的脉宽和相对较低的功率密度（$<10^6$ W/cm^2），正是这种长脉宽会产生额外的、显著的加热效应，这种加热效应会引起一些表面反应，可以用来控制纳米结构的形貌[43]。应用 μs-LAL 方法合成纳米材料的研究近年来才开始[43]，因此，一些基本物理和化学问题仍未阐明。例如，μs-LAL 过程的温度和压强能达到多高？μs-LAL 实际反应过程是怎样的？这个过程可以用高速光谱法表征吗？有些固体靶如难熔金属等是很难被 μs-LAL 过程烧蚀的，这是什么原因？我们能否实现 μs-LAL 对它们的烧蚀？此外，μs-LAL 合成产物产额与激光和液体参数的关系仍然未知。因此，要完全理解 μs-LAL 还有很长的路要走。

最后，我们将注意力转向 LAL 中的环境液体。传统上，LAL 纳米制备技术中的环境液体是水或可以在环境温度和压强下流动的有机溶剂。超临界 CO_2 流体作为一种新型环境液体用于 LAL 技术[44, 45]，这种液体与传统液体完全不同，其物理性质如导热率、热容量、介电常数和扩散常数很容易通过调节压强来控制。此外，超低温液氦也用于 LAL 技术[46]。液氦的热导率在温度为 2.17 K 以上时为 0.02 W/(m·K) (He I)，而超流氦中的热流在温度为 2.17 K 以下具有非线性的温度梯度，比 He I 高几个数量级。显然，这些独特的环境液体具有许多有趣的物理和化学特性。那么，我们能否在 LAL 技术中使用它们来合成更多新颖的纳米结构呢？在 LAL 过程中，这些液体的行为与传统液体相似吗？它们会强烈影响 LAL 过程中等离子体羽或空泡的形成吗？我们希望进一步的工作能够回答这些值得思考的问题。

11.3.2 应用领域拓展

从第 7～10 章对 LAL 技术及其在纳米材料制备中的应用介绍可以看出，经过几十年的发展，LAL 合成的纳米材料已经广泛应用于光电子学、微电子学、磁学、新能源与环境科学、生物医学等诸多领域。然而，我们希望通过分析 LAL 纳米制备技术存在的优缺点来重新审视 LAL 方法可能应用的更广阔领域和未来的发展趋势。

尽管近些年研究人员实现了产额达到每小时克级的 LAL 技术，但是，这种产额仍然远低于湿化学法等其他纳米材料合成方法所达到的水平。在大多数情况下，LAL 合成纳米材料的产额仍仅限于每小时毫克级。此外，LAL 合成的初始产物总是以胶体溶液而不是粉末形式获得。因此，如果一些应用需求是以克为单位或以胶体溶液为应用形式的话，那么 LAL 纳米制备技术毫无疑问是最适合的制备方法之一。此外，我们应该在拓展 LAL 技术应用领域时，充分利用 LAL 合成纳米材料的清洁表面。考虑到这些因素，我们认为 LAL 生产的纳米材料最适合用于生物医学和催化领域。

例如，我们认为 LAL 合成的纳米材料非常适合用于纳米毒理学研究。近年来，纳米毒理学已成为一门独立的新学科。纳米毒理学的研究表明，纳米材料与细胞、动物（人类）的相互作用极其复杂。目前，研究人员仍在试图理解纳米材料的理化特性是如何影响这些相互作用的，从而阐明哪些因素对动物（人类）健康的影响最为关键。与此同时，纳米颗粒的潜在毒性也已经越来越受到公众关注[47, 48]。实际上，纳米颗粒的应用和毒性研究取决于我们是否能够量化纳米颗粒与细胞的相互作用。许多研究报告表明，纳米颗粒的细胞摄取取决于纳米颗粒的表面结构[49, 50]。因此，保证用于检测的纳米材料的纯度对于获得令人信服的数据非常重要。例如，金纳米颗粒因其优异的生物相容性、对酸和碱的高惰性及易于与生物活性小分子缀合而成为最常用的生物医用纳米材料之一。目前，合成金纳米颗粒最常见的方法是化学溶液法。然而，这种湿化学法合成的金纳米颗粒上残留的配体和表面活性剂使得它们的毒性难以表征。换句话说，如果配体共轭的金纳米颗粒显示出疗效，则很难确定其中金纳米颗粒起到什么作用。事实上，不仅是金纳米颗粒，任何一种纳米材料都会面临同样的问题，即其毒性或疗效究竟是源于材料本身还是其他配体或添加剂。Petersen 等[51]和 Taylor 等[52]系统地研究了 LAL 合成的无配体金纳米颗粒的毒性［图 11-2（a）］。他们发现，与编码重组融合蛋白的质粒共孵育的金纳米颗粒没有改变蛋白表达或功能，但确实增加了转染效率［图 11-2（b）］。此外，在体外标记细胞所需的浓度水平下与这些金纳米颗粒共孵育时，没有观察到细胞毒性［图 11-2（c）］。因此，这些研究结果支持了 LAL 合成的无配体金纳米颗粒无毒的观点。我们甚至可以得出这样的结论：如果材料本身纯度很高的话，金纳米颗粒本身是无毒的。

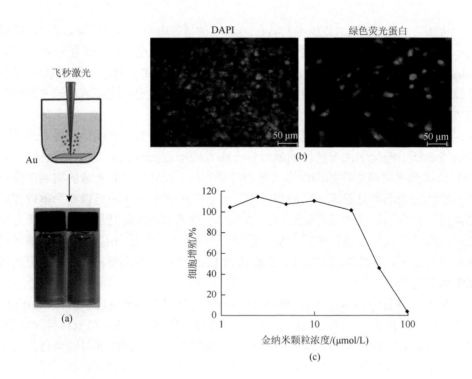

图 11-2 （a）fs-LAL 合成的金纳米颗粒胶体；（b）LAL 合成的金纳米颗粒在质粒 DNA 转染过程中不会干扰融合蛋白的生物活性；（c）当金纳米颗粒剂量低于 25 μmol/L 时，非内摄对细胞增殖没有影响（后附彩图）

除了上述研究方向外，还有两种方法可有效拓展 LAL 纳米制备技术的应用领域，也就是与其他纳米合成方法的有机结合和复合纳米结构的组装。我们知道，LAL 合成的纳米材料通常以胶体溶液的形式存在，如贵金属纳米颗粒胶体和半导体纳米颗粒胶体。而水热合成法也是在液体环境中发生反应。因此，将 LAL 纳米制备技术与水热合成法结合能够产生强大的功能，如使用 LAL 纳米制备技术为水热合成法制备特殊需求的前驱体，或将水热合成产物作为 LAL 的环境液体。例如，在非平衡环境中，LAL 合成的金属氧化物纳米材料（例如，氧化钛）的结晶度可能较差[53]。在这种情况下，水热处理可以有效地提高结晶度。当晶体生长发生在相对温和的环境中时，可以通过简单地改变水热处理的温度和时间来调节晶体形态和结构。

此外，电泳合成法和电化学合成法也非常适合与 LAL 纳米制备技术结合。由于 LAL 合成的胶体带有电荷，如果在 LAL 过程中设置电极，那么带电粒子将在外加电场的作用下沉积在电极表面。此外，如果在 LAL 过程中电极可以在电场作用下发生反应，那么靶材将会和电极发生反应。有趣的是，这两种技术已用于制造 SERS 基底和简单的 POM 纳米结构[54, 55]，并有望在未来实现更多新型纳米结构的组装。

复合纳米结构的组装也是拓展 LAL 纳米制备技术应用领域的方法之一。如前所述，LAL 纳米制备技术曾用来组装无机纳米颗粒-聚合物复合材料，并且基本做到没有任何界面污染，也不需要基质黏合剂。重要的是，这些复合纳米材料可以很好地将聚合物的生物相容性和柔韧性与无机纳米材料的导电和催化性能结合起来，充分显现了其在生物医学和化学催化方面的应用潜力，从而极大地拓展了应用领域。然而，目前这类复合纳米材料中的无机纳米颗粒仍然主要是贵金属如金和银等纳米晶[56]。因此，我们可以利用 LAL 合成的高活性纳米材料作为组装复合纳米材料的构成单元，来组装具有更大潜在应用价值的新型无机纳米颗粒-聚合物复合材料。

除了与聚合物形成复合纳米材料外，Liu 等[57]还成功地将纳米胶体簇（例如，Ge、Si 和 TiO_x）掺杂到赤铁矿纳米晶体中以提高其光电化学性能。同时，Lin 等[58]证明了纳米金刚石可以通过形成复合纳米材料，极大地改善 Cu_2O 纳米晶光催化剂的光催化析氢反应。此外，LAL 合成的纳米胶体溶液具有很高的化学活性，可以被用作湿化学法或其他纳米合成方法的前驱体。因为湿化学法常常使用含有各种离子的溶液作为前驱体，因此这种方法合成的溶液产物也可以被认为是亚稳态团簇。考虑到此类胶体溶液独特的反应活性，我们完全可以期待看到许多有趣的化学现象和独特的应用场景。

本书全面地介绍了 LAL 纳米制备技术的发展历程，阐述了 LAL 过程的基本物理与化学过程并描述了它的热力学和动力学特征，梳理了基于 LAL 机制发展的一系列 LAL 的纳米材料制备与纳米结构组装新技术，总结了 LAL 合成纳米材料的种类及所合成纳米材料的应用领域。我们可以看到，LAL 纳米制备技术有自己鲜明的特色和独到的地方，是其他纳米合成方法无法替代的，如最"干净"的纳米胶体溶液组装、高化学活性表面制备、新亚稳纳米相捕获等。同时，我们也能发现，研究人员可以针对具有不同功能需求的纳米材料与纳米结构而开发出不同的 LAL 纳米制备技术，这充分说明了 LAL 方法用于纳米材料制备领域的普适性。而 LAL 合成的纳米材料在诸多科学和技术领域的广泛应用则彰显了它的巨大应用潜力。为了使 LAL 纳米制备技术在未来科学、技术、工业领域发挥更大作用，我们需要在物理、化学、材料科学、生物医学等之间建立更紧密的跨学科联系，发展更强大、更实用的 LAL 纳米制备技术，并开发更多具有颠覆性的新型纳米材料。

参 考 文 献

[1] Yang G W. Laser ablation in liquids: Applications in the synthesis of nanocrystals[J]. Progress in Materials Science, 2007, 52 (4): 648-698.

[2] Patil P P, Phase D M, Kulkarni S A, et al. Pulsed-laser-induced reactive quenching at liquid-solid interface: Aqueous oxidation of iron[J]. Physical Review Letters, 1987, 58 (3): 238.

[3] Sylvestre J P, Poulin S, Kabashin A V, et al. Surface chemistry of gold nanoparticles produced by laser ablation in aqueous media[J]. The Journal of Physical Chemistry B, 2004, 108 (43): 16864-16869.

[4] Hebie S, Holade Y, Maximova K, et al. Advanced electrocatalysts on the basis of bare Au nanomaterials for biofuel cell applications[J]. ACS Catalysis, 2015, 5 (11): 6489-6496.

[5] Liu J, Liang C H, Tian Z F, et al. Spontaneous growth and chemical reduction ability of Ge nanoparticles[J]. Scientific Reports, 2013, 3: 1741.

[6] Li L H, Xiao J, Liu P, et al. Super adsorption capability from amorphousization of metal oxide nanoparticles for dye removal[J]. Scientific Reports, 2015, 5: 9028.

[7] Wagener P, Schwenke A, Barcikowski S. How citrate ligands affect nanoparticle adsorption to microparticle supports[J]. Langmuir, 2012, 28 (14): 6132-6140.

[8] Xiao J, Tian X M, Yang C, et al. Ultrahigh relaxivity and safe probes of manganese oxide nanoparticles for *in vivo* imaging[J]. Scientific Reports, 2013, 3: 3424.

[9] Petersen S, Barcikowski S. Conjugation efficiency of laser-based bioconjugation of gold nanoparticles with nucleic acids[J]. The Journal of Physical Chemistry C, 2009, 113 (46): 19830-19835.

[10] Blakemore J D, Gray H B, Winkler J R, et al. Co_3O_4 nanoparticle water-oxidation catalysts made by pulsed-laser ablation in liquids[J]. ACS Catalysis, 2013, 3 (11): 2497-2500.

[11] Yang S K, Cai W P, Liu G Q, et al. Optical study of redox behavior of silicon nanoparticles induced by laser ablation in liquid[J]. The Journal of Physical Chemistry C, 2009, 113 (16): 6480-6484.

[12] Liu J, Liang C H, Tian Z F, et al. Spontaneous growth and chemical reduction ability of Ge nanoparticles[J]. Scientific Reports, 2013, 3: 1741.

[13] Yang G W. Laser ablation in liquids: Applications in the synthesis of nanocrystals[J]. Progress in Materials Science, 2007, 52 (4): 648-698.

[14] Xiao J, Ouyang G, Liu P, et al. Reversible nanodiamond-carbon onion phase transformations[J]. Nano Letters, 2014, 14 (6): 3645-3652.

[15] Xiao J, Li J L, Liu P, et al. A new phase transformation path from nanodiamond to new-diamond via an intermediate carbon onion[J]. Nanoscale, 2014, 6 (24): 15098-15106.

[16] Pan B T, Xiao J, Li J L, et al. Carbyne with finite length: The one-dimensional sp carbon[J]. Science Advances, 2015, 1 (9): e1500857.

[17] Xiao J, Liu P, Liang Y, et al. Super-stable ultrafine beta-tungsten nanocrystals with metastable phase and related magnetism[J]. Nanoscale, 2013, 5 (3): 899-903.

[18] Chen X Y, Cui H, Liu P, et al. Double-layer hexagonal Fe nanocrystals and magnetism[J]. Chemistry of Materials, 2008, 20 (5): 2035-2038.

[19] Du X W, Qin W J, Lu Y W, et al. Face-centered-cubic Si nanocrystals prepared by microsecond pulsed laser ablation[J]. Journal of Applied Physics, 2007, 102 (1): 013518.

[20] Liu P, Cao Y L, Chen X Y, et al. Trapping high-pressure nanophase of Ge upon laser ablation in liquid[J]. Crystal Growth and Design, 2009, 9 (3): 1390-1393.

[21] Liu P, Cao Y L, Cui H, et al. Synthesis of GaN nanocrystals through phase transition from hexagonal to cubic structures upon laser ablation in liquid[J]. Crystal Growth and Design, 2008, 8 (2): 559-563.

[22] Wang J D, Yang G W, Zhang C Y, et al. Cubic-BN nanocrystals synthesis by pulsed laser induced liquid-solid interfacial reaction[J]. Chemical Physics Letters, 2003, 367 (1-2): 10-14.

[23] Yang G W, Wang J B. Carbon nitride nanocrystals having cubic structure using pulsed laser induced liquid-solid interfacial reaction[J]. Applied Physics A, 2000, 71: 343-344.

[24] Tan D Z, Lin G, Liu Y, et al. Synthesis of nanocrystalline cubic zirconia using femtosecond laser ablation[J]. Journal of Nanoparticle Research, 2011, 13: 1183-1190.

[25] Liu P, Cui H, Yang G W. Synthesis of body-centered cubic carbon nanocrystals[J]. Crystal Growth and Design, 2008, 8 (2): 581-586.

[26] Liu P, Cao Y L, Wang C X, et al. Micro-and nanocubes of carbon with C_8-like and blue luminescence[J]. Nano Letters, 2008, 8 (8): 2570-2575.

[27] Amendola V, Meneghetti M. Laser ablation synthesis in solution and size manipulation of noble metal nanoparticles[J]. Physical Chemistry Chemical Physics, 2009, 11 (20): 3805-3821.

[28] Zeng H B, Du X W, Singh S C, et al. Nanomaterials via laser ablation/irradiation in liquid: A review[J]. Advanced Functional Materials, 2012, 22 (7): 1333-1353.

[29] Messina G C, Wagener P, Streubel R, et al. Pulsed laser ablation of a continuously-fed wire in liquid flow for high-yield production of silver nanoparticles[J]. Physical Chemistry Chemical Physics, 2013, 15 (9): 3093-3098.

[30] Sajti C L, Sattari R, Chichkov B, et al. Ablation efficiency of α-Al_2O_3 in liquid phase and ambient air by nanosecond laser irradiation[J]. Applied Physics A, 2010, 100: 203-206.

[31] Barcikowski S, Menéndez-Manjón A, Chichkov B, et al. Generation of nanoparticle colloids by picosecond and femtosecond laser ablations in liquid flow[J]. Applied Physics Letters, 2007, 91 (8): 083113.

[32] Sajti C L, Sattari R, Chichkov B N, et al. Gram scale synthesis of pure ceramic nanoparticles by laser ablation in liquid[J]. The Journal of Physical Chemistry C, 2010, 114 (6): 2421-2427.

[33] Jiang Y, Liu P, Liang Y, et al. Promoting the yield of nanoparticles from laser ablation in liquid[J]. Applied Physics A, 2011, 105: 903-907.

[34] Kohsakowski S, Santagata A, Dell'Aglio M, et al. High productive and continuous nanoparticle fabrication by laser ablation of a wire-target in a liquid jet[J]. Applied Surface Science, 2017, 403: 487-499.

[35] Resano-Garcia A, Champmartin S, Battie Y, et al. Highly-repeatable generation of very small nanoparticles by

pulsed-laser ablation in liquids of a high-speed rotating target[J]. Physical Chemistry Chemical Physics, 2016, 18 (48): 32868-32875.

[36] Streubel R, Barcikowski S, Gökce B. Continuous multigram nanoparticle synthesis by high-power, high-repetition-rate ultrafast laser ablation in liquids[J]. Optics Letters, 2016, 41 (7): 1486-1489.

[37] Streubel R, Bendt G, Gökce B. Pilot-scale synthesis of metal nanoparticles by high-speed pulsed laser ablation in liquids[J]. Nanotechnology, 2016, 27 (20): 205602.

[38] Mafuné F, Kohno J, Takeda Y, et al. Formation and size control of silver nanoparticles by laser ablation in aqueous solution[J]. The Journal of Physical Chemistry B, 2000, 104 (39): 9111-9117.

[39] Wang H Q, Pyatenko A, Kawaguchi K, et al. Selective pulsed heating for the synthesis of semiconductor and metal submicrometer spheres[J]. Angewandte Chemie International Edition, 2010, 49 (36): 6361-6364.

[40] Yang J, Ling T, Wu W T, et al. A top-down strategy towards monodisperse colloidal lead sulphide quantum dots[J]. Nature Communications, 2013, 4: 1695.

[41] Rehbock C, Jakobi J, Gamrad L, et al. Current state of laser synthesis of metal and alloy nanoparticles as ligand-free reference materials for nano-toxicological assays[J]. Beilstein Journal of Nanotechnology, 2014, 5: 1523-1541.

[42] De Giacomo A, Dell'Aglio M, Santagata A, et al. Cavitation dynamics of laser ablation of bulk and wire-shaped metals in water during nanoparticles production[J]. Physical Chemistry Chemical Physics, 2013, 15 (9): 3083-3092.

[43] Niu K Y, Yang J, Kulinich S A, et al. Morphology control of nanostructures via surface reaction of metal nanodroplets[J]. Journal of the American Chemical Society, 2010, 132 (28): 9814-9819.

[44] Kato T, Stauss S, Kato S, et al. Pulsed laser ablation plasmas generated in CO_2 under high-pressure conditions up to supercritical fluid[J]. Applied Physics Letters, 2012, 101 (22): 224103.

[45] Saitow K, Yamamura T, Minami T. Gold nanospheres and nanonecklaces generated by laser ablation in supercritical fluid[J]. The Journal of Physical Chemistry C, 2008, 112 (47): 18340-18349.

[46] Lebedev V, Moroshkin P, Grobety B, et al. Formation of metallic nanowires by laser ablation in liquid helium[J]. Journal of Low Temperature Physics, 2011, 165: 166-176.

[47] Nel A, Xia T, Madler L, et al. Toxic potential of materials at the nanolevel[J]. Science, 2006, 311 (5761): 622-627.

[48] Teeguarden J G, Hinderliter P M, Orr G, et al. Particokinetics *in vitro*: Dosimetry considerations for *in vitro* nanoparticle toxicity assessments[J]. Toxicological Sciences, 2007, 95 (2): 300-312.

[49] Cho E C, Xie J W, Wurm P A, et al. Understanding the role of surface charges in cellular adsorption versus internalization by selectively removing gold nanoparticles on the cell surface with a I_2/KI etchant[J]. Nano Letters, 2009, 9 (3): 1080-1084.

[50] Verma A, Stellacci F. Effect of surface properties on nanoparticle: Cell interactions[J]. Small, 2010, 6 (1): 12-21.

[51] Petersen S, Soller J T, Wagner S, et al. Co-transfection of plasmid DNA and laser-generated gold nanoparticles does not disturb the bioactivity of GFP-HMGB1 fusion protein[J]. Journal of Nanobiotechnology, 2009, 7: 1-6.

[52] Taylor U, Klein S, Petersen S, et al. Nonendosomal cellular uptake of ligand-free, positively charged gold nanoparticles[J]. Cytometry Part A, 2010, 77A (5): 439-446.

[53] Zhang H M, Liang C H, Tian Z F, et al. Hydrothermal treatment of colloids induced via liquid-phase laser ablation: A new approach for hierarchical titanate nanostructures with enhanced photodegradation performance[J]. CrystEngComm, 2011, 13 (14): 4676-4682.

[54] Liu P, Liang Y, Lin X Z, et al. A general strategy to fabricate simple polyoxometalate nanostructures: Electrochemistry-assisted laser ablation in liquid[J]. ACS Nano, 2011, 5 (6): 4748-4755.

[55] Yang S K, Cai W P, Kong L C, et al. Surface nanometer-scale patterning in realizing large-scale ordered arrays of metallic nanoshells with well-defined structures and controllable properties[J]. Advanced Functional Materials, 2010, 20 (15): 2527-2533.

[56] Sowa-Söhle E N, Schwenke A, Wagener P, et al. Antimicrobial efficacy, cytotoxicity, and ion release of mixed metal (Ag, Cu, Zn, Mg) nanoparticle polymer composite implant material[J]. BioNanoMaterials, 2013, 14 (3-4): 217-227.

[57] Liu J, Liang C H, Zhang H M, et al. General strategy for doping impurities (Ge, Si, Mn, Sn, Ti) in hematite nanocrystals[J]. The Journal of Physical Chemistry C, 2012, 116 (8): 4986-4992.

[58] Lin Z Y, Xiao J, Li L H, et al. Nanodiamond-embedded p-type copper (I) oxide nanocrystals for broad-spectrum photocatalytic hydrogen evolution[J]. Advanced Energy Materials, 2015, 6 (4): 1501865.

附录　杨国伟研究组发表 LAL 纳米制备技术论文目录

[1] Yang G W，Wang J B，Liu Q X. Preparation of nano-crystalline diamonds using pulsed laser induced reactive quenching[J]. Journal of Physics：Condensed Matter，1998，10（35）：7923.

[2] Wang J B，Yang G W. Phase transformation between diamond and graphite in preparation of diamonds by pulsed-laser induced liquid-solid interface reaction[J]. Journal of Physics：Condensed Matter，1999，11（37）：7089.

[3] Yang G W，Wang J B. Carbon nitride nanocrystals having cubic structure using pulsed laser induced liquid-solid interfacial reaction[J]. Applied Physics A，2000，71：343-344.

[4] Yang G W，Wang J B. Pulsed-laser-induced transformation path of graphite to diamond via an intermediate rhombohedral graphite[J]. Applied Physics A，2001，72：475-479.

[5] Wang J B，Zhang C Y，Zhong X L，et al. Cubic and hexagonal structures of diamond nanocrystals formed upon pulsed laser induced liquid-solid interfacial reaction[J]. Chemical Physics Letters，2002，361（1-2）：86-90.

[6] Wang J B，Zhong X L，Zhang C Y，et al. Explosion phase formation of nanocrystalline boron nitrides upon pulsed-laser-induced liquid/solid interfacial reaction[J]. Journal of Materials Research，2003，18（12）：2774-2778.

[7] Wang J B，Yang G W，Zhang C Y，et al. Cubic-BN nanocrystals synthesis by pulsed laser induced liquid-solid interfacial reaction[J]. Chemical Physics Letters，2003，367（1-2）：10-14.

[8] Liu Q X，Yang G W，Zhang J X. Phase transition between cubic-BN and hexagonal BN upon pulsed laser induced liquid-solid interfacial reaction[J]. Chemical Physics Letters，2003，373（1-2）：57-61.

[9] Liu Q X，Wang C X，Zhang W，et al. Immiscible silver-nickel alloying nanorods growth upon pulsed-laser induced liquid/solid interfacial reaction[J]. Chemical Physics Letters，2003，382（1-2）：1-5.

[10] Liu Q X，Wang C X，Yang G W. Formation of silver particles and silver oxide plume nanocomposites upon pulsed-laser induced liquid-solid interface reaction[J]. The European Physical Journal B：Condensed Matter and Complex Systems，2004，41：479-483.

[11] Wang C X，Yang Y H，Liu Q X，et al. Phase stability of diamond nanocrystals upon pulsed-laser-induced liquid-solid interfacial reaction：Experiments and ab initio calculations[J]. Applied Physics Letters，2004，84（9）：1471-1473.

[12] Wang C X，Yang Y H，Yang G W. Thermodynamical predictions of nanodiamonds synthesized by pulsed-laser ablation in liquid[J]. Journal of Applied Physics，2005，97（6）：066104.

[13] Wang C X, Liu P, Cui H, et al. Nucleation and growth kinetics of nanocrystals formed upon pulsed-laser ablation in liquid[J]. Applied Physics Letters, 2005, 87 (20): 201913.

[14] Chen X Y, Cui H, Liu P, et al. Shape-induced ultraviolet absorption of CuO shuttlelike nanoparticles[J]. Applied Physics Letters, 2007, 90 (18): 183118.

[15] Yang G W. Laser ablation in liquids: Applications in the synthesis of nanocrystals[J]. Progress in Materials Science, 2007, 52 (4): 648-698.

[16] Liu P, Wang C X, Chen X Y, et al. Controllable fabrication and cathodoluminescence performance of high-index facets GeO_2 micro-and nanocubes and spindles upon electrical-field-assisted laser ablation in liquid[J]. The Journal of Physical Chemistry C, 2008, 112 (35): 13450-13456.

[17] Liu P, Cui H, Yang G W. Synthesis of body-centered cubic carbon nanocrystals[J]. Crystal Growth & Design, 2008, 8 (2): 581-586.

[18] Liu P, Cao Y L, Cui H, et al. Synthesis of GaN nanocrystals through phase transition from hexagonal to cubic structures upon laser ablation in liquid[J]. Crystal Growth & Design, 2008, 8 (2): 559-563.

[19] Liu P, Cao Y L, Cui H, et al. Micro-and nanocubes of silicon with zinc-blende structure[J]. Chemistry of Materials, 2008, 20 (2): 494-502.

[20] Chen X Y, Cui H, Liu P, et al. Double-layer hexagonal Fe nanocrystals and magnetism[J]. Chemistry of Materials, 2008, 20 (5): 2035-2038.

[21] Liu P, Cao Y L, Wang C X, et al. Micro-and nanocubes of carbon with C_8-like and blue luminescence[J]. Nano Letters, 2008, 8 (8): 2570-2575.

[22] Liu P, Wang C X, Chen J, et al. Localized nanodiamond crystallization and field emission performance improvement of amorphous carbon upon laser irradiation in liquid[J]. The Journal of Physical Chemistry C, 2009, 113 (28): 12154-12161.

[23] Lin X Z, Liu P, Yu J M, et al. Synthesis of CuO nanocrystals and sequential assembly of nanostructures with shape-dependent optical absorption upon laser ablation in liquid[J]. The Journal of Physical Chemistry C, 2009, 113 (40): 17543-17547.

[24] Liu P, Cao Y L, Chen X Y, et al. Trapping high-pressure nanophase of Ge upon laser ablation in liquid[J]. Crystal Growth & Design, 2009, 9 (3): 1390-1393.

[25] Liu P, Cui H, Wang C X, et al. From nanocrystal synthesis to functional nanostructure fabrication: Laser ablation in liquid[J]. Physical Chemistry Chemical Physics, 2010, 12 (16): 3942-3952.

[26] Jiang Y, Liu P, Liang Y, et al. Promoting the yield of nanoparticles from laser ablation in liquid[J]. Applied Physics A, 2011, 105: 903-907.

[27] Liu P, Liang Y, Lin X Z, et al. A general strategy to fabricate simple polyoxometalate nanostructures: Electrochemistry-assisted laser ablation in liquid[J]. ACS Nano, 2011, 5 (6): 4748-4755.

[28] Liang Y, Liu P, Li H B, et al. Synthesis and characterization of copper vanadate nanostructures via electrochemistry assisted laser ablation in liquid and the optical multi-absorptions performance[J].

CrystEngComm, 2012, 14 (9): 3291-3296.

[29] Liang Y, Liu P, Li H B, et al. ZnMoO$_4$ micro-and nanostructures synthesized by electrochemistry-assisted laser ablation in liquids and their optical properties[J]. Crystal Growth & Design, 2012, 12 (9): 4487-4493.

[30] Xiao J, Liu P, Liang Y, et al. Porous tungsten oxide nanoflakes for highly alcohol sensitive performance[J]. Nanoscale, 2012, 4 (22): 7078-7083.

[31] Liu P, Liang Y, Li H B, et al. Violet-blue photoluminescence from Si nanoparticles with zinc-blende structure synthesized by laser ablation in liquids[J]. AIP Advances, 2013, 3 (2): 022127.

[32] Xiao J, Liu P, Liang Y, et al. High aspect ratio β-MnO$_2$ nanowires and sensor performance for explosive gases[J]. Journal of Applied Physics, 2013, 114 (7): 073513.

[33] Liang Y, Zhu L F, Liu P, et al. Ag$_2$V$_4$O$_{11}$ nanobrushes for highly ethanol sensitive performance[J]. CrystEngComm, 2013, 15 (31): 6131-6135.

[34] Xiao J, Liu P, Liang Y, et al. Super-stable ultrafine beta-tungsten nanocrystals with metastable phase and related magnetism[J]. Nanoscale, 2013, 5 (3): 899-903.

[35] Liang Y, Liu P, Xiao J, et al. A microfibre assembly of an iron-carbon composite with giant magnetisation[J]. Scientific Reports, 2013, 3: 3051.

[36] Xiao J, Tian X M, Yang C, et al. Ultrahigh relaxivity and safe probes of manganese oxide nanoparticles for *in vivo* imaging[J]. Scientific Reports, 2013, 3: 3424.

[37] Liang Y, Liu P, Xiao J, et al. A general strategy for one-step fabrication of one-dimensional magnetic nanoparticle chains based on laser ablation in liquid[J]. Laser Physics Letters, 2014, 11 (5): 056001.

[38] Xiao J, Wu Q L, Liu P, et al. Highly stable sub-5 nm Sn$_6$O$_4$(OH)$_4$ nanocrystals with ultrahigh activity as advanced photocatalytic materials for photodegradation of methyl orange[J]. Nanotechnology, 2014, 25 (13): 135702.

[39] Liu J, Tian X M, Luo N Q, et al. Sub-10 nm monoclinic Gd$_2$O$_3$: Eu^{3+} nanoparticles as dual-modal nanoprobes for magnetic resonance and fluorescence imaging[J]. Langmuir, 2014, 30 (43): 13005-13013.

[40] Liang Y, Liu P, Yang G W. Fabrication of one-dimensional chain of iron-based bimetallic alloying nanoparticles with unique magnetizations[J]. Crystal Growth & Design, 2014, 14 (11): 5847-5855.

[41] Luo N Q, Yang C, Tian X M, et al. A general top-down approach to synthesize rare earth doped-Gd$_2$O$_3$ nanocrystals as dualmodal contrast agents[J]. Journal of Materials Chemistry B, 2014, 2 (35): 5891-5897.

[42] Xiao J, Li J L, Liu P, et al. A new phase transformation path from nanodiamond to new-diamond via an intermediate carbon onion[J]. Nanoscale, 2014, 6 (24): 15098-15106.

[43] Xiao J, Ouyang G, Liu P, et al. Reversible nanodiamond-carbon onion phase transformations[J]. Nano Letters, 2014, 14 (6): 3645-3652.

[44] Liu J, Tian X M, Chen H P, et al. Near-infrared to visible and near-infrared upconversion of monoclinic Gd$_2$O$_3$: Yb^{3+}/Tm^{3+} nanoparticles prepared by laser ablation in liquid for fluorescence imaging[J]. Applied Surface Science, 2015, 348: 60-65.

[45] Li L H, Deng Z X, Xiao J X, et al. A metallic metal oxide (Ti$_5$O$_9$) -metal oxide (TiO$_2$) nanocomposite as the

[46] Li L H, Xiao J, Yang G W. Amorphization of cobalt monoxide nanocrystals and related explosive gas sensing applications[J]. Nanotechnology, 2015, 26 (41): 415501.

[47] Chen F, Chen M, Yang C, et al. Terbium-doped gadolinium oxide nanoparticles prepared by laser ablation in liquid for use as a fluorescence and magnetic resonance imaging dual-modal contrast agent[J]. Physical Chemistry Chemical Physics, 2015, 17 (2): 1189-1196.

[48] Liu P, Chen H J, Wang H, et al. Fabrication of Si/Au core/shell nanoplasmonic structures with ultrasensitive surface-enhanced Raman scattering for monolayer molecule detection[J]. The Journal of Physical Chemistry C, 2015, 119 (2): 1234-1246.

[49] Xiao J, Liu P, Li L H, et al. Fluorescence origin of nanodiamonds[J]. The Journal of Physical Chemistry C, 2015, 119 (4): 2239-2248.

[50] Li L H, Xiao J, Liu P, et al. Super adsorption capability from amorphousization of metal oxide nanoparticles for dye removal[J]. Scientific Reports, 2015, 5: 9028.

[51] Lin Z Y, Xiao J, Yan J H, et al. Ag/AgCl plasmonic cubes with ultrahigh activity as advanced visible-light photocatalysts for photodegrading dyes[J]. Journal of Materials Chemistry A, 2015, 3 (14): 7649-7658.

[52] Lin Z Y, Liu P, Yan J H, et al. Matching energy levels between TiO_2 and α-Fe_2O_3 in a core-shell nanoparticle for visible-light photocatalysis[J]. Journal of Materials Chemistry A, 2015, 3 (28): 14853-14863.

[53] Xiao J, Liu P, Yang G W. Nanodiamonds from coal under ambient conditions[J]. Nanoscale, 2015, 7 (14): 6114-6125.

[54] Yan J H, Liu P, Lin Z Y, et al. Directional Fano resonance in a silicon nanosphere dimer[J]. ACS Nano, 2015, 9 (3): 2968-2980.

[55] Lin Z Y, Li J L, Zheng Z Q, et al. Electronic reconstruction of α-Ag_2WO_4 nanorods for visible-light photocatalysis[J]. ACS Nano, 2015, 9 (7): 7256-7265.

[56] Yan J H, Liu P, Lin Z Y, et al. Magnetically induced forward scattering at visible wavelengths in silicon nanosphere oligomers[J]. Nature Communications, 2015, 6: 7042.

[57] Pan B, Xiao J, Li J L, et al. Carbyne with finite length: The one-dimensional sp carbon[J]. Science Advances, 2015, 1 (9): e1500857.

[58] Ma C R, Yan J H, Wei Y M, et al. Second harmonic generation from an individual amorphous selenium nanosphere[J]. Nanotechnology, 2016, 27 (42): 425206.

[59] Ma C R, Xiao J, Yang G W. Giant nonlinear optical responses of carbyne[J]. Journal of Materials Chemistry C, 2016, 4 (21): 4692-4698.

[60] Ma C R, Yan J H, Liu P, et al. Second harmonic generation from an individual all-dielectric nanoparticle: Resonance enhancement versus particle geometry[J]. Journal of Materials Chemistry C, 2016, 4 (25): 6063-6069.

[61] Li L H, Yu L L, Lin Z Y, et al. Reduced TiO_2-graphene oxide heterostructure as broad spectrum-driven efficient water-splitting photocatalysts[J]. ACS Applied Materials & Interfaces, 2016, 8 (13): 8536-8545.

[62] Liu P, Yan J H, Ma C R, et al. Midrefractive dielectric modulator for broadband unidirectional scattering and effective radiative tailoring in the visible region[J]. ACS Applied Materials & Interfaces, 2016, 8 (34): 22468-22476.

[63] Liu X Y, Gao Y Q, Yang G W. A flexible, transparent and super-long-life supercapacitor based on ultrafine Co_3O_4 nanocrystal electrodes[J]. Nanoscale, 2016, 8 (7): 4227-4235.

[64] Yan J H, Liu P, Lin Z Y, et al. New type high-index dielectric nanosensors based on the scattering intensity shift[J]. Nanoscale, 2016, 8 (11): 5996-6007.

[65] Yan J H, Liu P, Ma C R, et al. Plasmonic near-touching titanium oxide nanoparticles to realize solar energy harvesting and effective local heating[J]. Nanoscale, 2016, 8 (16): 8826-8838.

[66] Yan J H, Lin Z Y, Ma C R, et al. Plasmon resonances in semiconductor materials for detecting photocatalysis at the single-particle level[J]. Nanoscale, 2016, 8 (32): 15001-15007.

[67] Lin Z Y, Xiao J, Li L H, et al. Nanodiamond-embedded p-type copper (I) oxide nanocrystals for broad-spectrum photocatalytic hydrogen evolution[J]. Advanced Energy Materials, 2016, 6 (4): 1501865.

[68] Lin Z Y, Li L H, Yu L L, et al. Dual-functional photocatalysis for hydrogen evolution from industrial wastewaters[J]. Physical Chemistry Chemical Physics, 2017, 19 (12): 8356-8362.

[69] Ma C R, Yan J H, Wei Y M, et al. Enhanced second harmonic generation in individual barium titanate nanoparticles driven by Mie resonances[J]. Journal of Materials Chemistry C, 2017, 5 (19): 4810-4819.

[70] Ma C R, Yan J H, Huang Y C, et al. Directional scattering in a germanium nanosphere in the visible light region[J]. Advanced Optical Materials, 2017, 5 (24): 1700761.

[71] Yan J H, Ma C R, Liu P, et al. Plasmon-induced energy transfer and photoluminescence manipulation in MoS_2 with a different number of layers[J]. ACS Photonics, 2017, 4 (5): 1092-1100.

[72] Chen T M, Tian X M, Huang L, et al. Nanodiamonds as pH-switchable oxidation and reduction catalysts with enzyme-like activities for immunoassay and antioxidant applications[J]. Nanoscale, 2017, 9 (40): 15673-15684.

[73] Lin Z Y, Li J L, Li L H, et al. Manipulating the hydrogen evolution pathway on composition-tunable CuNi nanoalloys[J]. Journal of Materials Chemistry A, 2017, 5 (2): 773-781.

[74] Lin Z Y, Li L H, Yu L L, et al. Modifying photocatalysts for solar hydrogen evolution based on the electron behavior[J]. Journal of Materials Chemistry A, 2017, 5 (11): 5235-5259.

[75] Xiao J, Li J L, Yang G W. Molecular luminescence of white carbon[J]. Small, 2017, 13 (12): 1603495.

[76] Yan J H, Ma C R, Liu P, et al. Electrically controlled scattering in a hybrid dielectric-plasmonic nanoantenna[J]. Nano Letters, 2017, 17 (8): 4793-4800.

[77] Xiao J, Liu P, Wang C X, et al. External field-assisted laser ablation in liquid: An efficient strategy for nanocrystal synthesis and nanostructure assembly[J]. Progress in Materials Science, 2017, 87: 140-220.

[78] Wu X J, Tian X M, Chen T M, et al. Inorganic fullerene-like molybdenum selenide with good biocompatibility synthesized by laser ablation in liquids[J]. Nanotechnology, 2018, 29 (29): 295604.

[79] Liu X Y, Wang J X, Yang G W. Amorphous nickel oxide and crystalline manganese oxide nanocomposite electrode

for transparent and flexible supercapacitor[J]. Chemical Engineering Journal, 2018, 347: 101-110.

[80] Lin Z Y, Li W J, Yang G W. Hydrogen-interstitial $CuWO_4$ nanomesh: A single-component full spectrum-active photocatalyst for hydrogen evolution[J]. Applied Catalysis B: Environmental, 2018, 227: 35-43.

[81] Ma C R, Yan J H, Huang Y C, et al. The optical duality of tellurium nanoparticles for broadband solar energy harvesting and efficient photothermal conversion[J]. Science Advances, 2018, 4 (8): eaas9894.

[82] Lin Z Y, Du C, Yan B, et al. Two-dimensional amorphous NiO as a plasmonic photocatalyst for solar H_2 evolution[J]. Nature Communications, 2018, 9: 4036.

[83] Chen T M, Zou H, Wu X J, et al. Fullerene-like MoS_2 nanoparticles as cascade catalysts improving lubricant and antioxidant abilities of artificial synovial fluid[J]. ACS Biomaterials Science & Engineering, 2019, 5 (6): 3079-3088.

[84] Lin Z Y, Du C, Yan B, et al. Amorphous Fe_2O_3 for photocatalytic hydrogen evolution[J]. Catalysis Science & Technology, 2019, 9 (20): 5582-5592.

[85] Liu Y, Liu P, Qin W, et al. Laser modification-induced $NiCo_2O_{4-\delta}$ with high exterior Ni^{3+}/Ni^{2+} ratio and substantial oxygen vacancies for electrocatalysis[J]. Electrochimica Acta, 2019, 297: 623-632.

[86] Lin Z Y, Du C, Yan B, et al. Two-dimensional amorphous CoO photocatalyst for efficient overall water splitting with high stability[J]. Journal of Catalysis, 2019, 372: 299-310.

[87] Liu Y, Yang F, Qin W, et al. $Co_2P@NiCo_2O_4$ bi-functional electrocatalyst with low overpotential for water splitting in wide range pH electrolytes[J]. Journal of Colloid and Interface Science, 2019, 534: 55-63.

[88] Huang Y C, Yan J H, Ma C R, et al. Trapping and filtering of light by single Si nanospheres in a GaAs nanocavity[J]. Nanoscale, 2019, 11 (35): 16299-16307.

[89] Yan J H, Ma C R, Huang Y C, et al. Dynamic radiative tailoring based on mid-refractive dielectric nanoantennas[J]. Nanoscale Horizons, 2019, 4 (3): 712-719.

[90] Huang Y C, Yan J H, Ma C R, et al. Active tuning of the Fano resonance from a Si nanosphere dimer by the substrate effect[J]. Nanoscale Horizons, 2019, 4 (1): 148-157.

[91] Ma C R, Yan J H, Huang Y C, et al. Photoluminescence manipulation of WS_2 flakes by an individual Si nanoparticle[J]. Materials Horizons, 2019, 6 (1): 97-106.

[92] Yan J H, Yu P, Ma C R, et al. Directional radiation and photothermal effect enhanced control of 2D excitonic emission based on germanium nanoparticles[J]. Nanotechnology, 2020, 31 (38): 385201.

[93] Ma C R, Yan J H, Huang Y C, et al. Direct-indirect bandgap transition in monolayer MoS_2 induced by an individual Si nanoparticle[J]. Nanotechnology, 2020, 31 (6): 065204.

[94] Chen T M, Yang F, Liu P, et al. General top-down strategy for generating single-digit nanodiamonds for bioimaging[J]. Nanotechnology, 2020, 31 (48): 485601.

[95] Ma C R, Liu Y, Zhao F, et al. Loss-favored ultrasensitive refractive index sensor based on directional scattering from a single all-dielectric nanosphere[J]. Journal of Materials Chemistry C, 2020, 8 (19): 6350-6357.

[96] Yang F, Zheng Z Q, Lin Z Q, et al. Visible-light-driven room-temperature gas sensor based on carbyne

[97] Chen T M, Yang F, Wu X J, et al. A fluorescent and colorimetric probe of carbyne nanocrystals coated Au nanoparticles for selective and sensitive detection of ferrous ions[J]. Carbon, 2020, 167: 196-201.

[98] Yan J H, Li Y C, Lou Z Z, et al. Active tuning of Mie resonances to realize sensitive photothermal measurement of single nanoparticles[J]. Materials Horizons, 2020, 7 (6): 1542-1551.

[99] Yan J H, Liu X Y, Ma C R, et al. All-dielectric materials and related nanophotonic applications[J]. Materials Science and Engineering: R: Reports, 2020, 141: 100563.

[100] Huang Y C, Yang G W. Light-matter interactions between germanium nanocavities and quantum dots at visible wavelengths[J]. The Journal of Physical Chemistry C, 2021, 125 (1): 812-818.

[101] Yang F, Liu P, Wu C W, et al. Paramagnetism of carbyne nanocrystals[J]. Materialstoday Communications, 2021, 26: 102152.

[102] Yan J H, Liu X Y, Mao B J, et al. Individual Si nanospheres wrapped in a suspended monolayer WS_2 for electromechanically controlled Mie-type nanopixels[J]. Advanced Optical Materials, 2021, 9 (5): 2001954.

[103] Yang F, Chen T M, Wu X J, et al. A hybrid gold-carbyne nanocrystals platform for light-induced crossover of redox enzyme-like activities[J]. Chemical Engineering Journal, 2021, 408: 127244.

[104] Yang F, Li C, Li J L, et al. Carbyne nanocrystal: One-dimensional van der Waals crystal[J]. ACS Nano, 2021, 15 (10): 16769-16776.

[105] Yang F, Zheng Z Q, He Y, et al. A new wide bandgap semiconductor: Carbyne nanocrystals[J]. Advanced Functional Materials, 2021, 31 (36): 2104254.

[106] Cai L K, Yan B, Xue Q, et al. Electronic modulation of NiO by constructing an amorphous/crystalline heterophase to improve photocatalytic hydrogen evolution[J]. Journal of Materials Chemistry A, 2022, 10 (36): 18939-18949.

[107] Yan B, Li Y W, Cao W W, et al. Highly efficient and highly selective CO_2 reduction to CO driven by laser[J]. Joule, 2022, 6 (12): 2735-2744.

[108] Yang G W. Synthesis, properties, and applications of carbyne nanocrystals[J]. Materials Science and Engineering: R: Reports, 2022, 151: 100692.

[109] Cao W W, Xu H K, Liu P, et al. The kinked structure and interchain van der Waals interaction of carbyne nanocrystals[J]. Chemical Science, 2023, 14 (2): 338-344.

后　　记

2024 年 11 月，中山大学将迎来世纪华诞。与此同时，中山大学材料科学与工程学院也将迎来十周年庆。中山大学材料科学学科有着悠久的历史，孙中山先生创办中山大学之初，在理学院的物理系就设有金属物理方向，自此开启了中山大学材料学科之门。学校于 2014 年 1 月成立材料科学与工程学院（筹），我被任命为执行院长；经过近两年的筹建，学院于 2015 年 12 月正式挂牌，我被任命为院长。秉承"立德树人，以德为先"的育人理念，学院凝集了中山大学在材料科学及相关学科领域的优势资源，致力于培养具有国际视野和创新精神、具备扎实广博理论基础和突出精深专业技能的学术精英及社会栋梁。作为学院的创始院长，我为学院的筹办和建设可谓殚精竭虑，倾注了大量心血，看着学院从无到有、从小到大、从大到强。如今，一座雄伟壮观的材料大楼坐落在郁郁葱葱、繁花似锦的中山大学广州校区东校园，学院各项事业已经步入了高质量发展的快车道。2023 年初的时候，我开始考虑应该做些什么来纪念一下即将到来的中山大学百年校庆和学院十周年庆，很自然地就想到了将自己数十年的研究积累，通过系统整理、凝练为学术专著出版，向中大世纪华诞和学院十周年庆献礼。机缘巧合，2023 年春节前夕，科学出版社的常诗尧编辑给我发送了节日问候的邮件并询问近期是否有出版学术书籍的意向。我欣然答应了常编辑撰写学术专著。

撰写中文学术专著对我来说是一件新鲜的事，因为我们发表的学术论文基本上都是英文的，需要将它们翻译成中文后再进行整理。我的博士生刘宁在进行博士论文研究的同时，抽出宝贵的时间，不辞辛苦地在论文翻译、数据和图片整理，以及文稿修订等方面为本书的撰写提供了极大的帮助，令人感激涕零！科学出版社的常诗尧编辑热情而专业，从专著题目的选定、章节结构的优化，到最后的成文定稿都给予了耐心、细致的指导和帮助，尤其是她在学术上的严谨给我留下了极深刻的印象！可以说，没有常编辑的努力，本书可能不会如此顺利地出版。所以，我本人对此表示由衷的感谢。

这本学术专著是以我个人名义出版的，但是它所涵盖的研究成果是我和我的学生们共同取得的。在本书出版之际，我向为本书内容做出重要贡献的王金斌博士、刘璞博士、肖俊博士等致以诚挚的感谢。

彩 图

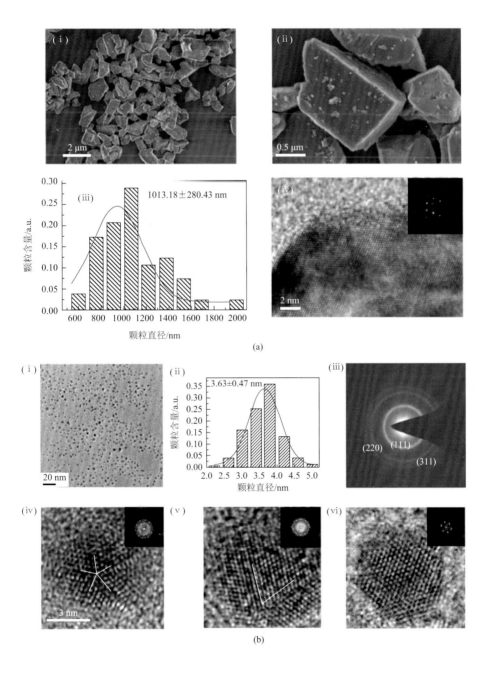

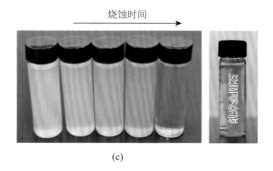

(c)

图 5-7　LAL 转化金刚石微晶为 ND

（a）初始金刚石微晶 SEM 图像（i）和（ii），分布直方图及其高斯拟合曲线（iii）显示颗粒尺寸为 1000 nm 左右，多分散度约 28%，以及相应高分辨透射电镜（HRTEM）（iv）和 SAED 图像（插图）；（b）个位数 ND 的低放大率 TEM 图像（i）和分布直方图及其高斯拟合曲线（ii）表明尺寸为 4 nm 左右，多分散度约 13%，以及相应 SAED 图像（iii），（iv）～（vi）为 ND 各种结构的 HRTEM 图像，包括五重孪晶、三重孪晶和单晶结构，插图显示了相应的傅里叶变换衍射图，表明金刚石结晶良好；（c）随着 LAL 过程的持续，合成胶体溶液颜色发生变化（左）和第五个瓶子的侧视图（右）

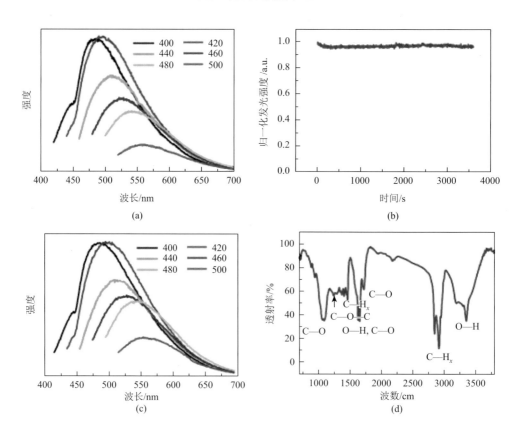

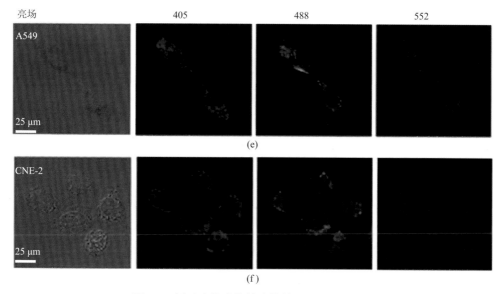

(e)

(f)

图 5-9 用于生物成像的个位数 ND 的各种光谱

(a) 不同波长下的发光光谱,从 400 nm 开始,以 20 nm 为增量;(b) 450 W 氙灯照射的 ND 发光稳定性;(c) 用 PEG$_{200N}$ 钝化后的 ND 的荧光光谱;(d) ND 的傅里叶变换红外光谱仪(FTIR)光谱,显示出各种表面基团的信号;(e) 和 (f) 在 405 nm、488 nm 和 552 nm 的激光辐照下,用荧光 ND 标记的 A549 和 CNE-2 细胞的激光扫描共聚焦显微镜图像,以及相应的亮场图像

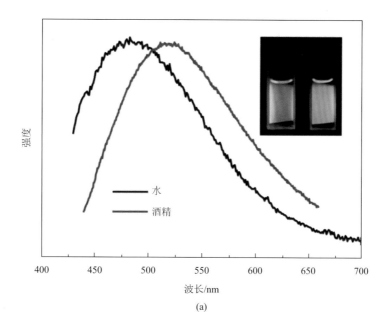

(a)

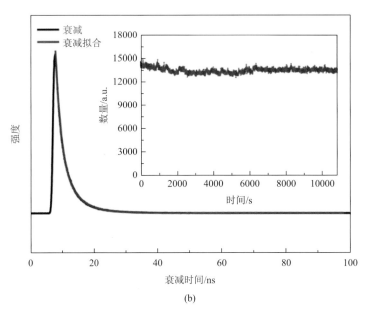

(b)

图 5-13 （a）波长与强度的关系，插图为不同环境液体合成的纳米金刚石的荧光颜色；（b）衰减时间与强度的关系，插图为荧光稳定性

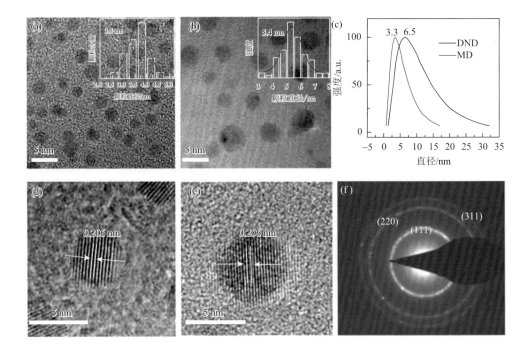

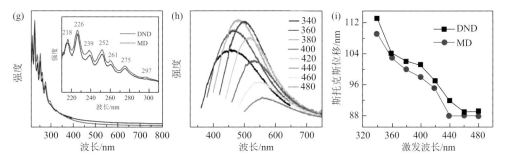

图 5-14 LAL 纳米金刚石的尺寸无关发光

(a) MD 尺寸 3.5 nm 和 (b) DND 尺寸 5.4 nm 的良好分散纳米金刚石的典型形态；(c) 测量的纳米金刚石的有效直径；(d) MD 和 (e) DND 的 HRTEM 图像；(f) 相应 SAED 图像；(g) 纳米金刚石胶体紫外可见吸收光谱，插图是 210～310 nm 的放大图；(h) 不同激发波长的发光光谱，两种类型的纳米金刚石表现出相似发光行为；(i) DND 和 MD 的斯托克斯位移与激发波长的关系

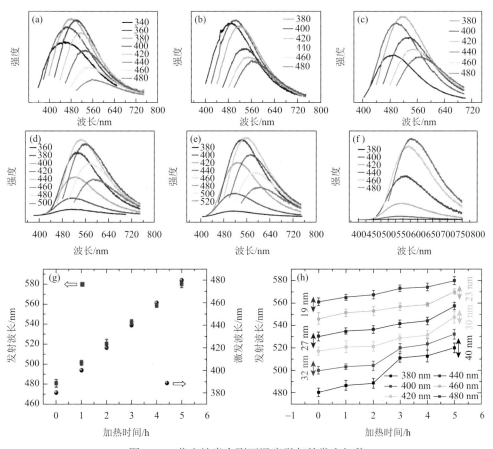

图 5-15 荧光纳米金刚石温度引起的发光红移

(a) LAL 合成纳米金刚石荧光光谱；(b) ～ (f) 合成的纳米金刚石胶体在 65℃下加热 1～5 h 后的荧光光谱；(g) 由最佳激发波长激发的最强峰的变化；(h) 在加热不同时间后，在特定发射波长激发下每个发射峰的红移

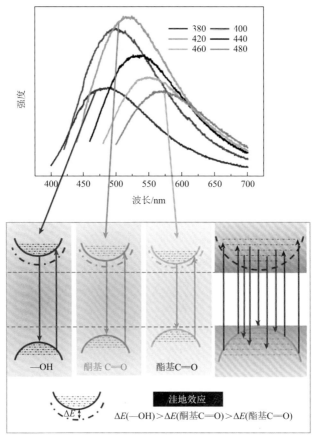

图 5-17 纳米金刚石的荧光起源

—OH、酮基 C＝O 和酯基 C＝O 分别由蓝色、绿色和黄色发光表示，依赖于激发的荧光的本质在于这三组的相对强度，并且洼地效应是不同的：蓝色的 LUMO 比绿色和黄色的 LUMO 变化更大

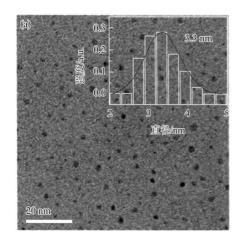
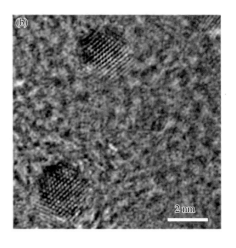

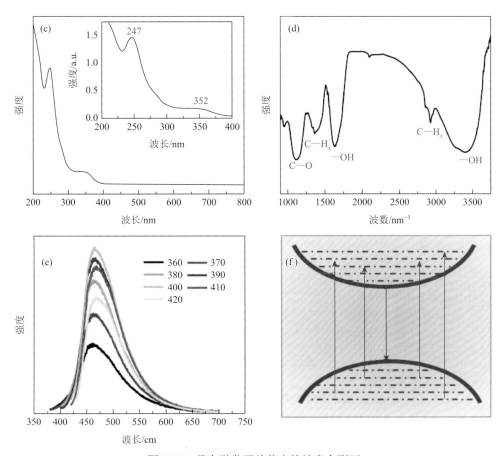

图 5-18 具有激发无关荧光的纳米金刚石

(a) 以水为环境液体 LAL 合成的纳米金刚石, 尺寸约为 3.3 nm, 具有良好分散性; (b) 相应的 HRTEM 图像; (c) 纳米金刚石的紫外可见吸收光谱, 显示出比在酒精中合成的纳米金刚石少得多的吸收峰; (d) 纳米金刚石的 MFTIR 光谱, 表明—OH 基团在样品中占主导地位; (e) 具有激发无关荧光的纳米金刚石发光光谱和 (f) 荧光机理示意图

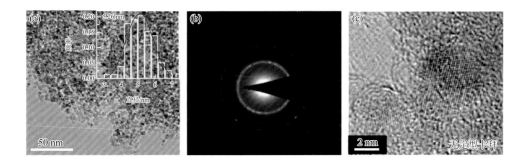

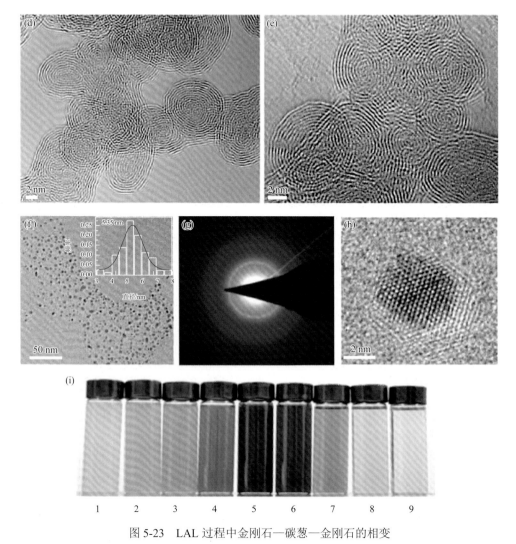

图 5-23 LAL 过程中金刚石—碳葱—金刚石的相变

(a) 初始原料爆炸法纳米金刚石 TEM 图像显示这些纳米颗粒是团聚的，插图为粒径分布，显示平均尺寸为 5.26 nm；(b) 相应 SAED 图像；(c) HRTEM 图像；(d)、(e) 碳葱 TEM 图像；(f) LAL 合成纳米金刚石 TEM 图像，分散性得到改善，插图为粒径分布；(g) 相应 SAED 图像；(h) HRTEM 图像；(i) LAL 作用过程中合成胶体溶液颜色的变化

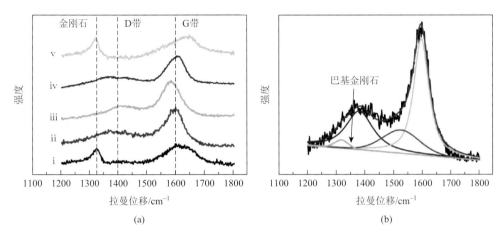

图 5-25 从初始原料纳米金刚石到 LAL 合成的纳米新金刚石相变的拉曼光谱分析

（a）初始原料纳米金刚石（曲线 i）显示 1325 cm^{-1} 的金刚石峰和 1600 cm^{-1} 的加宽峰；巴基金刚石（曲线 ii）显示峰值移向 1360 cm^{-1}；碳葱（曲线 iii）显示金刚石峰不断向 1400 cm^{-1} 移动（D 带），同时 G 带也明显下降到 1584 cm^{-1}；巴基金刚石（曲线 iv）显示 D 带波峰下降而 G 带向上移动到更高的波数；最终的纳米新金刚石（曲线 v）显示金刚石峰（1325 cm^{-1}）再次生成，G 带移动到 1650 cm^{-1}，高于初始原料纳米金刚石。（b）巴基金刚石曲线的详细拟合 [（a）中的曲线 ii]，绿色拟合曲线位于 1324 cm^{-1}，可以归属于纳米金刚石

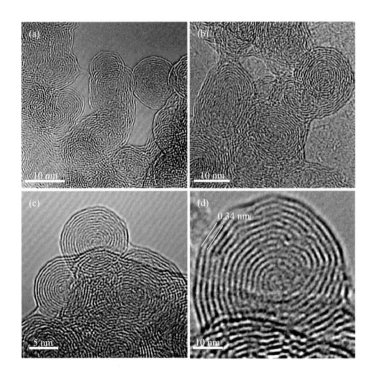

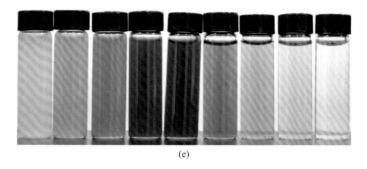

(e)

图 5-30 作为相变中间相的碳葱

(a)～(d) 聚集的球形碳葱；(e) 合成液体颜色随 LAL 作用时间的变化

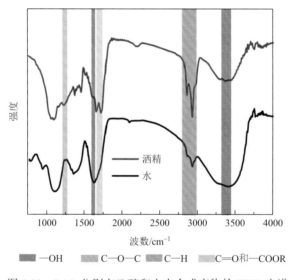

图 5-33 LAL 分别在乙醇和水中合成产物的 FTIR 光谱

图 5-46 carbyne 纳米晶胶体（左）和粉末（右）照片

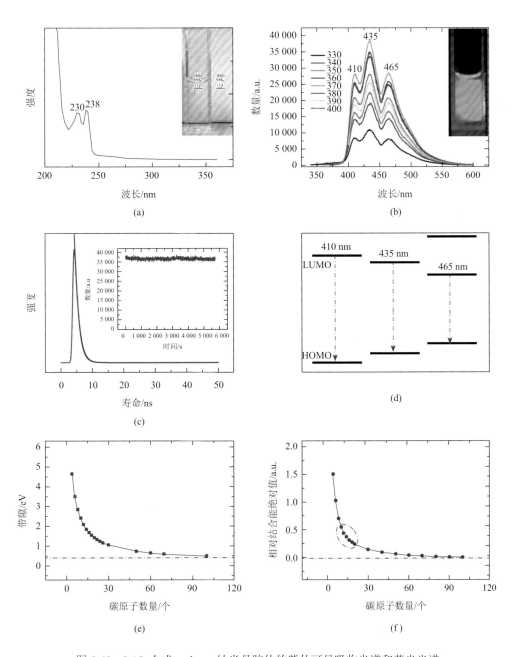

图 5-48 LAL 合成 carbyne 纳米晶胶体的紫外可见吸收光谱和荧光光谱

(a) 紫外可见吸收光谱,插图为无色透明溶液的光学照片;(b) 荧光光谱,插图是用 370 nm 的光激发的紫蓝色荧光;(c) 发光寿命测量为 1.3 ns,插图表示使用 450 W 氙灯辐照 1.5 h 未观察到光漂白现象;(d) 不同长度碳链的三种荧光行为;(e) 带隙对碳原子数量的依赖性;(f) 作为比较,垂直坐标表示相对结合能的绝对值,负结合能对应于稳定的构型,绝对值越大,碳链的稳定性就越高,结合能的绝对值随着碳原子数量的增加而降低,但降低速度逐渐变慢,最终达到特定值

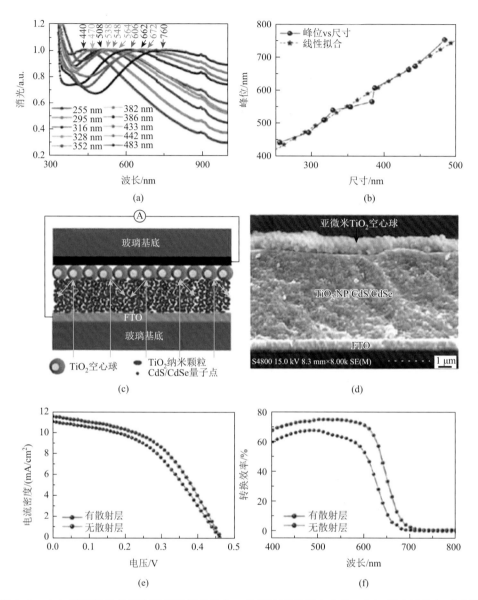

图 9-5 （a）不同激光能量密度辐照样品的归一化紫外可见消光光谱；（b）消光峰值与球形颗粒平均尺寸的关系；（c）QDSSC 的 TiO_2 空心球颗粒散射层示意图；（d）TiO_2 空心球颗粒覆盖的量子点敏化太阳电池介孔电极的横截面 SEM 图像；（e）电流密度与电压的关系；（f）有和没有 TiO_2 空心球颗粒散射层的 QDSSC 的入射光子-电子转换效率

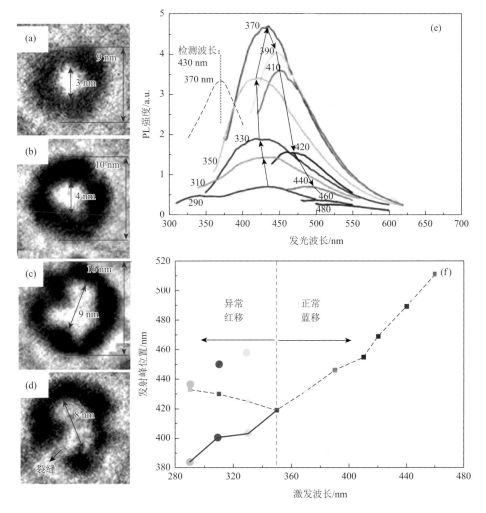

图 10-1　(a)～(d) 各种形态的 SiC 纳米环；(e) 不同波长激光激发下的 SiC 纳米环的光致发光光谱；(f) 发射峰位置与激发波长的函数关系

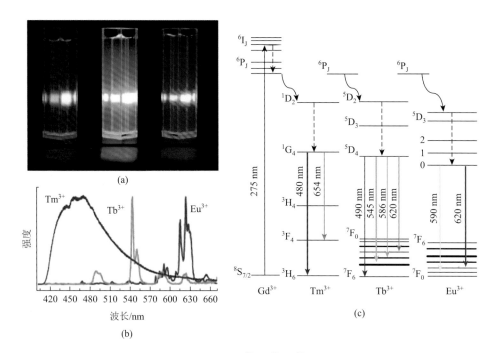

图 10-2 LAL 合成 Gd_2O_3∶$Tm^{3+}/Tb^{3+}/Eu^{3+}$ 纳米颗粒的荧光特性

（a）纳米颗粒的荧光图像，从左到右分别为 Gd_2O_3∶$Tm^{3+}/Tb^{3+}/Eu^{3+}$，激光波长为 275 nm；（b）相应的发射光谱；（c）光敏化剂 Gd^{3+} 与不同激活剂 Tm^{3+}、Tb^{3+} 和 Eu^{3+} 之间的能量转移示意图

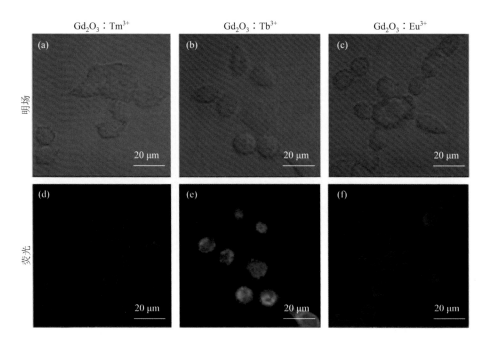

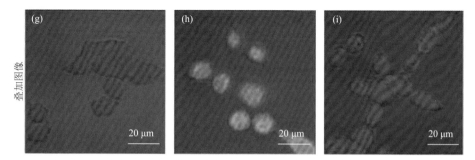

图 10-3　分别与 Gd_2O_3：Tm^{3+}/Tb^{3+}/Eu^{3+} 纳米颗粒一起孵育的细胞的共焦显微镜图像
（a）～（c）明场图像；（d）～（f）405 nm 激光激发下的荧光图像；（g）～（i）叠加图像

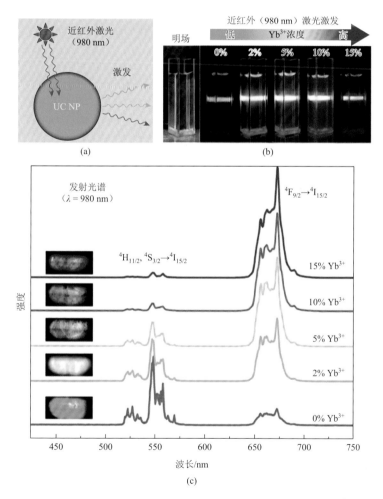

图 10-4　（a）近红外激光激发下纳米颗粒上转换过程的示意图；（b）Gd_2O_3：Yb^{3+}（0～15%）胶体在 980 nm 激光激发下的荧光图像；（c）相应的发射光谱

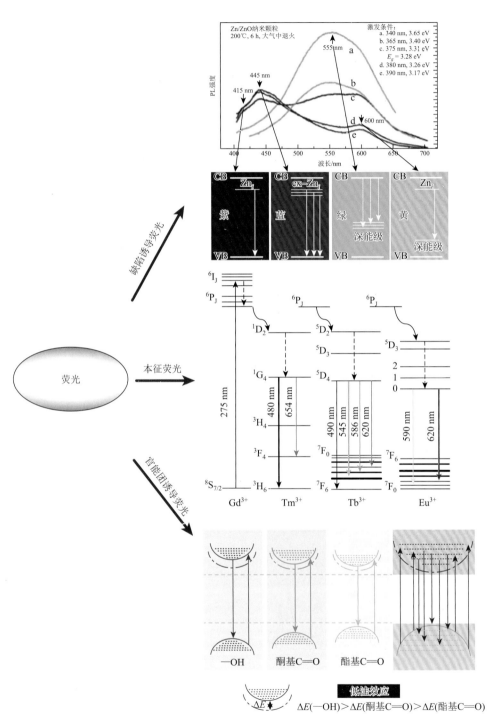

图 10-6　LAL 合成的荧光纳米颗粒的三种荧光发光来源

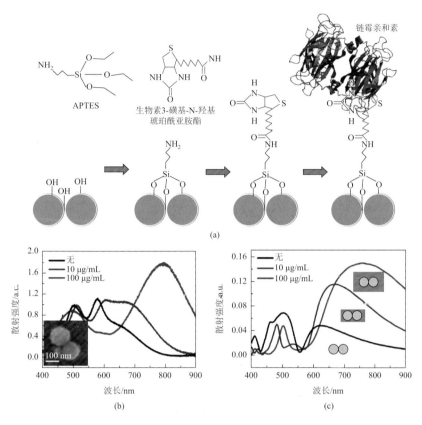

图 10-9 LAL 组装的硅纳米球二聚体传感器检测示意图

(a) 传感器用于生物大分子检测的表面功能化的示意图;(b) 实验测量了使用不同浓度的链霉亲和素的背向散射光谱,插图是典型二聚体的 SEM 图像;(c) 模拟背向散射光谱,以揭示当增加与表面结合的链霉亲和素分子数量时的强度变化

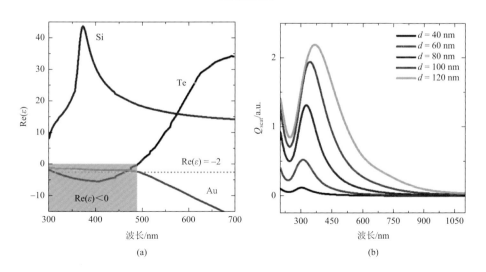

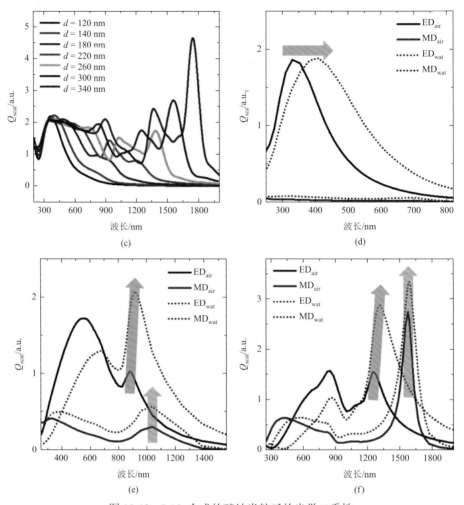

图 10-12 LAL 合成的碲纳米粒子的光学二重性

（a）与 Au（等离子体材料）和 Si（全介质材料）相比，碲的介电常数的实部；（b）直径小于等于 120 nm 的碲纳米颗粒的类等离子体行为；（c）直径在 120~340 nm 的碲纳米颗粒的全介质材料行为；（d）~（f）电偶极子（ED）和磁偶极子（MD）对于在空气（air）和水中（wat）直径分别为 100 nm、200 nm 和 300 nm 的碲纳米颗粒的散射效率的贡献

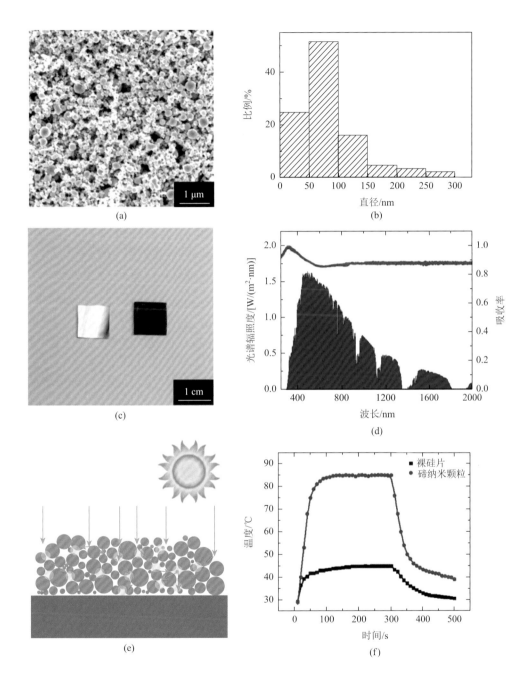

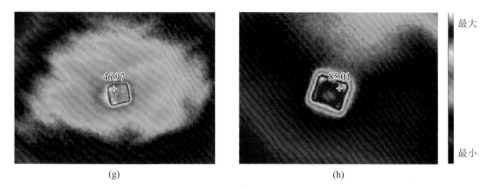

图 10-13 沉积在硅衬底上的碲纳米颗粒层的光热效应

(a) 碲纳米颗粒自组装的 SEM 图像；(b) 基于 (a) 中 SEM 图像的尺寸分布 (来自 500 个颗粒的统计数据)；(c) 裸硅片 (左) 和沉积在硅衬底上的碲纳米颗粒层 (右) 的照片；(d) 碲纳米颗粒吸收剂的吸收光谱 (红色曲线)，蓝色区域是太阳光谱；(e) 太阳光照射下的碲纳米颗粒层的示意图；(f) 裸硅片 (黑色曲线) 和碲纳米颗粒吸收体 (红色曲线) 的随时间变化的温度变化；(g) 和 (h) 分别为硅晶片和碲纳米颗粒吸收体的稳态热图像

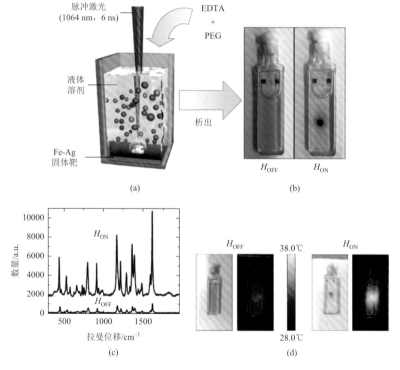

图 10-15 (a) LAL 纳米制备示意图和 (b) Fe-Ag 纳米颗粒溶液的光学照片，黄绿色为 Ag 等离子体的典型颜色，黑色区域为通过在比色皿一侧放置小型 NdFeB 磁铁施加磁场所积累的磁性纳米颗粒；(c) 磁聚焦之前 ($H_{OFF}=0$，黑线) 和磁聚焦之后 ($H_{ON}>0$，蓝线)，MG/Fe-Ag 纳米颗粒分散液的拉曼光谱；(d) 在磁聚焦前后，使用波长 785 nm 连续激光辐照 Fe-Ag 纳米颗粒分散液，进行光热加热实验

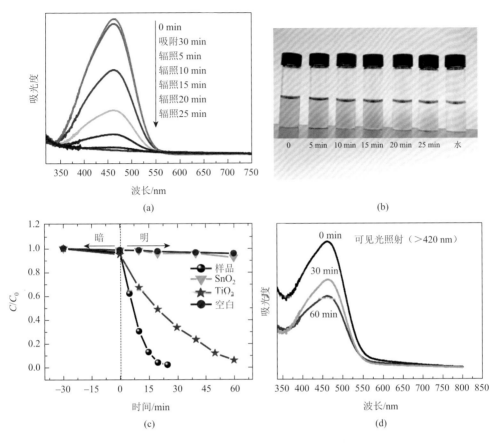

图 10-21 （a）$Sn_6O_4(OH)_4$ 纳米晶降解甲基橙的光谱图；（b）对应的降解后溶液光学照片图；
（c）$Sn_6O_4(OH)_4$ 纳米晶与 SnO_2 纳米晶、TiO_2 纳米晶及空白对照的降解效果对比图；
（d）$Sn_6O_4(OH)_4$ 纳米晶可见光降解的效果

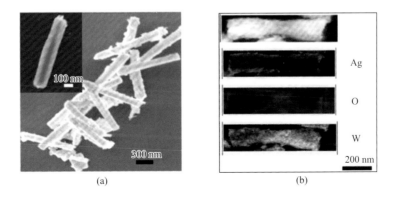

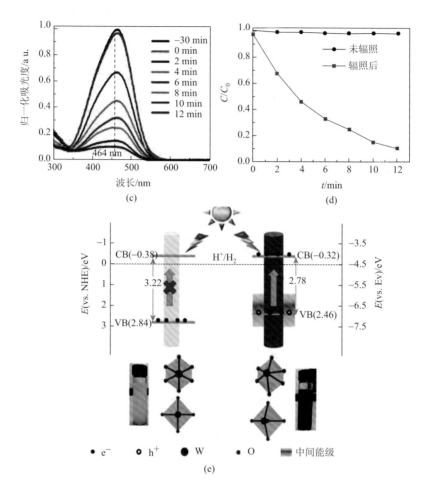

图 10-22 （a）LAL 处理后样品低倍和（插图）高倍放大的 SEM 图像；（b）EDX 元素分析图；（c）MO 溶液被降解的典型紫外可见吸收光谱变化（–30 min 表示样品配好后静止 30 min，再用光照）；（d）未 LAL 处理和 LAL 处理的 Ag_2WO_4 样品降解的 MO 溶液的相对浓度变化图；（e）基于团簇扭曲和中间能级导致带隙变窄的可见光光催化机制示意图

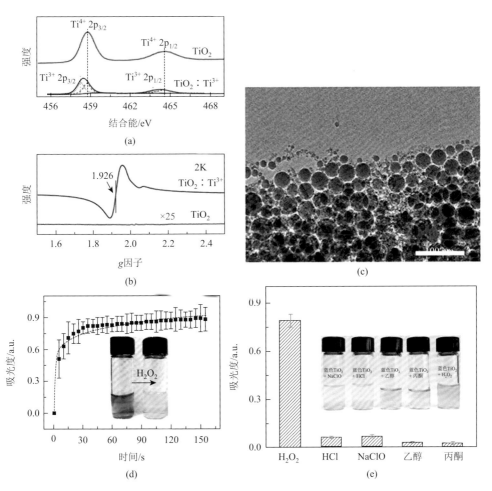

图 10-23 （a）Ti 2p XPS；（b）355 nm 脉冲激光辐照前后的金红石粉末在 2 K 下的 EPRS；（c）TiO_2：Ti^{3+} 纳米颗粒的 TEM 图像；（d）TiO_2：Ti^{3+} 纳米颗粒和 H_2O_2 在 450 nm 处的典型吸光度动力学，插图为未添加（左）和添加（右）H_2O_2 溶液的 TiO_2：Ti^{3+} 纳米颗粒颜色变化；（e）通过监测相对吸光度进行 H_2O_2 检测的选择性分析，插图的分析物浓度如下：1 mol/L 次氯酸钠（NaClO）、2 mol/L 盐酸（HCl）、1 mol/L 乙醇、1 mol/L 丙酮、0.5 mol/L 过氧化氢（H_2O_2）

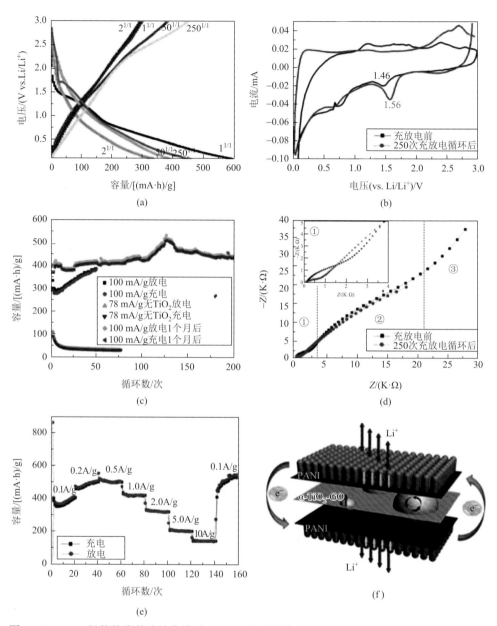

图 10-29　LAL 组装的聚苯胺纳米棒/非晶 TiO_2 纳米颗粒-氧化石墨烯复合纳米片/聚苯胺纳米棒三明治纳米结构作为锂离子电池正极材料的电化学性能

(a) 放电/充电容量；(b) CV 测试；(c) 循环稳定性；(d) EIS 测试，插图为①区放大图，横纵表示实部，纵横表示虚部；(e) 通过改变电流密度来提高三明治纳米结构的稳定性；(f) 三明治纳米结构中 Li^+ 和 e^- 的传输示意图

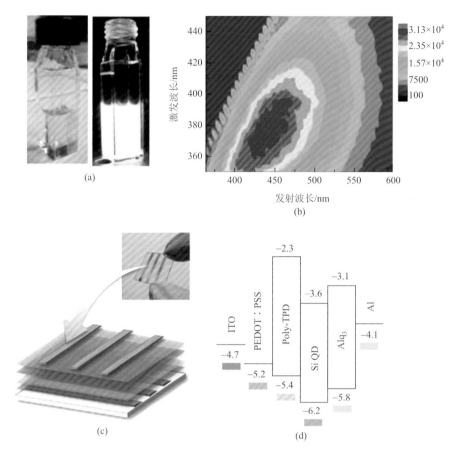

图 10-32 （a）在环境光下分散在异丙醇中的 Si QD 的照片（左）和波长 365 nm 的 UV LED 激发下胶体溶液的照片（右）；（b）分散在异丙醇中的 Si QD 的激发-发射光谱三维图；（c）组装的夹层结构 Si QD 混合式 LED 的照片和示意图；（d）Si QD 混合式 LED 的能级图（单位：eV）

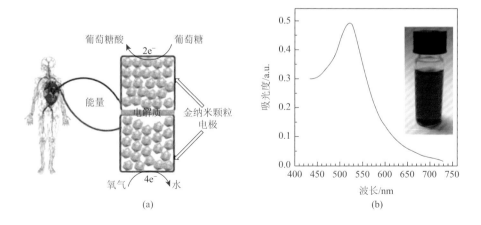

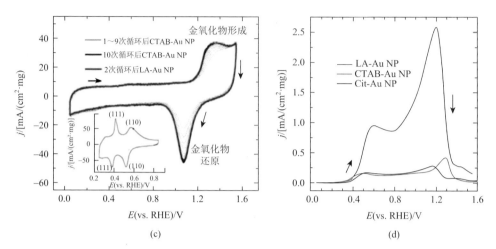

图10-34 （a）用于植入式生物设备的混合生物燃料电池的示意图；（b）LAL合成的Au纳米颗粒的典型消光光谱，插图为LAL合成Au纳米颗粒的去离子水溶液的玻璃比色皿；（c）LAL合成Au纳米颗粒（LA-Au纳米颗粒，红色）和化学溶液法合成Au纳米颗粒（CTAB-Au纳米颗粒）形成的Au电极的不同循环次数的循环伏安图，插图为局部放大图；（d）CTAB-Au纳米颗粒（红色）、Cit-Au纳米颗粒（蓝色）和LA-Au纳米颗粒（黑色）电极的伏安图

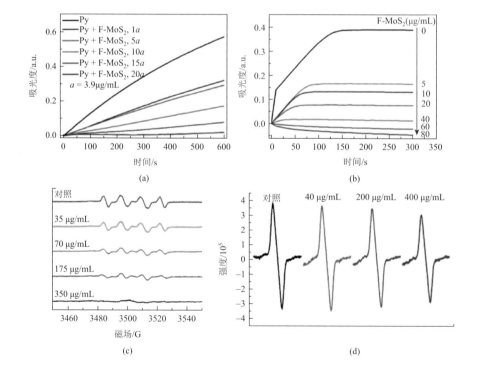

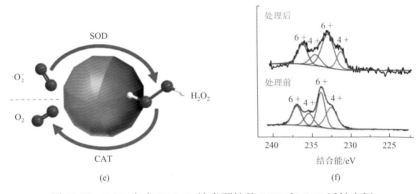

图 10-39 LAL 合成 F-MoS$_2$ 纳米颗粒的 SOD 和 CAT 活性表征

(a) 不同浓度的 F-MoS$_2$ 纳米颗粒与邻苯三酚（Py）共孵育体系在 318 nm 处的吸光度随时间的变化曲线；(b) 与黄嘌呤氧化酶、黄嘌呤、DTPA 共孵育体系在 550 nm 处吸光度随时间的变化曲线；(c) 不同浓度的 F-MoS$_2$ 纳米颗粒与黄嘌呤氧化酶、黄嘌呤、DTPA、BMPO 共孵育体系的 ESR 谱；(d) 与 CTPO、H$_2$O$_2$ 共孵育体系的 ESR 谱；(e) 基于 F-MoS$_2$ 纳米颗粒的级联催化过程的示意图；(f) 处理前后的 F-MoS$_2$ 纳米颗粒高分辨 XPS 的 Mo 3d 峰

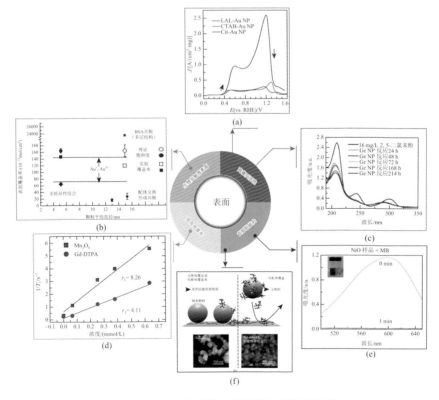

图 11-1 LAL 合成的高性能纳米颗粒催化剂

(a) 具有高催化活性的 Au 纳米颗粒；(b) 具有高共轭效率的无配体 Au 纳米颗粒；(c) 具有高催化活性的 Ge 纳米颗粒；(d) 具有高弛豫率的 Mn$_3$O$_4$ 纳米颗粒；(e) 具有高吸附能力的 NiO 纳米颗粒；(f) 具有高吸附能力的无配体纳米颗粒

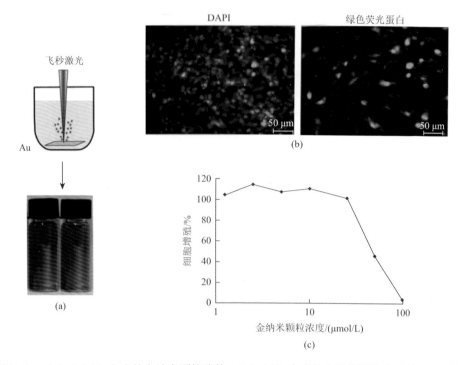

图 11-2 （a）fs-LAL 合成的金纳米颗粒胶体；（b）LAL 合成的金纳米颗粒在质粒 DNA 转染过程中不会干扰融合蛋白的生物活性；（c）当金纳米颗粒剂量低于 25 μmol/L 时，非内摄对细胞增殖没有影响